DIGITAL COMMUNICATIONS WITH FIBER OPTICS AND SATELLITE APPLICATIONS

DIGITAL COMMUNICATIONS WITH FIBER OPTICS AND SATELLITE APPLICATIONS

HARROLD B. KILLEN, Ph.D.

University of Houston

PRENTICE HALL

Englewood Cliffs, New Jersey 07632

Library of Congress Cataloging-in-Publication Data

Killen, Harold B.
Digital communications with fiber optics and satellite applications / by Harold B. Killen.
p. cm.
Includes index.
ISBN 0-13-213018-1
1. Digital communications. I. Title.
TK5103.7.K55 1987
621.38'041--dc19 87-21152
CIP

Editorial/production supervision
and interior design: *Theresa A. Soler/Anne Kenney*
Cover design: *Diane Saxe*
Manufacturing buyer: *Pete Havens*
Cover photograph courtesy of NASA

A Division of Simon & Schuster
Englewood Cliffs, New Jersey 07632

Printed in the United States of America
10 9 8 7 6 5 4 3 2 1

ISBN 0-13-213018-1

Prentice-Hall International (UK) Limited, *London*
Prentice-Hall of Australia Pty. Limited, *Sydney*
Prentice-Hall Canada Inc., *Toronto*
Prentice-Hall Hispanoamericana, S.A., *Mexico*
Prentice-Hall of India Private Limited, *New Delhi*
Prentice-Hall of Japan, Inc., *Tokyo*
Simon & Schuster Asia Pte. Ltd., *Singapore*
Editora Prentice-Hall do Brasil, Ltda., *Rio de Janeiro*

In memory of my father,
Harrell Baxter Killen

CONTENTS

PREFACE xi

ACKNOWLEDGMENTS xv

1. INTRODUCTION 3

1.1 Digital Data Transmission Systems 3

1.2 Coding and Error Control 5

1.3 Digital System Classifications 8

1.4 Summary 12

2. DIGITAL COMMUNICATION CONCEPTS 13

2.0 Introduction 15

2.1 Digital Communication 15

2.2 Baseband Pulse Transmission 23

2.3 Delta Modulation 26

3. DIGITAL SIGNALING TECHNIQUES 33

3.0 Introduction 35

3.1 Receiver Evaluation Criteria 35

3.2 Digital Signal-encoding Formats 38

3.3 Digital Modulation Techniques 42
3.4 Bandwidth Efficiency Plane 71
3.5 Bandwidth Compression 73

4. ERROR-CONTROL CODING **81**

4.0 Introduction 83
4.1 Theoretical Basis for Channel Coding 83
4.2 Channel Encoding 87
4.3 Coding Principles 88
4.4 Types of Codes 91
4.5 Convolutional Codes 100
4.6 Channel Coding with Multi-Phase Signals 105

5. SPREAD-SPECTRUM TECHNIQUES **111**

5.0 Introduction 113
5.1 Principles of Spread Spectrum 113
5.2 Generation of Pseudorandom Sequences 117
5.3 Spread-spectrum Modulation 126

6. PHASE-LOCKED LOOPS **133**

6.0 Introduction 135
6.1 PLL Characteristics 135

7. DIGITAL SYNCHRONIZATION CIRCUITS **151**

7.0 Introduction 153
7.1 Carrier Synchronization 153
7.2 Bit Synchronization 164
7.3 Bit Synchronization for Random Baseband Signals 171
7.4 Frame Synchronization Techniques 182

8. DIGITAL RADIO SYSTEMS **189**

8.0 Introduction 191
8.1 Comparison of Digital Radio Modulation Techniques 192

8.2 Bandlimited Digital Radio System Design 199
8.3 System Requirements 208
8.4 Digital Radio Spectrum Measurements 210
8.5 Receiving System Design Parameters 212
8.6 Communication Link Analysis 218
8.7 Channel Simulation Test for Digital Radios 225

9. SPREAD-SPECTRUM COMMUNICATION SYSTEMS 233

9.0 Introduction 235
9.1 Processing Gain and Jamming Margin 235
9.2 Spread-spectrum Signal Reception 237
9.3 Communication System Performance 250
9.4 CDMA 256

10. OPTICAL COMMUNICATIONS 263

10.0 Introduction 265
10.1 Optical Fiber Characteristics 268
10.2 Optical Transmitters and Receivers 274
10.3 Digital Modulation and Detection Techniques 281
10.4 Pulse Modulation Techniques 285

11. SATELLITE COMMUNICATION NETWORKS 297

11.0 Introduction 299
11.1 The INTELSAT Network 300
11.2 Satellite Classes and Stationkeeping 300
11.3 TDMA Synchronization Techniques 306
11.4 System Parameters 312
11.5 System Considerations 318
11.6 Locating the Satellites 324

APPENDIX 329

INDEX 343

PREFACE

Digital Communications with Fiber Optics and Satellite Applications is intended as a one-semester introductory text in the art of digital communications. The primary objective of this text is twofold. We wish to provide both an introduction to the basic principles of digital communications and also a rather detailed discussion of the system theory that leads to the design of digital systems. The emphasis is always on the practical with applications of the theory being emphasized more than the theoretical. Students enrolled in a course in digital communications are likely to have diverse backgrounds. Consequently, the text does not require exposure to the theory of probability and random processes. We choose primarily to concentrate on the results of these fields. Detailed theoretical treatment can be added at the instructor's option.

Chapters 1 and 2 provide an overview of digital communications and an introduction to the concepts that lead to the implementation of digital systems. This discussion leads to a rather detailed discussion of digital signaling techniques, which is the subject of Chapter 3. This chapter with a discussion of modulation and demodulation techniques introduces many of the topics to which succeeding chapters are devoted. Among these are the need for error-control coding and system synchronization. Thus Chapter 4 is devoted to the subject of error-control coding. The theoretical basis for channel coding is presented. Coding principles are discussed and common codes such as block codes, cyclic codes, and convolutional codes are introduced. Decoding techniques are discussed. The generation of pseudonoise codes is introduced as a prelude to the discussion of spread-spectrum codes. Secure communications is steadily becoming a more common requirement, and Chapter 5 is devoted to spread-spectrum techniques. Techniques such as pseudonoise spreading and frequency hopping are introduced, and the generation and correlation properties of spread-spectrum codes such as the Gold codes are discussed. Chapters 6 and 7 are devoted to synchronization techniques. Since phase-locked loops play a particularly important role in system synchronization, Chapter 6 is devoted entirely to the understanding and application of phase-locked loops. Chapter 7 extends the material in Chapter 6 by introducing the hierarchy of synchronization problems encountered in digital systems. A detailed discussion of carrier and bit synchronization is presented. Techniques for implementing devices are included. This material, by necessity, is closely related to the digital modulation techniques and codes discussed in earlier chapters. The effects of noise and noise-induced design trade-offs are discussed and the complete design of a bit synchronizer is presented.

Chapters 8 through 11 are systems chapters. The previous chapters, in presenting fundamentals, also introduced a considerable amount of system material. In these chapters, this material is combined to discuss and analyze the design of complete digital systems. The design of a digital communication system begins with a choice of modulation technique. Chapter 8 addresses the design of digital radio systems. Modulation techniques are compared, and systems aspects such as fading are presented. This is followed with the design of bandlimited digital radio systems. The determination of the composite characteristics of the transmitter and receiver filters is discussed. This, in turn, is followed by a discussion of digital radio spectrum measurements needed to meet the interference and fidelity requirements as specified by the FCC. The calculation of receiver sensitivity is presented and system design trade-offs are discussed. Design techniques for the digital link are included and techniques for hardware simulation of the digital channel are presented.

Spread-spectrum techniques including code generation was introduced in Chapter 5. Chapter 9 presents a discussion of spread-spectrum systems. Systems calculations for processing gain and jamming margin are presented. Techniques for code synchronization and tracking are discussed and the performance of spread-spectrum systems is analyzed. This chapter concludes with a discussion of code division multiple access systems.

Chapters 10 and 11 are devoted to a discussion of optical communications and satellite communications, respectively. Digital techniques discussed in the text are applied extensively in these two systems. A discussion of the design of fiber optics links is presented, and digital modulation and demodulation techniques are considered. PCM intensity modulated systems are discussed as well as pulse frequency intensity modulation. To some extent, fiber optic systems and satellite communication systems are competitors. Digital communications over satellite networks is discussed in Chapter 11. In particular, details of the INTELSAT system are presented and a discussion of time-division multiple access is included. System parameters are introduced, and the design of a complete C-band satellite link is presented.

An enormous variety of specialized techniques have been developed over the years for digital communication systems. In this sense, one book cannot provide a complete understanding of the overall problem of digital system design. End-of-chapter problems are included to test and extend the student's knowledge of the subject matter. An effort was made to blend the best of industrial practice with an academic emphasis so that the text can be carried into the field by students and used as a practical reference.

The author wishes to express his appreciation to the associates and reviewers who contributed to the development of the book and to the organizations that gave permission to use technical material. Special thanks is also due to the Prentice-Hall staff for their work in producing the text.

Harold B. Killen

ACKNOWLEDGMENTS

The following figures are reprinted with permission from the September 1982 issue of *Computer Design*, Copyright 1982, PennWell Publishing Company, Advanced Technology Group: Figures 8-24 and 8-25.

The following figures are reprinted with permission from *Electronic Design*, Vol. 27, No. 25; copyright Hayden Publishing Co., Inc., 1986: Figures 5-13, 5-14, and 9-2.

The following figures are reprinted with permission from The Institute of Electrical and Electronics Engineers, Inc., E. W. Dusio et al, "Communications Satellite Software: A Tutorial Computer" April 1983; © copyright 1983 IEEE: Figures 11-1 and 11-2.

The following figure is reprinted with permission from The Institute of Electrical and Electronics Engineers, Inc., Q. K. Bennett, "INTELSAT VI—A Continuing Evolution"; *Proc. of the IEEE*; Vol. 72; No. 11; November, 1984 © copyright 1984 IEEE: Figure 11-5.

The following figure is reprinted with permission from the Institute of Electrical and Electronics Engineers, Inc., R. W. Duesing, "Software Models for the Intelsat System," *Computer*, April 1983, © copyright 1983 IEEE: Figure 11-6.

The following figures are reprinted with permission from The Institute of Electrical and Electronics Engineers, Inc., S. J. Campanella et al, "Satellite Communication Networks," *Proc. of the IEEE*, Vol. 72, No. 11, November 1984, © copyright 1984 IEEE: Figures 11-8, 11-9, 11-10, and 11-11.

The following figure is reprinted with permission from The Institute of Electrical and Electronics Engineers, Inc., C. M. Siperko, LaserNet—A Fiber Optic Intrastate Network," *IEEE Communications Magazine*, Vol. 25, No. 5, May 1985, © copyright 1985 IEEE: Figure 1-12.

The appendix is reprinted with permission from The Institute of Electrical and Electronics Engineers, Inc., W. L. Morgan, "Satellite Locations—1984," *Proc. of the IEEE*, Vol. 72, No. 11, November 1984, © copyright 1984 IEEE.

The following figures are reprinted with permission from The Institute of Electrical and Electronics Engineers, Inc., W. Stallings, "Digital Signaling Techniques," *IEEE Communications Magazine*, Vol. 22, No. 12, December 1984, © copyright 1984 IEEE: Figures 3-2 and 3-4.

The following figure is reprinted with permission from The Institute of Electrical and Electronics Engineers, Inc., R. A. Scholz, "Frame Sychronization Techniques," *Trans on Communications*, Vol. Com-28, No. 8, August 1980, © copyright 1980 IEEE: Figure 7-23.

The following figure is reprinted with permission from The Institute of Electrical and Electronics Engineers, Inc., John D. Oetting, "A Comparison of Modulation Techniques for Digital Radio," *IEEE Trans on Communications*, Vol. Com-27, No. 12, December 1979, © copyright 1979 IEEE: Figure 8-5.

The following figures are reprinted with permission from The Institute of Electrical and Electronics Engineers, Inc., H. D. Morais et al, "Bandwidth Efficiency and Probability of Error Performance of MSK and Offset QPSK Systems," *IEEE Trans on Communications*, Vol. Com-27, No. 12, December 1979, © copyright 1979 IEEE: Figures 8-6, 8-8, and 8-12.

The following figures are reprinted with permission from *Microwave Journal*, R. Moser and J. Stover, "Generation of Pseudo Random Sequences for Spread Spectrum Systems," *Microwave Journal*, May 1985, © copyright 1985 *Microwave Journal*: Figures 5-11 and 5-12.

The following is reprinted with permission from Prentice-Hall, Inc., Harold B. Killen, *Telecommunications & Data Communication System Design with Troubleshooting*, © copyright 1986, pp. 156–157, Prentice-Hall, Inc.: Pages 343 and 344.

The following is reprinted with permission from Prentice-Hall, Inc., Harold B. Killen, *Telecommunications & Data Communication System Design with Troubleshooting*, © copyright 1986, pp. 110–113, Prentice-Hall, Inc.: Pages 281, 282 and 283.

DIGITAL COMMUNICATIONS WITH FIBER OPTICS AND SATELLITE APPLICATIONS

1

INTRODUCTION

1.0 INTRODUCTION

Knowledge may be considered to be the end product of thought and information. We may or may not be able to measure the increase in human thinking, but the increase in the production, dissemination, and processing of information is a highly visible fact. In fact, in the interval since World War II this increase has been so rapid we are witness to a virtual information explosion. Central to this phenomenon is the fact that information attains its maximum utility only when it moves. In organized societies, this has resulted in a communication explosion. The need to transmit data and the development of integrated circuits proved to be a resoundingly successful combination, each spurred on by the other. With microprocessors and digital transmission systems multiplying, the growth of digital communications technology is undergoing revolutionary changes. These changes are being brought on by advances in technology such as high-speed analog-to-digital convertors and in digital processing theory. Coupled to this is a requirement for both data security and reliability. The end result is that future communication systems will become more and more digital. In this text we examine the principles and techniques of digital communication systems. Fundamental principles are presented first. This material leads to system applications along with techniques for implementing complete communication systems.

1.1 DIGITAL DATA TRANSMISSION SYSTEMS

The elements of a generalized communication system are shown in Figure 1.1. Note that a communication system consists of an information source and a receiver connected by a data transmission system. Information generated by the source is directed to the transmitting terminal, where it is encoded into a form suitable for transmission over the transmission path. The encoding process can take many forms and is treated in detail in the text. At this point, we simply note that an encoding process is required.

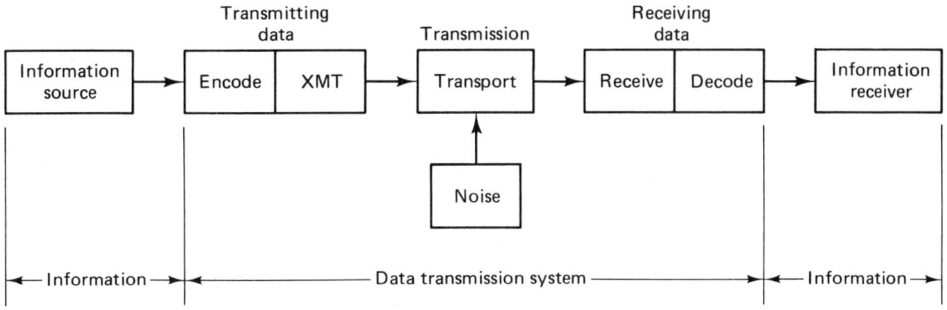

Figure 1.1 Generalized communication system.

The transmitter next generates the appropriate electrical or electromagnetic signal and applies it to the transmission path. The function of the transmission path is to carry this signal from the transmitter to the receiver. It may consist of a pair of wires, coaxial line, free space, and so on. During transmission, the transmitted signal has noise added to it. Thus at the input to the receiver, both the signal and noise exists. One of the primary functions of the receiver is to read the incoming signal correctly in the presence of the noise. The signal that is read is decoded, and the output is a representation of the original information sent.

Obviously, from the preceding discussion, we are interested in encoding and decoding techniques. A little thought will allow us to generate encoding schemes. For example, a voltage level may be used to represent a letter of the alphabet. Or a particular voltage level might represent a group of letters such as a word. Transmitting a symbol that represents an entire word would appear to be more efficient than transmitting a symbol that represents only a single letter. We begin to see the possibilities of encoding. It is instructive to approach the problem on a more scientific basis.

We noted earlier that the receiver must decode the signal to extract the original information. That is, it must continually choose the correct answer from a number of possible answers. Consider, for example, a teletypewriter. If we neglect numbers and punctuation symbols, the correct answer is always 1 letter out of the 26-letter alphabet or a space, for a choice of 1 of 27 possibilities.

We can assign to each letter a number that represents its information content. A logical basis is the number of choices that the receiver must make. In this case, the information value of each letter is 27. Suppose, however, that we choose to transmit 2-letter combinations. The number of possible choices is now 27^2. For 3-letter combinations, the number rises to 27^3. We see that this approach rapidly leads to large values. Also, a little thought will convince us that the actual information content is not increasing in such an exponential fashion. It seems more reasonable, for example, that transmitting 2 letters is equivalent to twice the information transmitted by a single letter; so in our example, the information content is 54 rather than 27^2. This indicates that the information content varies according to a logarithmic scale rather than an arithmetic scale. Based on these concepts, the formal definition of the quantity of information, H, transmitted is defined to be

$$H = \log N \tag{1.1}$$

where N is the number of equally probable choices.

Note that in this equation we have not assigned specific units for the information content, H. Equation (1.1) was arrived at through the concept of choice. The most simple and basic choice that can be made is a choice of 1 out of 2. That is, something either exists or it does not. The value of N is then fixed to be equal to 2. If a value of 1 is desired as our basic unit of information, the base of the logarithm must also be 2. We can rewrite equation 1.1 as

$$H = \log_2 N \quad \text{bits} \tag{1.2}$$

TABLE 1.1 INFORMATION VALUES OF LETTERS AND LETTER COMBINATIONS FOR FOUR ALPHABETS

Letter or combination	Information (Bits) for 2-letter alphabet	3-letter alphabet	4-letter alphabet	27-letter alphabet
1	1	1.58	2	4.76
2	2	3.16	4	9.52
3	3	4.75	6	14.28

where the quantity of information, H, is expressed in bits and N is again the number of equally probable choices. Using equation (1.2), the information values for different letter alphabets is tabulated in Table 1.1.

Referring to this table, we note that the information content of letter combinations now vary in a logical fashion. It is interesting to note that the 1-out-of-2 choice used as the basic unit represents a method of counting that is based on 2, that is, a binary system. In fact the term *bit* is an abbreviation for binary digit. Hence this relates directly to digital electronics.

1.2 CODING AND ERROR CONTROL

Refer to Figure 1.1. We can define coding as the rearrangement of the incident data into some other format at the transmitter in accordance with a set of predetermined conventions. The receiver performs the inverse of this to recover the original data. Coding is undertaken for three primary reasons:

- To increase the information capacity of the system.
- To enable error detection and possible error correction to be performed.
- To alleviate transmission problems.

Based on a purely theoretical approach, all three reasons should lead to the same coding scheme. This one ideal coding scheme, however, has not been realized (or even approached for that matter) and, consequently, many different coding schemes exist.

It does not require a great amount of imagination to realize that the information capacity of the digital system must be interrelated with both the coding and the transmission path. Obviously, the speed of signaling that can be realized over the path is a key factor in the information capacity of the system. The transmission-path characteristics required to transmit enough information to the receiver for it to perform its required regeneration function for an incident signal of N elements per second of varying amplitudes was investigated by H. Nyquist in 1928. Nyquist was interested primarily in telegraph transmission. His results, however, were quite general, and still influence transmission thinking of today. In his analysis, he postulated certain known distortions of an incident rectangular signal that would still allow it to be recognized at the receiver. He then evaluated the bandwidth and band shaping required.

Nyquist evaluated the transmission requirements for three criteria. Many others have been evaluated since, particularly for free space paths. The one of interest to us at present is illustrated in Figure 1.2. With this particular criteria, the transmission path is arranged for correct reproduction of the incident signal amplitude at the exact center (sampling time) of each element, t_1, t_2, t_3, and so on. Note in Figure 1.2(b), the transmission path has amplitude, a, from 0 frequency up to a sharp cutoff at $f_c = 1/2T_1$. The phase shift, b, is linear over the same frequency range. If the pulse repetition rate, T_1, is 1 ms, we require a transmission path bandwidth, f_c, of 500 Hz. In words, Nyquist proved that to transmit N signal elements per second, a bandwidth of $N/2$ Hz was theoretically sufficient. This bandwidth is called the *Nyquist bandwidth* in his honor.

The transmission path illustrated cannot necessarily be realized. Transmit-

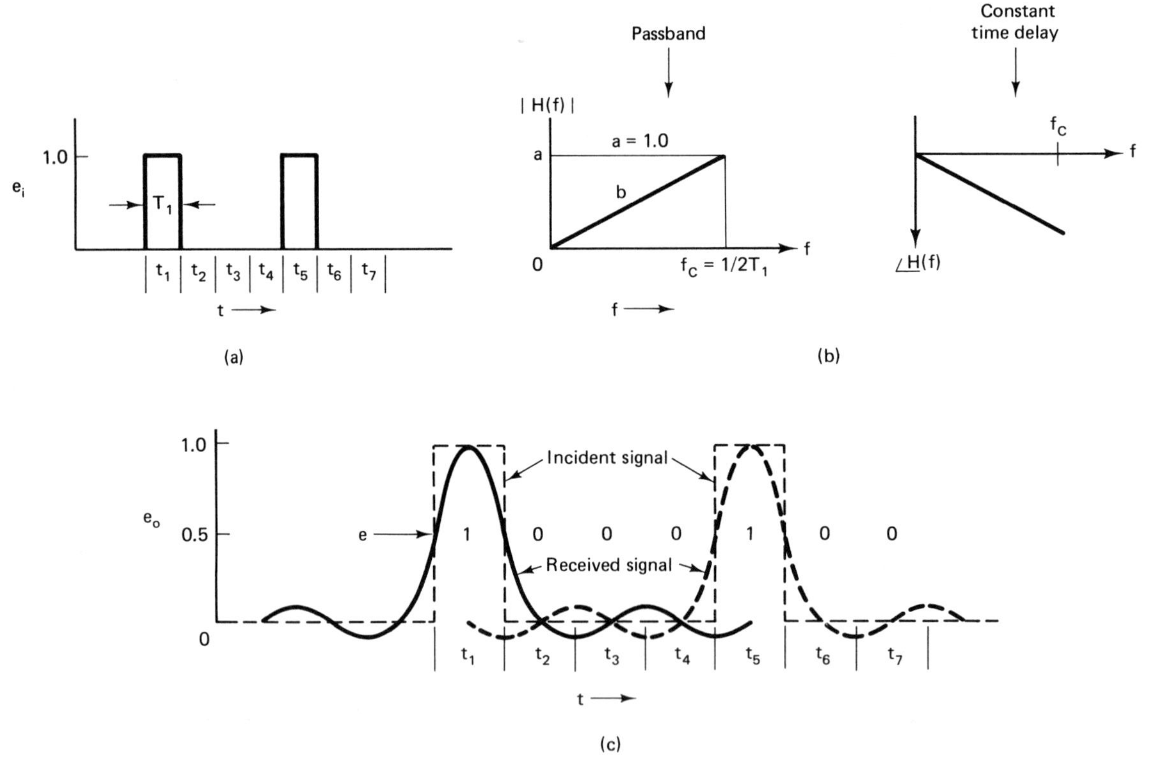

Figure 1.2 The problem of "how to get there." (a) Input data signal. (b) Transmission path. (c) Output data signal.

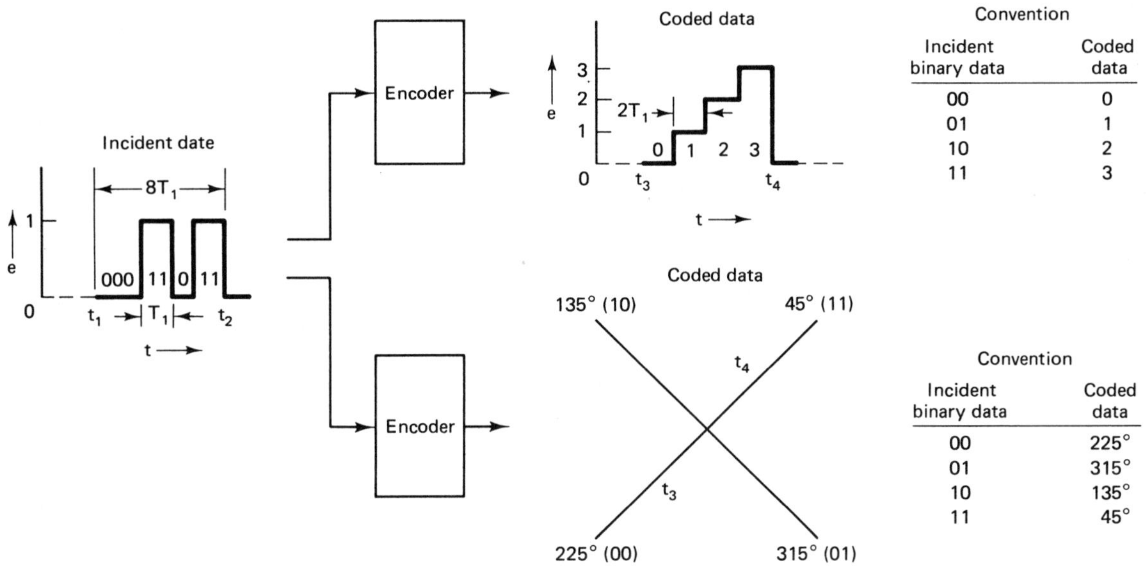

Figure 1.3 Representation of two coding schemes and their conventions.

ting digital signals over physical lines results in considerably poorer performance. A bandwidth on the order of the pulse repetition rate is needed. If we are to improve the system's information capacity, we must, out of necessity, try to make each signal element carry more information. The amount of information can be increased by coding.

A simple coding scheme is illustrated in Figure 1.3. An 8-bit sequence of digital data is shown incident to the encoders. The encoders examine each pair of bits, starting at t_1, and generate a signal in accordance with the conventions at the immediate right. Note that these conventions transform from a radix 2 (binary) numbering system to a radix 10 (decimal) system. The uppermost encoder output is 0123. The actual symbol transmitted is the multilevel voltage waveform. Each voltage level represents 2 binary bits. We can represent the output of the lower encoder as a sinusoidal waveform such as $E \sin(\omega t + \phi_i)$, where ϕ_i takes on values of 225°, 315°, 135°, and 45° for the incident digital data.

The multilevel-amplitude baseband signal is not common in present-day systems, but the phase-shift method or a combined carrier-amplitude–phase-shift method is widely used in today's digital systems and is referred to as quadraphase modulation (QAM). If we increase the number of possible phases, we can increase the system's information capacity. For example, in an 8-phase system, each phase represents 3 bits. Systems employing 64 phases are in use today. This increased information capacity, however, is not gained without paying a price. It is obvious from Figure 1.3 that as the number of phases increase (become closer together), the receiver must make a more difficult choice. Noise introduced into the system results in phase jitter. The received phase varies in a statistical manner. To reduce this effect, the signal must be made even greater than the noise as the number of phases increase. That is, the ratio of signal to noise must become larger.

In 1948, C. E. Shannon developed a comprehensive theory that indicated the information capacity that could be obtained with optimum coding in a band-limited channel with noise. This result expressed in equation form is

$$H = W_I \log_2 \left(\frac{S}{N} + 1 \right) \tag{1.3}$$

where

H = information capacity in bits per second
W_I = channel information bandwidth in hertz
S = average power of the signal
N = average power of the noise.

Shannon's optimum coding ensures that the errors in transmission can be made vanishingly small and also that the signal power will be as large as is practical, since it is uniformly distributed over the available bandwidth. This, then, is the ideal but unobtainable coding mentioned earlier. Equation (1.3) can be used to determine the penalty paid by not using the optimum code. We return to this subject in Chapter 3 when we compare digital modulation techniques and in Chapter 4 when we discuss the theoretical basis for coding.

Coding can also be used to permit the detection and, perhaps, the correction of errors at the receiver. The simplest and most widely used type of coding for error detection is parity coding. In this technique, messages are coded before transmission so that there is always an odd or even number of 1s in each character in the transmitted message. This error-detection scheme is referred to simply as *odd* or *even parity*. No means is available to correct the error. The system fails when multiple errors occur. With more complex coding arrangements, both error detection and correction are possible. In essence, sufficient information must be transmitted to permit reconstruction of the correct message. The disadvantage is that the additional information sent uses part of the available bandwidth. In modern-day digital communication systems coding theory and modulation theory are merging and becoming less recognizable as separate entities. The phase-shift scheme in Figure 1.3 is an illustration of this trend.

1.3 DIGITAL SYSTEM CLASSIFICATIONS

Consider once again the system in Figure 1.1. It is obvious that when transmitting digital data, the receiver and transmitter must, in some manner, be kept "in step." That is, the receiver must know when to look. The receiver must be synchronized with the transmitter. This problem is illustrated in Figure 1.4. Ideally, in a system

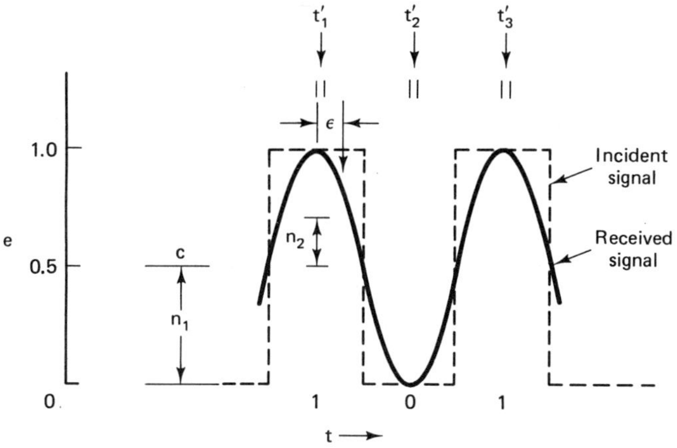

Figure 1.4 The problem of "when to look."

such as this, where the Nyquist bandwidth is provided, the receiver should sample at times t_1', t_2', t_3', and so on. An example illustrates the situation.

EXAMPLE 1.1

Refer to Figure 1.4. What signal-to-noise ratio (SNR) is required if the receiver is to make a correct decision (a) at time t_1 and (b) at time $t_1 + \epsilon$, $n_2 = 0.2$.

Solution

(a) If the incoming signal is above level C, a 1 is indicated. Below this level, a 0 is indicated. The receiver will work correctly as long as the noise, n_1, does not exceed this level. For $C = 0.5$, the SNR required at t_1' is

$$20 \log_{10} \frac{1}{n_1} = 20 \log \frac{1}{0.5} = +6 \text{ dB}$$

(b) $\text{SNR} = 20 \log \frac{1}{n_2} = 20 \log \frac{1}{0.2} = 14 \text{ dB} \qquad \text{at } t = t_1' + \epsilon$

An increase of 8 dB is required for the second situation. This is a penalty so severe that it cannot be allowed to happen. Sufficient information must be provided in the incoming data to allow the receiver to be synchronized with a time error smaller than the ϵ shown in Figure 1.4. This is not a trivial problem, and considerable design effort is centered around overcoming this problem. The timing feature typically results in digital systems being divided into two classes: synchronous and asynchronous. In the following sections, we discuss the characteristics of the two system classifications.

1.3.1 Asynchronous Systems

Generally speaking, asynchronous data comes in spurts. That is, the typical asynchronous system transmits data a character or a word at a time with no coherent clock between words. This contrasts with a synchronous system, which transmits data for longer periods in accordance with a reference clock. The most common asynchronous data format is shown in Figure 1.5. Note that an 8-bit word is preceded by a *start* bit and is then followed by one or two *stop* bits. Thus we see that each word's timing reference is independent of the reference for preceding or succeeding words. By convention the input signal is held in a *marking* condition until a data message is transmitted. The transition from the mark to the space state identifies the start bit. The locations of the data bits are referenced to this time.

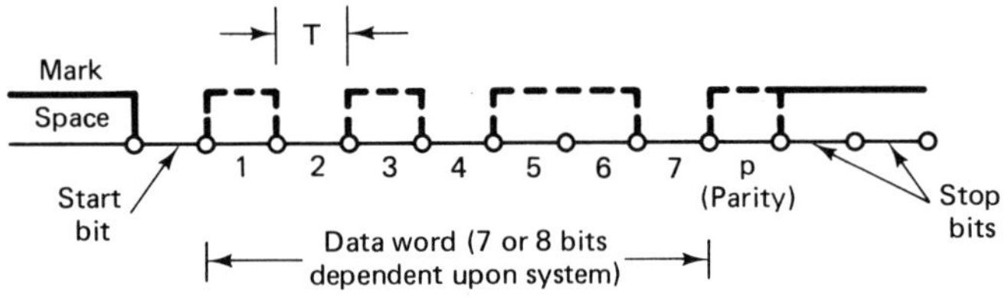

Figure 1.5 Asynchronous transmissions do not have a coherent clock. A start signal that prepares the receiver for reception precedes each group of code elements, whereas a stop signal follows the word to prepare the receiver for the next character.

Now that we have the timing diagram, how do we drive a clock from it? The clock circuit design for a 1-word message is shown in Figure 1.6. Refer to Figure 1.5. Note that signal transitions occur only when data bits change from one state to the other. The start bit must initiate a local clock reference. The design in Figure 1.6, then, must provide a local clock whose phase can be adjusted to match the signal (symbol) transition times.

In operation, the free-running astable multivibrator is tuned to oscillate at a frequency slightly lower than the symbol repetition rate. The mark-space detector produces an output pulse (sync) near the midbit period of the start bit. This

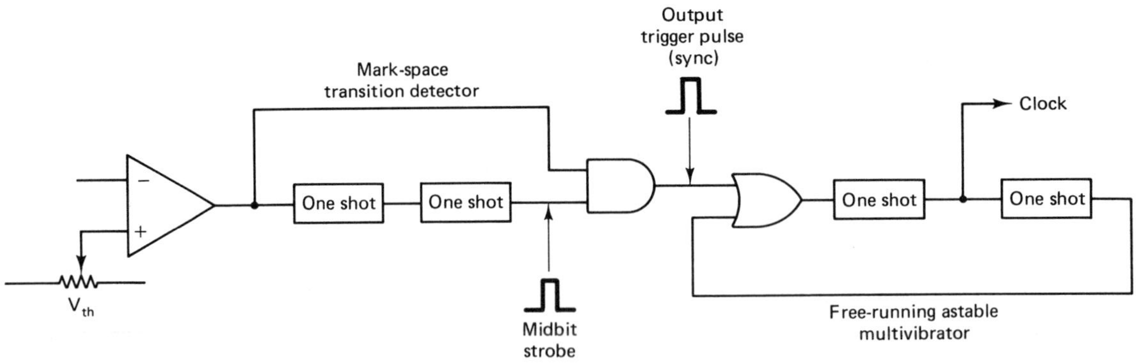

Figure 1.6 Each time a transition from mark to space occurs, the start-bit detection logic retriggers the multivibrator, thus keeping the local clock in step.

trigger pulse resets the multivibrator so that succeeding output clock pulses occur at approximately midbit time. Each time the data changes from mark to space, a sync pulse is generated that keeps the local clock in sync. Between transitions, the multivibrator runs free at its preset frequency. After receiving the first valid start bit, a counter may be enabled to produce output strobes corresponding to the stop bits (end-of-message detect). Obviously, this solution suffers from frequency drift between sync pulses. It is, however, inexpensive. The regeneration of data using synchronized systems is discussed extensively in Chapter 7. In the next section, we consider the elements of a synchronous system.

1.3.2 Synchronous Systems

A synchronous data transmission system can be defined as one that transmits data at a constant speed. As with the asynchronous system, the data must provide a means of extracting a local clock, that is, a self-clocking code may be used. Consider the circuit shown in Figure 1.7. A bipolar *return-to-zero* (RZ) code is used. The signal returns to zero at the midpoint for either a logic 1 or logic 0. A clock can be extracted by first full-wave rectifying the signal and then filtering to obtain the dc level. Refer to Figure 1.7. The dc level is used to establish a threshold for the comparitors. Applying the RZ signal to this circuit then produces

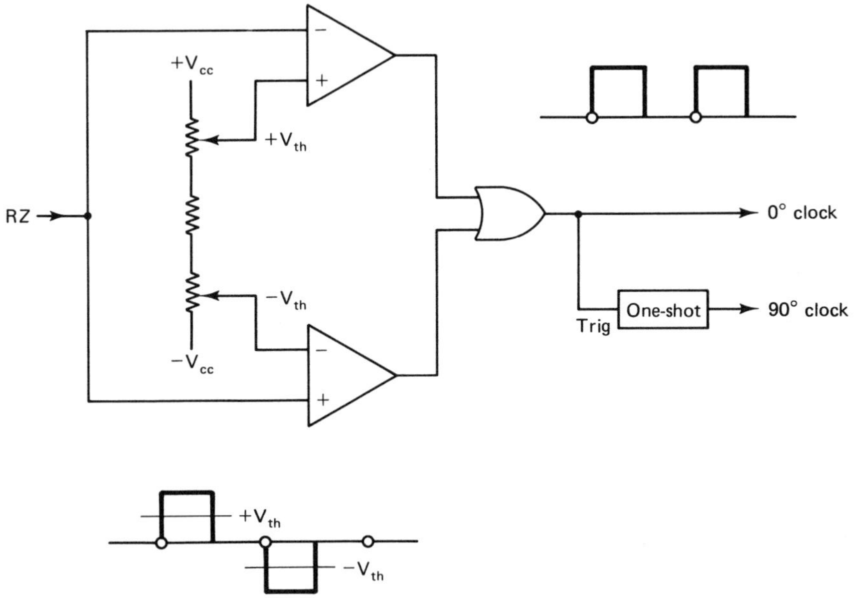

Figure 1.7 Synchronous systems can extract the timing clock directly from the data. The received signal is rectified and limited about a bias, thus producing a bit-rate clock signal.

a clock pulse every bit period. In the absence of noise, this clock can be used dirrectly.

The effect of a noisy signal is to produce a clock with jitter. This happens because the noise displaces the signal at the threshold point. Since we cannot eliminate noise in practical digital communication systems we must develop techniques to deal with it. In succeeding chapters the basics of noise processes are presented to provide the necessary background for discussing the synchronization problem in more detail.

1.4 SUMMARY

This chapter presented an overview of the elements of digital communication coding and transmission systems. This phase of communications provides the foundation for today's modern telecommunication systems. In the following chapters we will become familiar with topics such as digital radios, digital signal processing, and the newer encoding schemes such as spread spectrums. The emphasis throughout the text is on the application of techniques rather than purely theoretical. It is estimated that by the 1990s most communication systems will be digital. I invite you to join me in this study.

2

DIGITAL COMMUNICATION CONCEPTS

2.0 INTRODUCTION

When information is represented by a finite set of discrete values, the result is referred to as a digital signal. Digital data can occur in one of two ways. We can, for example, *sample* an analog signal and then digitize the resulting pulse so that it is represented by a digital word. This operation—converting to digital—is called *quantization*. Alternatively, the data may simply occur in digital form. An example is a sequence of numbers representing the time various events occurred. If we choose to replace this set of numbers by another set of symbols, we introduce an additional process, that of *encoding*.

At this point we have introduced three operations: sampling, quantizing, and encoding. These operations are illustrated in Figure 2.1 and together constitute a process known as pulse code modulation (PCM). Referring to this figure we see that the signal is consecutively sampled, quantized, and encoded before being applied to the channel for transmission. To complete the system, we need a receiver (Figure 2.2). The output of the decoder is the sampled signal and the low-

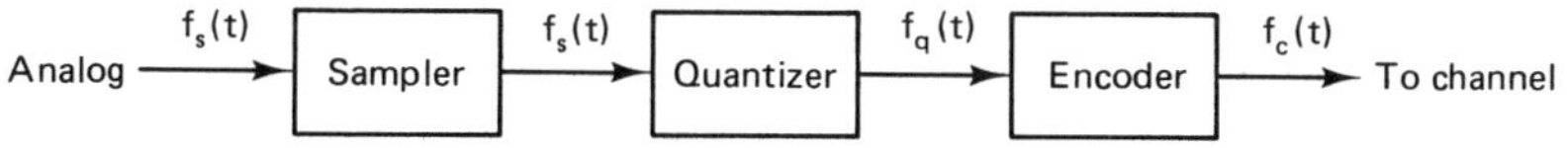

Figure 2.1 Block diagram of a PCM XMTR.

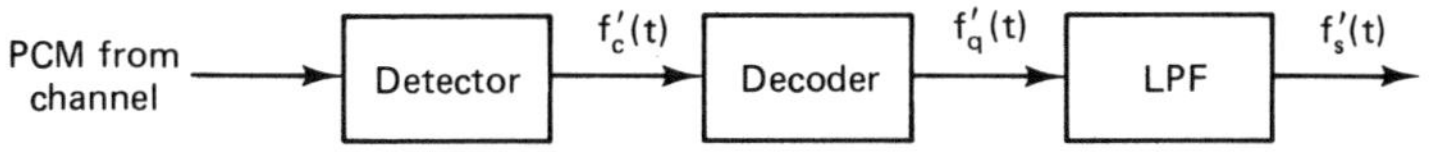

Figure 2.2 Block diagram of a PCM PCVR.

pass filter output is the original analog signal. The transmission of information in this fashion by PCM is generally referred to as *digital communication*. In the following sections we examine the basics of this process.

2.1 DIGITAL COMMUNICATION

Digital communication using PCM entered the field of electronic communications during World War II. Since then it has matured and, although challenged in some areas, enjoys supremacy in data communications. A code group $(A_0 \ldots A_n)_2$ of binary digits ($A_i = 1$ or 0) can be used to represent the number

$$N = A_0 + A_1 \cdot 2 + A_2 \cdot 2^2 + \cdots + A_n \cdot 2^n \tag{2.1}$$

We use this number in PCM to represent the value of a signal voltage at a given instant of time. An instant of time later, corresponding to one sampling interval, a second code group represents the new signal amplitude, and so on. These code

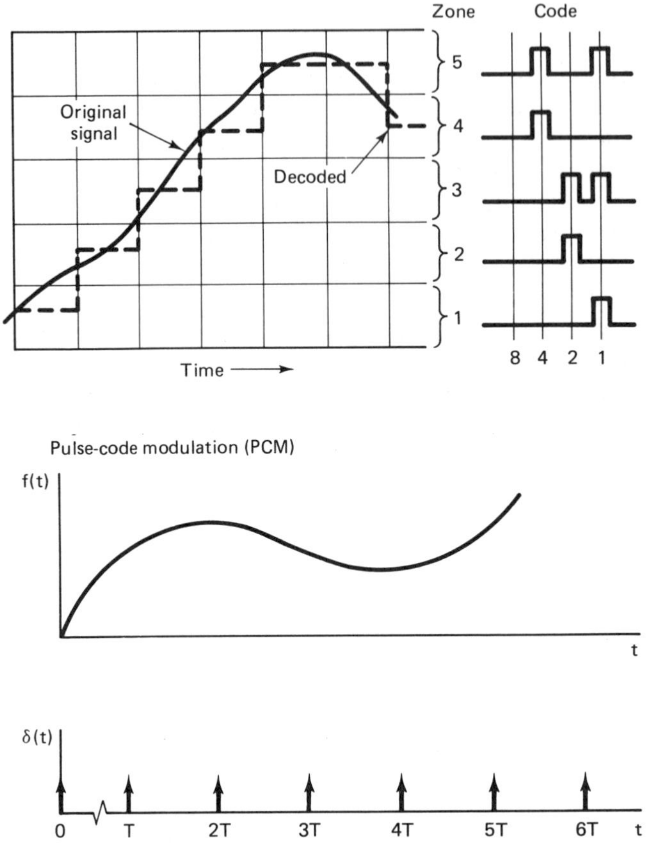

Figure 2.3 Sampling process.

groups are transmitted as a digital pulse train and at the receiver are decoded into a voltage proportional to the binary number.

This process is indicated in Figure 2.3. The continuous signal is sampled at the instant shown and coded into the binary number at the right. The number represents the signal amplitude to an accuracy of plus or minus $\frac{1}{2}$ of the least significant bit ($\pm\frac{1}{2}$ LSB). If the amplitude falls anywhere in a given zone, it is coded into the binary word assigned to that zone. Note that this results in a staircase appearance for the decoded signal. From inspection, we see that decreasing the sampling interval improves the accuracy of the decoded signal. The difference between the decoded signal and the original signal is referred to as the *quantizing error.*

2.1.1 Quantizing Noise

The quantizing error effectively introduces noise into the reconstructed signal. This happens because we are representing a continuous signal by a finite number of codes. The relative value of this noise may be approximated by examining the

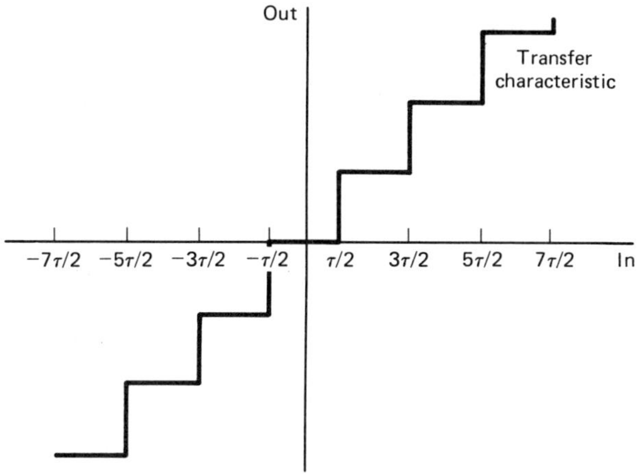

(a)

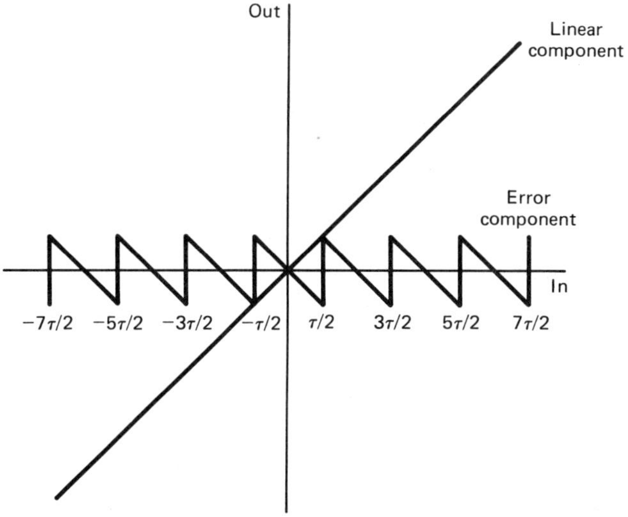

(b)

Figure 2.4 (a) Transfer characteristics. (b) Linear and error components.

transfer characteristics of a typical coder-decoder link (Figure 2.4). This figure divides the characteristics into linear and error components. The difference in amplitude between the sampled signal value and the quantized level is the *quantization error*. For an n-digit code there are 2^n levels; thus the peak error magnitude is

$$\frac{1}{2} \cdot \frac{1}{2^n - 1}$$

The rms error can be computed for the error level as follows:

$$V_{N,rms} = [\Delta V_s] \left[\frac{1}{\tau} \int_{-\tau/2}^{\tau/2} \left(\frac{1}{2^n - 1} \frac{1}{\tau} t \right)^2 dt \right]^{1/2} = [\Delta V_s] \frac{1}{\sqrt{12}} \frac{1}{2^n - 1} \tag{2.2}$$

where V_s is the total digitizing range.

EXAMPLE 2.1

A 5-V peak-to-peak signal is sampled and digitized to an accuracy of 5 bits. (a) What is the quantizing rms error voltage? (b) What is the signal-to-quantizing-noise ratio?

Solution

(a) $\text{rms error} = (5)\left(\frac{1}{\sqrt{12}} \frac{1}{2^5 - 1}\right) = 0.0451$

(b)

$$\frac{S}{N} = \begin{cases} \left[\dfrac{5/2\sqrt{2}}{0.0451}\right]^2 = 1536.4 \Rightarrow 31.865 \text{ dB}; & \text{sinusoidal signal, } V_{s,rms} = \dfrac{\Delta V_s}{2\sqrt{2}} \\ \left[\dfrac{\frac{5}{8}}{0.0451}\right] = 192.05 \Rightarrow 22.834 \text{ dB}; & \text{Gaussian signal, } V_{s,rms} = \dfrac{\Delta V_s}{8} \end{cases}$$

where $V_s = 8\sigma_s$.

2.1.2 Companding

We showed in the previous section that the rms quantizing error is on the order of $1/\sqrt{12}$ times the step size. Consequently, a weak signal that does not utilize the full range of codes will have a poorer signal-to-quantizing-noise ratio than the stronger signals. For speech, the result is an intolerable SNR for a person who speaks weakly. A technique known as *companding* may be used to solve this problem (see Figure 2.5). Compression takes place at the transmitter and the signal is expanded to its original form at the receiver.

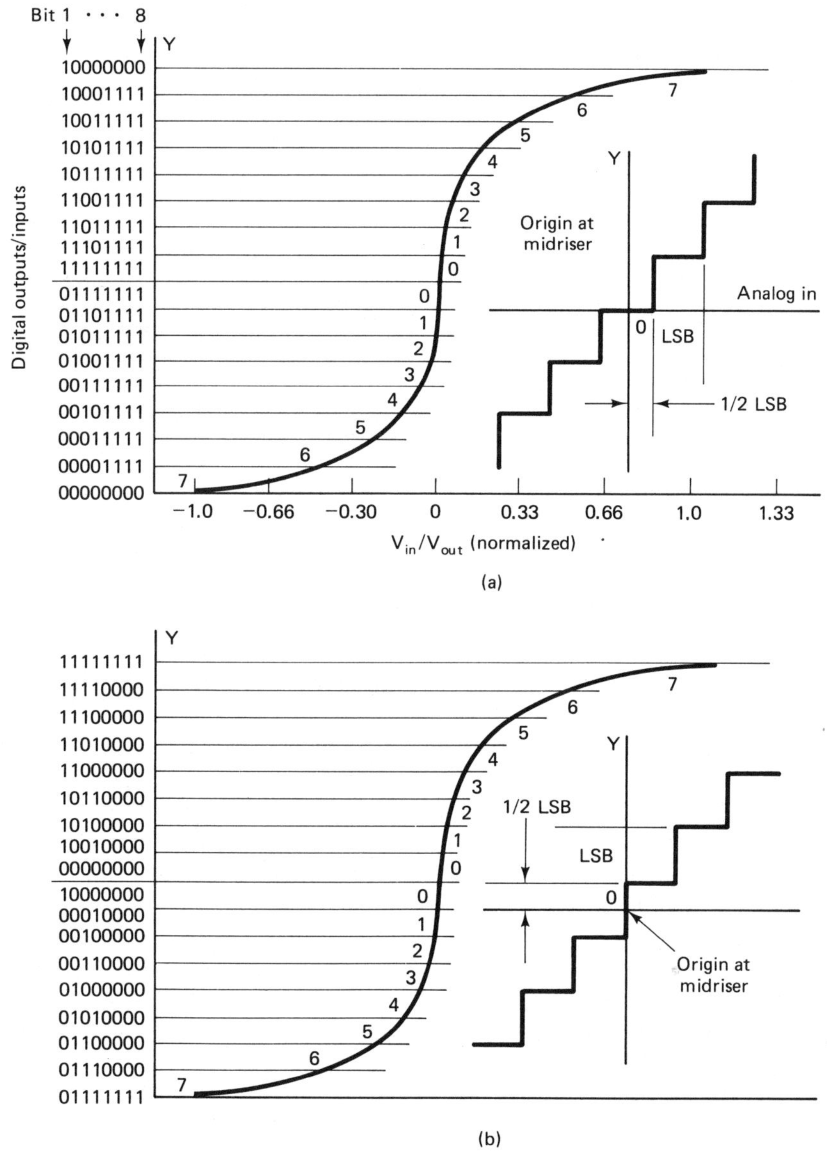

Figure 2.5 (a) μ-255 law. (b) A-law.

One of the most common compression characteristics has the form

$$E_{out} = \frac{\ln(1 + \mu E_{in})}{\ln(1 + \mu)} \tag{2.3}$$

where E_{in} is the normalized sample amplitude at the compressor input. E_{out} is the normalized compressed output. μ is a parameter chosen to improve the signal/

quantizing ratio of weaker signals at the expense of stronger ones. Telephone signals are compressed throughout the world. The Bell System compands with the μ-255 law, whereas much of the rest of the world uses the A-87.6 law. Both methods increase the dynamic range.

The μ-law, when normalized for a coding range of ± 1, is

$$Y(x) = \operatorname{sgn}(x) \frac{\ln(1 + \mu/|x|)}{\ln(1 + \mu)} \tag{2.4}$$

for x falling between ± 1. In this expression $\mu = 255$. This gives rise to the Bell System's μ-255 law. The European standard for companding is the A-law. It requires two equations to define this law. For the absolute value of x, $|x|$ greater than $1/A$,

$$Y(x) = \operatorname{sgn}(x) \frac{1 + \log A|x|}{1 + \log A} \tag{2.5}$$

For $|x|$ between 0 and $1/A$,

$$Y(x) = \operatorname{sgn}(x) \frac{A|x|}{1 + \log A} \tag{2.6}$$

The variable A is equal to 87.6.

Coders-decoders (codecs) approximate the smooth analog curves of Figure 2.5 with straight-line segments. In the most popular approximation for both laws, the first of each word's 8 bits defines its segment's polarity (sign bit). The following 3 bits determine one of eight major segments (chords). The last 4 bits divide each major segment into 32 steps. These steps are uniform within each chord, but the step size increases as the chords progress away from zero. In the longest two of the μ-law chords, the step sizes are 128 times greater than those in chord 0.

The differences between μ and A-law coding are most apparent for weak signals. The transfer characteristics show that the μ-law specifies two codes for 0 V, thereby giving the completely quiet state a midtread position on the amplitude staircase waveform. On the other hand, the A-law characteristic shows no quiet codes, and so the quiet state interpolates to a midrise position. This first A-law chord breaks into 64 separate divisions.

2.1.3 Pulse Regeneration

PCM reduces any given signal to a random binary pulse stream. The pulse pattern retains the original information. Consequently, the signals all look alike and are treated alike by the transmission path. Using time-domain multiplexing (TDM), the pulse train may contain voice, video, telemetry, and so on. These are advantages that relate to PCM. A much stronger advantage, however, lies in the principle of pulse regeneration.

We have the option of regenerating the pulse after it has passed through a noisy bandlimited medium (Figure 2.6). Note that the pulses out of the cable have

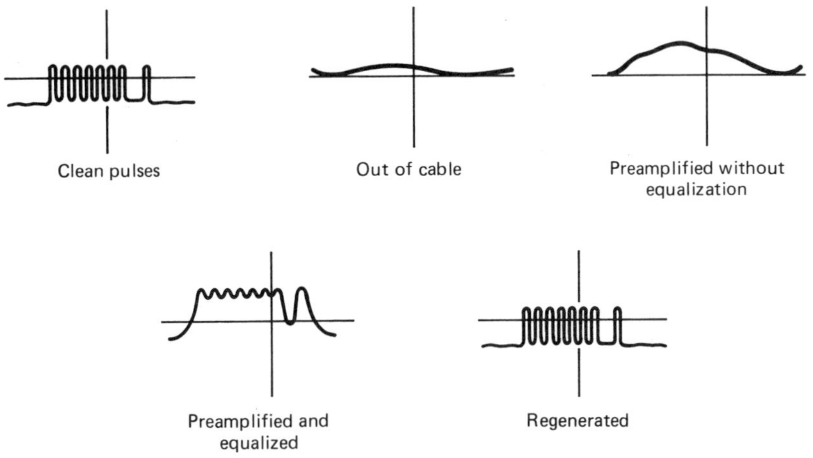

Figure 2.6 PCM regeneration principle.

little high-frequency content. Frequency-shaped amplification produces equalized and amplified pulses that are confined to their own time slots. A regenerator then looks at each time slot, and if it finds a voltage greater than half peak-pulse height, generates a new clean output pulse. Recall from Chapter 1 that this implies an SNR of at least 6 dB.

2.1.4 Sampling Theorem

In Chapter 1 we observed one of the basic results of Nyquist's work was that a signal of N-hertz bandwidth required a transmission path of at least $\frac{N}{2}$-hertz bandwidth for transmission. Nyquist further showed that if a signal is sampled, the sampling rate must be at least twice the highest frequency contained in the signal. This is necessary to ensure that the original signal can be reconstructed from the samples. Refer to Figure 2.7. In Figure 2.7a we see that with some smoothing the original signal could be reconstructed by first connecting the maximum values of the samples. In Figure 2.7b (see page 22), the signal is undersampled, and the recovered signal is a sine wave of lower frequency. That is, the original frequency has been folded over to a lower frequency. This result is referred to as *aliasing*. The sampling rate, then, is a primary system consideration in any data acquisition system.

Let the spectrum of the original analog signal be represented by $F(j\omega)$. If the original signal, $F(j\omega)$, is bandlimited so that it does not contain frequencies greater than ω_m, the spectrum $F^*(j\omega)$, of the sampled signal is as shown in Figure 2.8 (see page 22). We see that a sampled signal contains all the original frequencies and, in addition, contains modulation products wherein the original signal appears as sidebands about the sampling frequency and all harmonics of the sampling

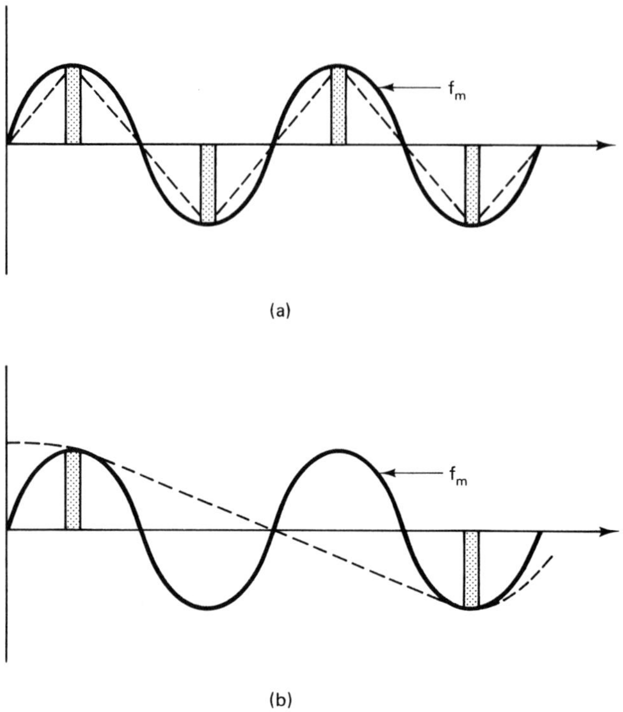

Figure 2.7 Illustration of sampling theorem.

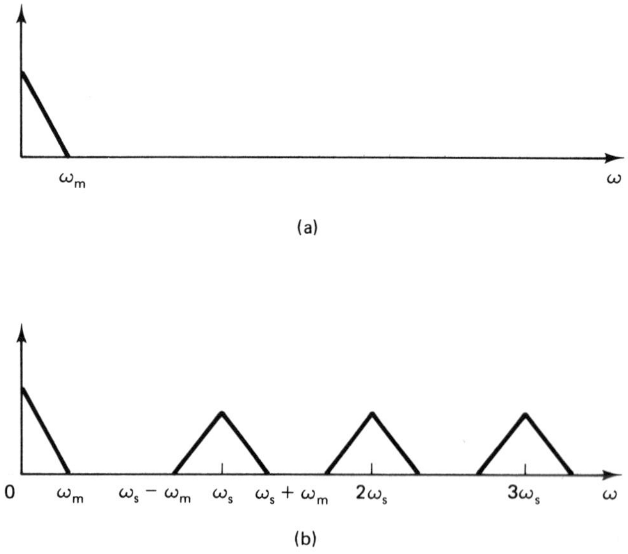

Figure 2.8 (a) Spectrum of original signal. (b) Spectrum of sampled signal.

frequency. The spectrum of the sampled signal actually possesses a sin x/x envelope and thus, the energy falls off at the higher harmonics.

Note from Figure 2.8 that if $\omega_s > 2\omega_m$, a *guard space* exists between the original signal component and the lower sideband of the sampling frequency, ω_s. If the guard space is preserved, the original signal may be recovered by a low-pass filter (see Figure 2.2). Thus, to the extent that a perfect filter can be built, the signal can be recovered with no degradation.

2.2 BASEBAND PULSE TRANSMISSION

The manner in which a pulse train is formed for transmission is determined by the characteristics of the medium. For example, if dc transmission or dc restoration is difficult, it is preferable to transmit a wave with no dc component. If high-frequency interchannel cross talk is a problem, a wave without excessive energy at the upper portion of the spectrum is preferable.

A number of pulse shapes are available for baseband transmission. A number of variations with their spectra are illustrated in Figure 2.9. Note that all these pulse trains represent the digital word 1011. The first two are unipolar with 50 percent and 100 percent duty cycle, respectively. Both have a continuous and dc component in the frequency spectrum. The spectrum of the 100 percent duty cycle pulse is zero at the pulse repetition frequency, f_r. Neither of the bipolar pulses have a dc component. The spectra of the pulses are important in designing pulse regeneration equipment. For example, a timing component may be added at the spectra nulls in (b) and (c). This component may be extracted by the regenerator to serve as a clock.

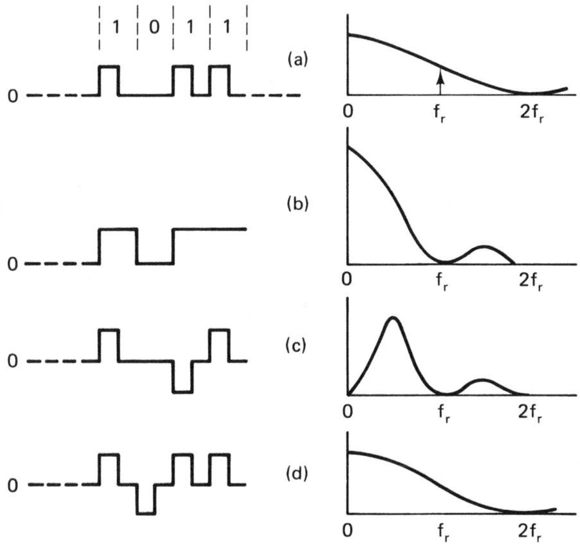

Figure 2.9 Spectra of various random pulse trains. (a) Unipolar 50 percent. (b) Unipolar 100 percent. (c) Bipolar 50 percent. (d) Bipolar 50 percent.

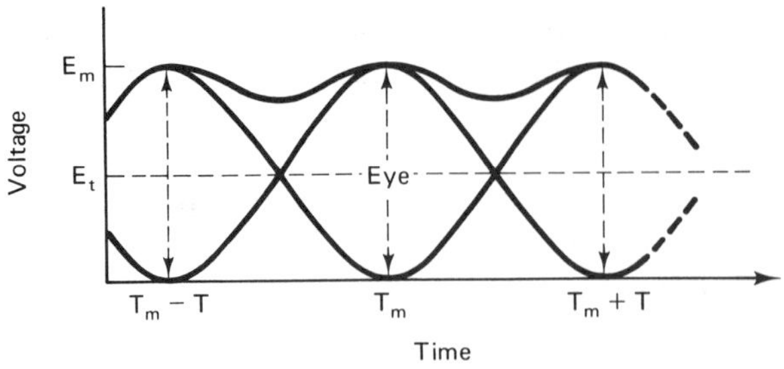

Figure 2.10 Eye diagram for a random pulse train.

The effect of intersymbol influence due to transmission over a line or through a filter may be evaluated using an *eye pattern*. An eye diagram for a random pulse train is sketched in Figure 2.10. This pattern may be obtained from an oscilloscope if the pulse output from the line is fed to the vertical input of the scope and the sampling clock $C(t - nT_m)$ is fed to the external trigger of the scope. The horizontal time base is set approximately to the symbol duration. The sampling clock occurs at the center of the pulse and thus the pulse should reach its maximum height at this time. Whether or not the distorted pulse from the line reaches its maximum height at the sampling time depends on the value of the previous bits. An imperfect filter on line will introduce intersymbol interference (ISI). If, for example, the pulse reaches only 90 percent of its height, the 90 percent eye opening corresponds to $[-20 \log_{10} 90/100] = 0.9$ dB ISI degradation.

2.2.1 Bandwidth Requirements

We noted in Chapter 1 that it is theoretically possible to transmit a pulse train of repetition frequency f_r over a channel bandwidth $\frac{1}{2}f_r$ (see Figure 1.2), which is the Nyquist bandwidth. The received pulse is of the form sin x/x, with zeros of the tails occuring at multiples of a time slot. This theoretically leads to no ISI but to an eye that requires perfect clock stability and zero clock sample width.

EXAMPLE 2.1

It is desired to transmit speech over a PCM channel with 8-bit accuracy. Assume the speech is baseband limited to 4 kHz. Determine the minimum bandwidth (BW_{min}) required for the PCM signal.

Solution

$$\text{Bit rate} = 2 \times 4000 \times 8 = 64{,}000 \text{ bits/s}$$

$$BW_{min} = \frac{1}{T} = \tfrac{1}{2}f_r = 32{,}000 \text{ Hz}$$

Example 2.1 illustrates the calculation of PCM bandwidth needed under the ideal conditions postulated by Nyquist. A more practical number is arrived at by

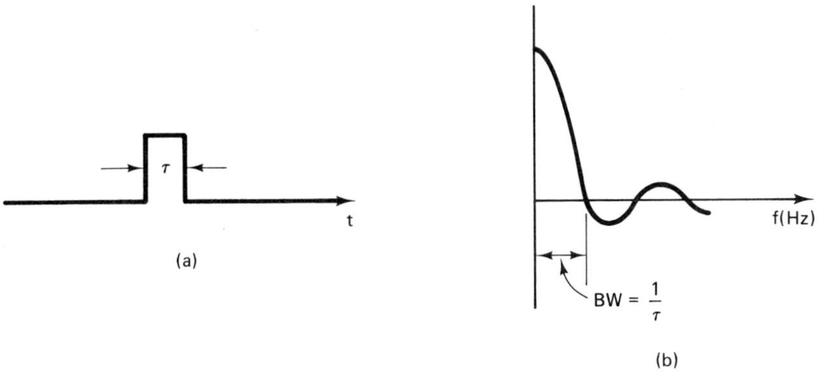

Figure 2.11 (a) Single pulse. (b) Frequency spectrum of the pulse.

considering the frequency spectrum of a single pulse (Figure 2.11). For design purposes, the bandwidth is often taken to be the distance to the first zero crossing. That is,

$$\mathrm{BW} = \frac{1}{\tau}\ \mathrm{Hz} \tag{2.7}$$

where τ is the width of the pulse. Now our PCM wave train is composed of a sequence of ones and zeros. Regardless of the sequence, the bandwidth can be no greater than that required for a single pulse, and the bandwidth is as expressed by equation (2.7). Let us suppose that we sample an analog signal at the Nyquist rate. The time, T, between pulses is then given by

$$T = \frac{1}{2f_m} \tag{2.8}$$

where f_m is the highest frequency in the signal. Extending this to n signals (multiplexing), the width of the sampled pulse will be

$$\tau_n = \frac{T}{n} = \frac{1}{2nf_m}\ . \tag{2.9}$$

If we wish to digitize each pulse, each code word can only last T/n seconds. For an m-bit PCM word, the width of each bit pulse is then T/mn. Using equation (2.7), the PCM bandwidth, $\mathrm{BW_{PCM}}$, is given by

$$\begin{aligned}\mathrm{BW_{PCM}} &= \frac{1}{\tau} = \frac{1(m^*n)}{T}\ \mathrm{Hz} \\ &= (m)\left(\frac{\omega_m}{\pi}\right)\ \mathrm{Hz}\end{aligned} \tag{2.10}$$

For the binary process, the number of levels, L, in the quantization process is given by

$$L = 2^m \tag{2.11}$$

Solving for m and substituting the result in equation (2.10) results in

$$BW_{PCM} = \left(\frac{n\omega_m}{\pi}\right)(\log_2 L)\text{Hz} \tag{2.12}$$

EXAMPLE 2.2

A PCM system transmits 24 time-multiplexed voice signals, each bandlimited to 8 KHz. What bandwidth is required if an 8-bit PCM word is used?

Solution

$$\begin{aligned} BW &= (2nf_m)(m) \\ &= (2 \times 24 \times 8000)(8) \\ &= 3.072 \text{ MHz} \end{aligned}$$

2.3 DELTA MODULATION

From the preceding discussion, it is evident that an increase in the number of bits (resolution) directly increases the bandwidth. A reduction of bandwidth while retaining the ability to transmit the information has been a subject of continuing research. Delta modulation is one result from this work. Using this technique, the knowledge of past information is used to simplify the coding technique. An example of a quantized signal for delta modulation is shown in Figure 2.12. As shown in Figure 2.12, the signal is first quantized into discrete levels with a constant step size Δ. Note that the quantized signal is constrained to move only one quantization level at each transition point, with this point corresponding to the instant of sampling. Also, the quantized signal must change at each sampling point. Transmission then becomes a matter of binary communication. That is, a 1 indicates a positive transition, whereas a 0 indicates a negative transition. The transmitted bit train for Figure 2.12 is

1 1 1 1 1 0 0 1 0 - - -

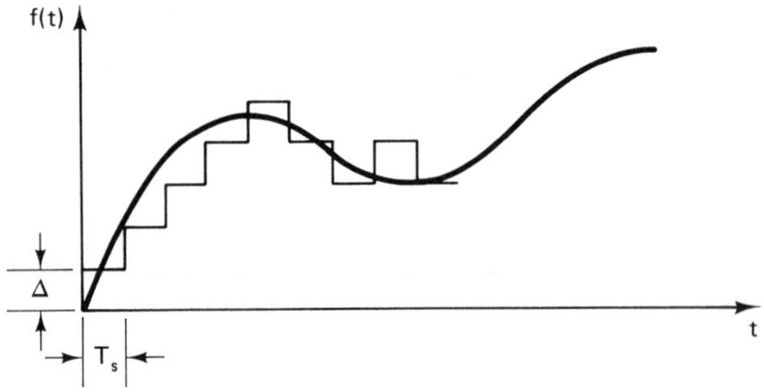

Figure 2.12 Quantized signal for delta modulation.

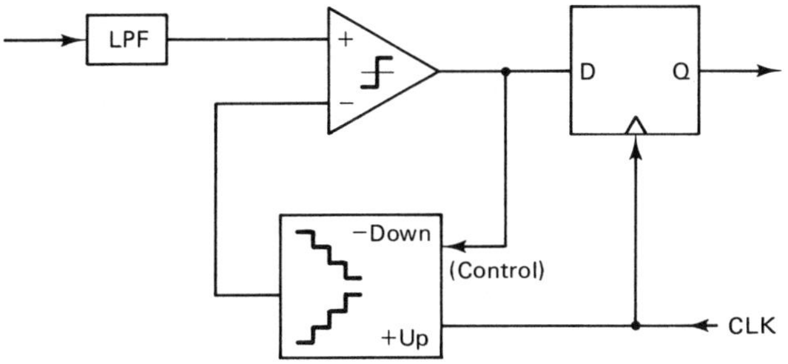

Figure 2.13 Delta modulator.

A delta modulator (encoder) can be constructed using a comparator along with a staircase generator (see Figure 2.13). Referring to this diagram, the analog input is referred to the staircase function at each sampling instant. If the input is greater than the staircase, a positive step is initiated. If it is less, a negative step is initiated. The low-pass filter eliminates potential aliasi ¸ problems due to high-frequency input-signal components.

The delta demodulator is simply a staircase generator (Figure 2.14). A staircase is generated using an integrator. If a 1 is received, the integrator increments positively. A 0 received results in a negative increment. The low-pass filter filters (smoothes) the output waveform.

Considering delta modulation on a systems level, we see that two key parameters are involved: *step size* and *sampling period*. The steps must be able to follow the signal; otherwise we have a situation known as *overloading*. Specifically, to enable the system to track a faster signal means that either the step size or the sampling frequency must be increased. Increasing the sampling frequency increases the bandwidth. On the other hand, increasing the step size increases the quanitization error. This results in quantizing noise defined in the same manner as it was for PCM.

As an example, suppose that the step size and the sampling interval have been chosen to avoid overloading. With this assumption, the magnitude of the quantization noise will never exceed the step size, Δ. Now, referring to Figure 2.12, the error can range between $+\Delta$ and $-\Delta$. If we assume, furthermore, that

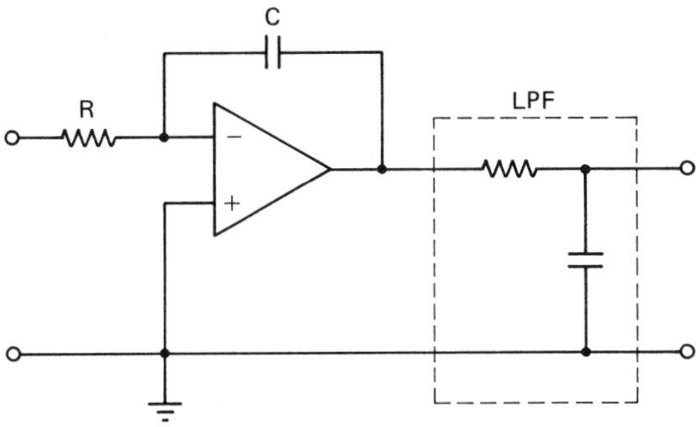

Figure 2.14 Delta demodulator.

the errors will be uniformly distributed between these two extremes, the mean square value (rms squared) value of the quantization noise may be expressed as

$$\overline{e}^2 = \int_{-\Delta}^{\Delta} \frac{1}{2\Delta} e^2 \, de = \frac{\Delta^2}{3} \tag{2.13}$$

If this noise is passed through the low-pass filter (Figure 2.15), the output noise power is given by

$$P_N = \frac{\Delta^2}{3} \frac{f_m}{f_s} \tag{2.14}$$

where

f_m = −3-dB filter cutoff frequency
f_s = sampling frequency

The reader is referred to an advanced text for the actual derivation of this useful result. An example illustrates the use of these concepts.

EXAMPLE 2.3

Using delta modulation, it is desired to transmit the following signal:

$$f(t) = 4 \cos 500t + 3 \cos 1500t$$

Find the signal-to-quantization-noise ratio.

Solution We must first determine the necessary sampling frequency and step size.

The Nyquist rate for the waveform is

$$T_s = \frac{1}{f_s} = \frac{2\pi}{2(1500))} = \frac{\pi}{1500} \text{ s}$$

The fastest rate at which the signal can change is ±3V in π/3000 s. The maximum change will occur when both sinusoids are approximately in phase. Suppose we choose a sampling frequency that is 40 times the Nyquist rate. This corresponds to 20 steps in π/3000 s. If we choose a step size of $\frac{1}{2}$ V, this is more than sufficient to prevent overloading since the maximum signal is 7 V. The quantization noise power is then

$$P_N = \frac{\Delta^2}{3} \frac{f_m}{f_s} = \frac{(\frac{1}{2})^2}{3} \frac{750/\pi}{60{,}000/\pi}$$

$$= \frac{1}{12} \frac{1}{80} = 1.04 \text{ MV}$$

The SNR is

$$\text{SNR} = \frac{(4^2 + 3^2)/2}{1.04 \times 10^{-3}} = 12.02 \times 10^3 = 40.79 \text{ dB}$$

2.3.1 Continuously Variable Slope Delta Modulator

The performance of the basic delta modulation scheme can be improved without increasing bandwidth. This improvement is realized if the step size adapts to the changing input rather than remaining constant. This technique is referred to as continuously variable slope delta (CVSD) modulation, or adaptive delta modulation.

A scheme for implementing this is shown in Figure 2.15. Referring to this figure, we see that the innermost control loop is a simple delta modulator (Figure 2.13). The additional dynamic range is achieved by varying the gain (step size) of this loop. In operation, the last few bits of the modulator are monitored continuously using the shift register. The control action depends on the shift register contents. One possible CVSD algorithm simply monitors the register to see if it contains all ones or zeros. This condition is called *coincidence*. The occurrence of this condition indicates that the gain is too small, that is, overloading is occurring. The resulting coincidence output drives a low-pass filter and the voltage output of this filter controls the integrator gain through a voltage-to-current (*V*/*I*) converter. The other input to the slope polarity switch is the sign bit or the up-down control of the delta modulator. The key ingredient of the ones-zeros algorithm is that it provides a measure of the average power level of the input signal. Other schemes are sometimes used that provide more instantaneous information about the shape of the input curve. Obviously, in the scheme we are considering, the gain of the integrator is minimum when the number of ones and zeros are equal. It is maximum when all 1s or all 0s occur.

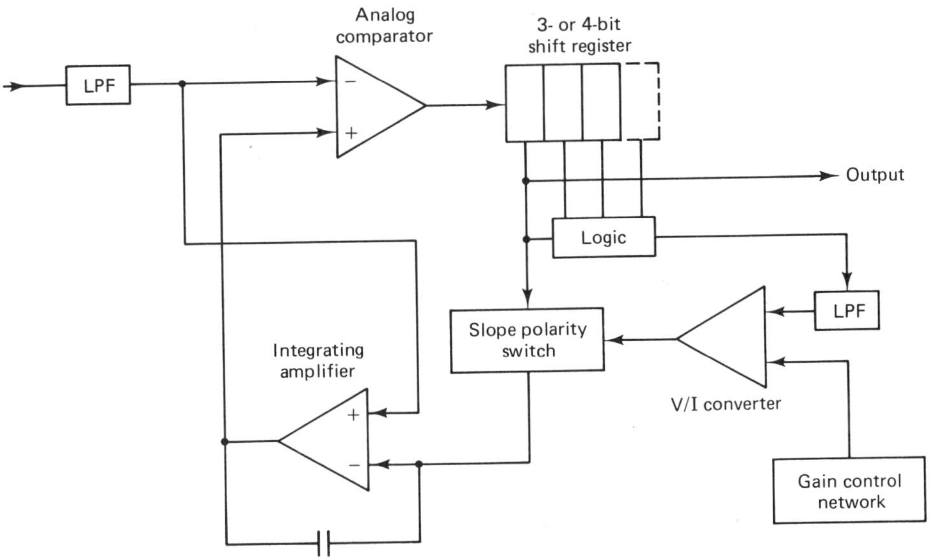

Figure 2.15 CVSD block diagram.

The algorithm is repeated in the receiver and thus the data can be recovered. Because the algorithm operates only on the past serial data, it changes the nature of the bit stream without changing the bit rate.

2.3.2 Differential PCM

Differential PCM is a variation of the delta system just considered. In fact, it is often referred to as delta PCM (DPCM). With this system, the modulator does not send the difference between adjacent samples but instead sends the difference between a sample and its predicted value (See Figure 2.16). An additional advantage of this system is that we now have the option of using the code of our choice.

The prediction $\hat{f}(nT_s)$ is made on the basis of previous samples. It is formed by adding the prediction to the difference term, $f(nT_s) - f(nT_s)$. In terms of only one sample, the output of the predictor is

$$\hat{f}(nT_s) = Af([n - 1]T_s) \tag{2.15}$$

where A is a weighting function added by the predictor. The prediction error, $e(nT_s)$, is then

$$\begin{aligned} e(nT_s) &= f(nT_s) - \hat{f}(nT_s) \\ &= f(nT_s) - Af([n - 1]T_s) \end{aligned} \tag{2.16}$$

The weighting function is chosen to make the error as small as possible.

In our study of digital communication systems thus far, we have already seen that fundamental choices must be made in the type of digital system to be designed or utilized. Facts without direction are of little use to us in resolving these problems. In succeeding chapters, we undertake a systematic study of principles and techniques that lead to systems design. Chapter 3 contains an extensive discussion of digital signaling techniques, which expands on the basics of this chapter and also introduces much of the remainder of the text.

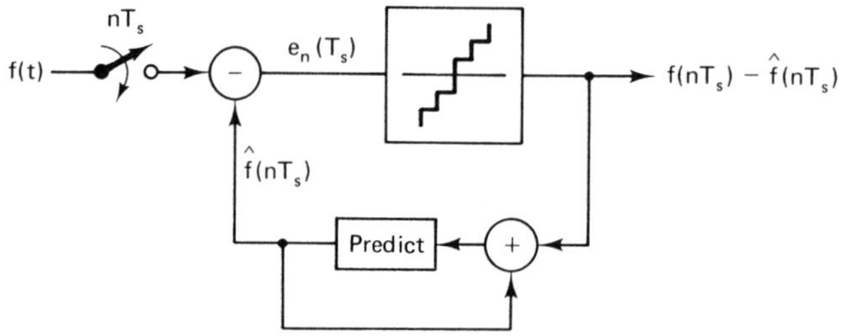

Figure 2.16 DPCM transmitter.

PROBLEMS

2.1. **(a)** In Example 2.1, how many quantization levels are there per sample?

(b) Determine the number of quantization levels that can be used if the number of bits per sample in a given PCM code is (a) 5, (b) 8, (c) x.

2.2. An analog signal is sampled at the Nyquist rate $1/T$ and quantized using M quantization levels.

(a) Show that the time duration τ of 1 bit of the binary encoded signal is $\tau \leq T/\log_2 M$.

(b) When is the equality sign valid?

2.3. Determine the minimum sampling rate necessary in order to sample and perfectly reconstruct the signal $f(t) = (\sin 6280\, t)/6280\, t$. Utilize the Fourier transform pair

$$f(t) = \left(\frac{W}{2\pi}\right) \frac{\sin Wt/2}{Wt/2} \qquad F(\omega) = \begin{cases} 1 & \text{for } |\omega| < \dfrac{W}{2} \\ 0 & \text{for } |\omega| > \dfrac{W}{2} \end{cases}$$

2.4. A telephone system encodes 96 voice channels (4 KHz) into 8-bit binary PCM. (a) Neglecting synchronizing pulses, what is the system bit rate? (b) Calculate the bandwidth required for the system.

2.5. Twenty-four telephone channels with bandwidths of 300 to 3500 Hz are to be transmitted. If the channels are TDM in a PAM (Pulse Amplitude Modulation) wave, determine: (a) the sampling rate, (b) the number of pulses per second to be transmitted (assume no synchronization pulses), and (c) the required bandwidth. State any assumptions about guard time between pulses.

2.6. A modulation scheme is employed whose amplitude can assume one of 64 different levels. A level is transmitted each 10^{-4} s through a channel.

(a) What is the baud rate?

(b) How many binary digits are necessary to specify a particular level?

(c) What is the bit rate through the channel if the binary code is transmitted?

2.7. A 1.5-MHz information signal with a dynamic range of 64 mV is sampled, quantized, and encoded using a direct binary code. The quantization is linear with 512 levels. Determine each of the following.

(a) The maximum possible bit duration

(b) The amplitude of one quantization level

2.8. A delta modulation system employs a step size of 0.1 V and a sampling interval of 10^{-4}. What is the maximum amplitude that a 1-KHz sinusoid can have and not produce slope overload? What is the maximum frequency that a 1-V sinusoid can have and not produce slope overload?

2.9. **(a)** If all letters are equally likely, calculate the information contents in bits of the English alphabet.

(b) Assume now that the vowels (a, e, i, o, u) are removed. How much is the information content reduced?

2.10. Assume that the bandwidth required to transmit a signal digitally equals the number of binary digits per second in the sampled and quantized waveform. What bandwidth would be required to transmit a speech signal (BW = 3.5 kHz) with a signal-to-quantization-noise ratio of 30 dB? A television video signal (BW = 4.0 MHz) with a signal-to-quantization-noise ratio of 40 dB?

2.11. We desire to encode the output of a 20-kHz music channel into PCM. What is the required bit rate if a signal-to-quantization-noise ratio greater than or equal to 45 dB is required?

2.12. A delta modulator has as input a signal described by $f(t) = 5t + 1$. Assuming a step size of 1 V and 10 samples/s, sketch the output of the delta modulator.

2.13. **(a)** Design a TDM that will accomodate 11 sources with the following specifications:

- Analog, 2-KHz bandwidth: source 1
- Analog, 4-KHz bandwidth: source 2
- Analog, 2-KHz bandwidth: source 3
- Digital, synchronous at 7200 bits/s: sources 4–11

The analog sources are to be converted to 4-bit PCM words. Draw a diagram showing one composite TDM PCM frame.

(b) What is the required bandwidth for the TDM PCM signal?

2.14. A baseband digital communication system operates at a rate of 9600 bit/s. If 8-bit words are encoded into each level for transmission over the system, find the minimum required channel bandwidth.

2.15. Assure that a digital system sends one of 32 possible levels over a channel every 0.8 s. What is the word length and data rate of the system?

2.16. The following sequence is transmitted by delta modulation: 1101110010111001. Sketch the resulting analog waveform at the output of the receiver.

3

DIGITAL SIGNALING TECHNIQUES

3.0 INTRODUCTION

A major concern in digital communications is how to both detect and control transmission errors caused by channel noise so that error-free data can be delivered to the user. An approach to this problem is the application of coding techniques (see Chapter 4). That is, we make use of error-detecting and error-correcting codes. Digital systems, however, do make errors, and this is usually described as the bit error rate (BER) of the system. We might say, for example, that 1 error out of 10^8, or a BER of 10^{-8}, is pretty good, whereas 10^{-3} is quite bad. The data is actually transmitted as digital signals. For example, binary data is transmitted by encoding each data bit into signal elements. If the signal elements all have the same algebraic sign—all positive or all negative—we say the signal is *unipolar*. In polar signaling, one logic state is represented by a positive voltage level and the other, by a negative voltage level. In recent years, the process of employing codes and then encoding the resulting bits into digital signals has to some extent merged. In the following we examine these techniques in modern digital communication systems.

3.1 RECEIVER EVALUATION CRITERIA

The generation of digital signals is illustrated in Figure 3.1. Upon receiving the signal, the receiver must accomplish two important tasks. These are (1) it must know when a bit begins and ends, and (2) it must decide if the received signal represents a 1 or a 0. A number of factors affect this process. Among these are the SNR, the data rate, and the bandwidth of the signal. Basically, we can make the following observation:

- An increase in data rate increases the BER (the probability that a received bit is in error).
- An increase in the SNR decreases the BER.
- An increase in bandwidth allows an increase in the data rate at the same BER.

In addition to these factors, the encoder also is an important factor in improving performance. Basically, the encoder performs the function of mapping from data bits to signal elements. Some of the more common encoding formats

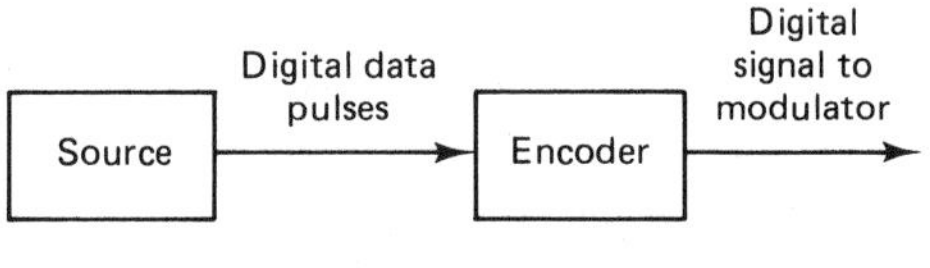

Figure 3.1 Digital signaling of digital data.

TABLE 3.1 DEFINITION OF DIGITAL SIGNAL ENCODING FORMATS

Nonreturn to zero–level (NRZ-L)	
1 =	high level
0 =	low level
Nonreturn to zero–mark (NRZ-M)	
1 =	transition at beginning of interval
0 =	no transition
Nonreturn to zero–space (NRZ-S)	
1 =	pulse in first half of bit interval
0 =	no pulse
Biphase–level (Manchester)	
1 =	transition from high to low in middle of interval
0 =	transition from low to high in middle of interval
Biphase–mark	
Always a transition at beginning of interval	
1 =	no transition in middle of interval
0 =	transition in middle of interval
Biphase–space	
Always a transition at beginning of interval	
1 =	no transition in middle of interval
0 =	transition in middle of interval
Differential Manchester	
1 =	no transition in middle of interval
0 =	transition at beginning of interval
Delay modulation (Miller)	
1 =	transition in middle of interval
0 =	no transition if followed by 1 Transition at end of interval if followed by 0
Bipolar	
1 =	pulse in first half of bit level, alternating polarity pulse to pulse
0 =	no pulse

are listed in Table 3.1 and are illustrated in Figure 3.2. We can group these encoding formats into the following categories:

- Nonreturn to zero (NRZ)
- Return to zero (RZ)
- Biphase
- Delay modulation
- Multilevel binary

Before describing these techniques, it is instructive first to devise ways of comparing the various encoding formats. Factors that are used to accomplish this may vary. In the following we suggest several:

- *Signal frequency spectrum.* Less bandwidth is required if there is no high frequency content. Conversely, if there is no dc component, ac coupling can be employed.

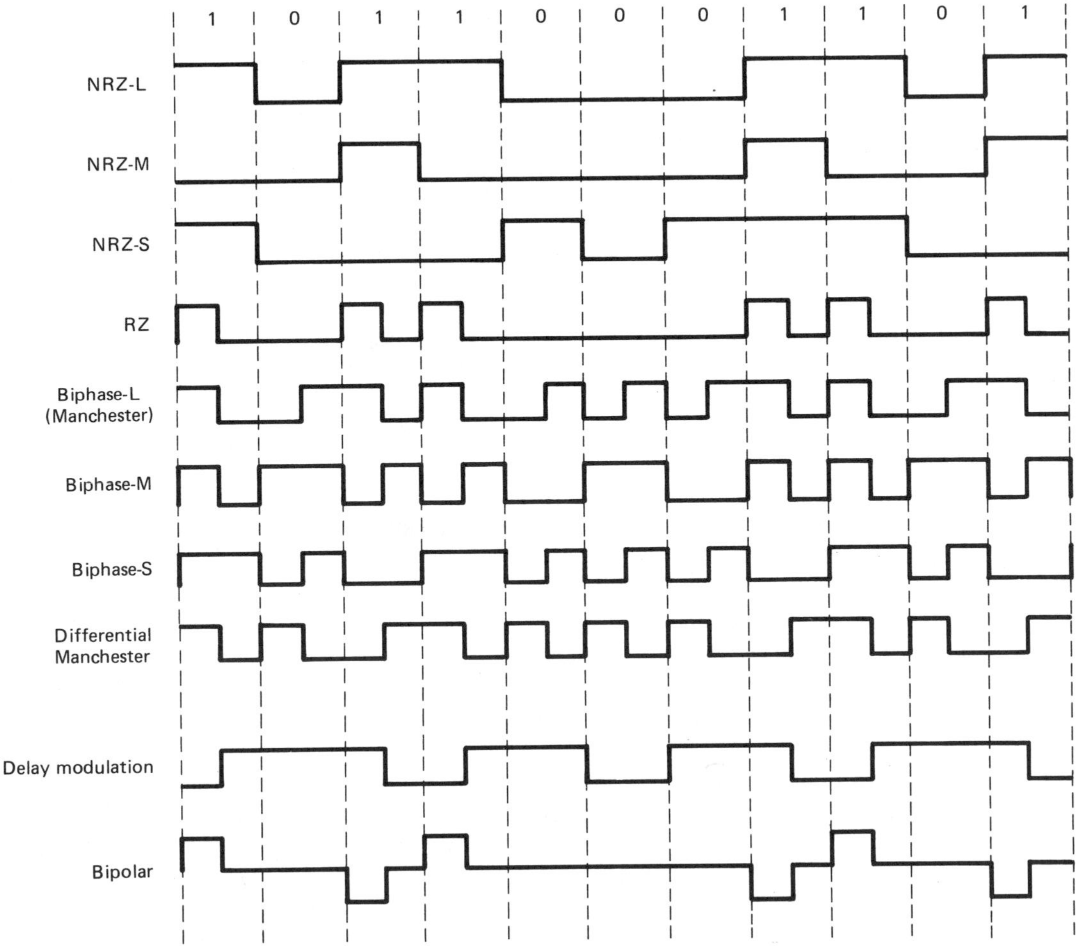

Figure 3.2 Digital signal-encoding formats.
(© 1984, IEEE)

- *Signal synchronization capability.* We previously mentioned the need to determine the beginning and end of each bit position. It is helpful if information is contained in the signal frequency spectrum that will permit the synchronization of a clock in the receiver. This is sometimes referred to as a self-clocking capabillity.
- *Signal interference and noise immunity.* Some codes exhibit superior performance in the presence of noise. This usually is reflected by a smaller BER rate for the same SNR.
- *Error-detection capability.* Some signaling schemes inherently possess this feature.
- *Cost and complexity.* A factor which must be considered in any design.

With these considerations in mind, we can now consider the encoding formats in Table 3.1.

3.2. DIGITAL SIGNAL-ENCODING FORMATS

The digital signal-encoding formats itemized in Table 3.1 are all of the baseband variety. The primary signal-processing advantage of digital encoding compared to analog schemes is that lower signal power can be used for a fixed signal quality. Digital signal quality (average signal power to average noise power) increases exponentially with bandwidth for elementary coding schemes, whereas the quality increases linearly for nonencoded schemes such as FM. A discussion of the various techniques follows.

3.2.1 Nonreturn to Zero (NRZ)

Referring to Figure 3.2, we see that NRZ codes share the property that the voltage level is constant during a bit interval. That is, there is no transition (no return to zero voltage level). In terms of complexity, these are the simplest codes to implement, with the simplest being NRZ-L. This code is generally used to generate or interpret digital data by data processing terminals and other devices. If we desire to transmit with another code, this code is typically generated from the NRZ-L signal (see Figure 3.2). Two other versions of this code are *mark* and *space*, that is, NRZ-M and NRZ-S. These codes have the advantage that they are differential codes and are less subject to dc line-bias voltage-shift effects.

We introduced the idea of differential encoding in Chapter 2. In differential encoding, the signal is decoded by comparing the polarity of adjacent signal elements rather than the absolute value of a signal element. Two benefits accrued from this: (1) It may be more reliable to detect a transition in noise than to compare a value to a threshold, and (2) it is easy to lose the sense of the polarity of a signal. For example, on a multidrop twisted pair line, the accidental reversal of the lines results in the inversion of 1s and 0s for NRZ-L. With differential encoding, this cannot happen.

NRZ codes are easiest to engineer and also make the most efficient use of bandwidth. This is illustrated in Figure 3.3 where the power spectral density of various encoding schemes are plotted. The frequency is normalized to the data rate. Note that most of the energy in an NRZ signal lies between dc and half the bit rate. That is, with a bit rate of 9600 bits/s, most of the signal energy is concentrated between dc and 4800 Hz. With the advantages mentioned we may well ask ourselves why we consider encoding schemes. Consider the disadvantages. The primary limitations of NRZ signals are the presence of a dc component and the lack of a self-synchronization capability. For example, with a long string of 1s for NRZ-L or NRZ-S, the output is a constant voltage. Based on the signal alone, any drift between the transmitter and receiver cannot be corrected. These

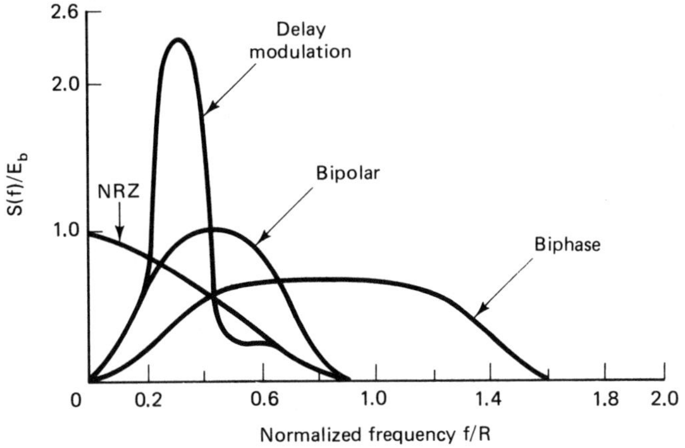

Figure 3.3 Spectral density of various digital signal-encoding schemes.

limitations make these codes unattractive for signal-transmission applications. However, because of their simplicity and relatively low frequency response characteristics, NRZ codes are commonly used for digital magnetic recording and for asynchronous coding, where synchronization is provided by the occurrence of the start pulses.

3.2.2 Return to Zero (RZ)

The RZ format actually provides no improvement over the previously discussed NRZ techniques. Referring to Figure 3.2, we note a distinction between data rate and modulation rate. The data rate is $1/t_B$, where t_B is the bit repetition period. The modulation rate is the average number of transitions that occur per second. Thus, for RZ, the maximum modulation rate, R, is given by

$$R = \frac{2}{t_B} \text{ baud} \tag{3.1}$$

This occurs during a string of 1s. The minimum naturally occurs during a string of 0s. Because the modulation rate is twice that for NRZ, the bandwidth of the signal is also twice as large. The limitations of NRZ apply to RZ as well. That is, there is a dc component and no synchronization (with a string of zeros). RZ is used in some applications because of its simplicity but is not a preferred choice for most applications.

3.2.3 Biphase

A number of biphase encoding formats exist. Acutally, the term *biphase* encompasses biphase-L (Manchester), biphase-M, biphase-S, and differential Manchester. These schemes are devised to overcome the limitations of NRZ and RZ signal-encoding techniques. A common characteristic of biphase schemes is that

TABLE 3.2 SIGNAL TRANSITION RATE

	Minimum	101010 . . .	Maximum
NRZ-L	0 (all 0s or 1s)	1.0	1.0 (1010 . . .)
NRZ-M	0 (all 0s)	0.5	1.0 (all 1s)
NRZ-S	0 (all 0s)	0.5	1.0 (all 0s)
RZ	0 (all 0s)	1.0	2.0 (all 1s)
Manchester	1.0 (1010 . . .)	1.0	2.0 (all 0s or 1s)
Biphase-M	1.0 (all 0s)	1.5	2.0 (all 1s)
Biphase-S	1.0 (all 1s)	1.5	2.0 (all 0s)
Differential Manchester	1.0 (all 1s)	1.5	2.0 (all 0s)
Delay modulation	0.5 (1010 . . .)	0.5	1.0 (all 0s or 1s)
Bipolar	0 (all 0s)	1.0	2.0 (all 1s)

they all require at least one transition per bit period. There may be as many as two transitions per bit period (see Table 3.2). We see then that the modulation rate, R, for biphase is twice that for NRZ with a corresponding greater bandwidth.

The question then, again, is why we use the biphase schemes. There are several advantages:

- *Synchronization.* Referring to Figure 3.2 we see that there is a predictable transition during each bit period. The receiver can synchronize on this transition. Specifically, for Manchester and differential Manchester, there is always a transition in the middle of the bit interval. For biphase-M and biphase-S, there is always a transition at the beginning of a bit time. Biphase codes, then, are known as self-clocking codes.
- *No dc component.* The guaranteed transition ensures that the average dc level is zero.
- *Error detection.* The absence of an expected transition can be used to detect errors. Noise would have to invert the signal both before and after the expected transition to cause an undetected error.

Referring to Figure 3.3, note that most of the energy in biphase codes is between one-half and one times the bit rate. There is no dc component. All the biphase codes are differential with the exception of Manchester.

NRZ is still the most widely used code in data communication systems. Biphase codes are also popular techniques for data transmission, and the Manchester code is gaining ground rapidly. This code is popular in magnetic tape recording and is used as an input signal for fiber-optic modulation systems. Local area network (LAN) standards specify both Manchester and differential Manchester. In fact, the Manchester code has been specified for the IEEE 802.3 LAN standard for baseband coaxial cable using *carrier sense multiple access/collision detection* (CSMA/CD), that is, *Ethernet.* Differential Manchester has been specified for the IEEE 802.5 standard for the token ring LAN, using either baseband coaxial cable or twisted pair. Differential Manchester is preferred for twisted pair because it uses differential encoding.

3.2.4 Delay Modulation

Delay modulation, also known as Miller encoding, is an interesting alternative to the biphase technique. With this technique there is at least one transition per two bit times. There is never more than one transition per bit. We see then that Miller coding requires less bandwidth than biphase and has a lower modulation rate. It also possesses some synchronization capability. Referring to Figure 3.3, we see that the bandwidth for Miller is considerably less than for either NRZ or biphase. However, the figure is misleading. For worst-case bit patterns, Miller can have a bandwidth greater than NRZ along with a significant dc component.

The BER for several digital encoding schemes is shown in Figure 3.4. Note that Manchester and NRZ exhibit identical performance with an improvement over Miller of 3 dB. The derivation of formula leading to these curves is based on the probability of error in white noise. Without entering into this analysis we can intuitively see the reason for the results. At the receiver a decision is made based on whether a received bit internal contains a 1 or 0. For both NRZ and Manchester, there are only two elementary pulse waveforms from which to choose. This statement also applies to RZ, but RZ makes inefficient use of signal time. There is either a half-pulse or no pulse. The decision is even more difficult for the Miller code because it makes use of four elementary codes.

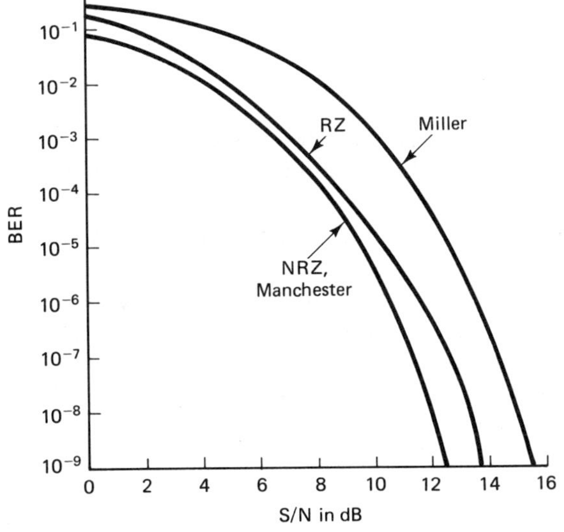

Figure 3.4 Theoretical BER for various digital encoding schemes. (© 1984, IEEE)

3.2.5 Multilevel Binary

Multilevel binary uses more than two signal levels. An example of this is bipolar (Figure 3.2). It is widely used by AT&T for T1-PCM carriers (See Chapter 7). From Figure 3.3, we see that it has a bandwidth centered on one-half the bit rate. In terms of characteristics, there is no dc component, and a lack of synchronization capability. Some error-detection capability does exist, since successive 1s must have opposite signs.

3.3 DIGITAL MODULATION TECHNIQUES

Modulation, in general, is a process by which one or more parameters of a waveform are varied in a manner directly related to an information-bearing signal. For example, a sinusoidal signal has three features that can be varied: amplitude, frequency, and phase (refer to equation 3.2):

$$E_c(t) = A \sin(\omega_c t + \phi) \tag{3.2}$$

Using the information signal to vary A results in amplitude modulation (AM) of $E_c(t)$. Varying ω_c or ϕ results in frequency or phase modulation. We refer to the unmodulated sinusoidal signal described by equation (3.2) as the carrier with frequency, f_c. The different types of modulation are illustrated by equation (3.3).

Modulated carrier

$$S(t) = A(t) \sin[2\pi f(t) + \phi(t)] \tag{3.3}$$

Amplitude

Frequency

Phase

If a continuum of values of the modulated parameter is allowed, the modulation is said to be *analog*. If only specific values are allowed, the modulation is said to be *digital*.

Thus digital modulation refers to techniques in which amplitude, frequency, and/or phase assume one of a set of specific values. Accordingly, these techniques are called amplitude-shift keying (ASK), frequency-shift keying (FSK), phase-shift keying (PSK), and QAM (quadrature-amplitude modulation). QAM is a combination of ASK and PSK and is discussed in Section 3.3.6.1. An example of ASK is shown in Figure 3.5. In this figure, the information is encoded as binary data. Mathematically, we can describe the modulated signal by

$$S_i(t) = E \cos \omega_0 t \cdot d(t) \tag{3.4}$$

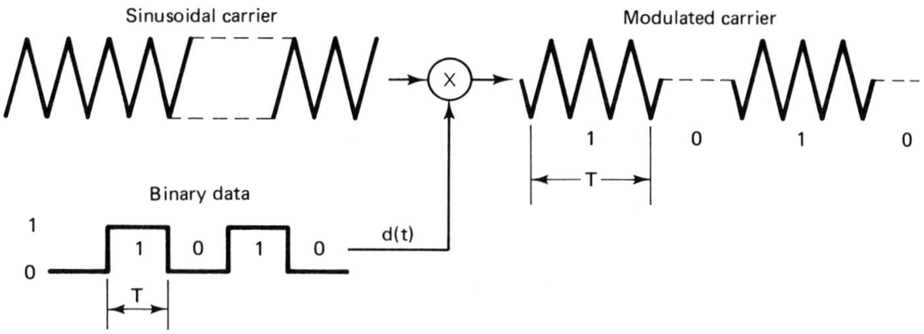

Figure 3.5 On/off amplitude-shift keying (ASK).

where

$$E = \text{peak value of the carrier}$$
$$d(t) = \text{binary data}$$

For the example shown, a sinusoidal signal of amplitude E is transmitted when the digital data signal is a logic 1; otherwise, no carrier is transmitted. This particular scheme has the advantage of simplicity but is not generally used for several reasons. For example, the available carrier power is used inefficiently. With random data, the carrier is on only half the time. Constant envelope signals are preferred for most advanced systems (see Section 5.3.1).

3.3.1 FSK Fundamentals

FSK is used much more than ASK (Figure 3.6). Referring to the figure, we see that implementation of an FSK modulator is conceptually simple. Two carriers of frequency $\omega_0 + \Delta\omega$ and $\omega_0 - \Delta\omega$ are generated. A logic 1 keys on one carrier, whereas a logic 0 keys on the other. Another approach is simply to use a classical analog frequency modulator, which will respond to the two voltage levels. The two approaches can produce slightly different signal spectra. FSK is often used because of the following advantages:

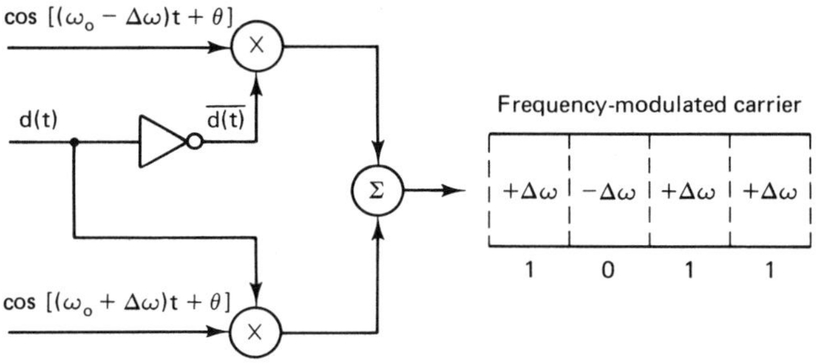

Figure 3.6 Digital frequency-shift keying.

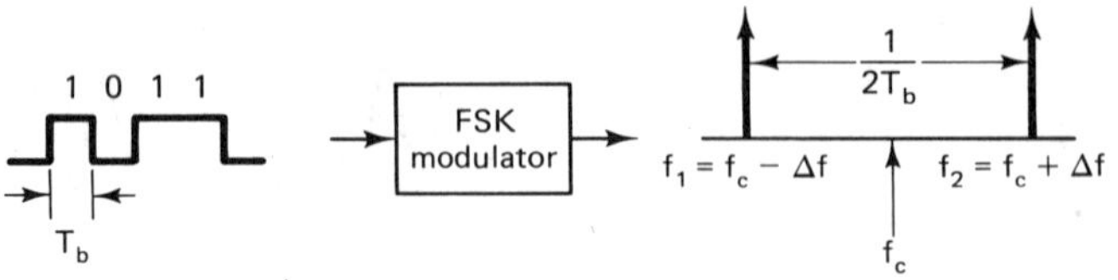

Figure 3.7 FSK principles.

1. Coding is simple and inexpensive.
2. FSK is an FM technique. The receiver can amplify the received signal without automatic gain control (AGC).
3. FSK is affected less by noise than AM-sensitive coding.
4. Hard-limiting can be employed to improve the SNR and reduce the effects of interfering signals.

The FSK output can be related to the digital input as shown in Figure 3.7. Note from Figure 3.7 that a logic 0 corresponds to frequency f_1 and a logic 1 corresponds to frequency f_2. The required peak frequency deviation, Δf, is given by

$$\Delta f = \frac{f_2 - f_1}{2} = \frac{1}{4T_b} \text{ Hz, minimum} \tag{3.5}$$

where T_b is the bit duration in seconds and f_2 and f_1 are expressed as

$$f_1 = f_c - \Delta f = f_c - \frac{1}{4T_b} \tag{3.6}$$

and

$$f_2 = f_c + \Delta f = f_c + \frac{1}{4T_b} \tag{3.7}$$

Using equations (3.6) and (3.7), we can express the FSK signal as

$$S_{FSK}(t) = A \cos[2(f_c \pm \Delta f)t] \tag{3.8}$$

or

$$S_{FSK}(t) = A \cos \pm 2\pi f t \cos 2\pi f_c t - A \sin(\pm 2\pi f t) \sin 2\pi f_c t \tag{3.9}$$

Finally, by using equation (3.5), the FSK signal can be related to the bit rate. That is,

$$S_{FSK}(t) = A \cos \frac{\pm \pi t}{2T_b} \cos 2\pi f_c t - A \sin \frac{\pm \pi t}{2T_b} \sin 2\pi f_c t \tag{3.10}$$

From Figure 3.7, we see that FSK can be considered as consisting of two pulsed sinusoidal waves of frequencies f_1 and f_2. Referring to Table A-1 in the appendix, the frequency spectrum of a pulsed sinusoidal wave is a sin x/x function. Thus we can indicate the spectrum for FSK as shown in Figure 3.8. If we assume that the zero crossings contain most of the energy, the bandwidth for FSK is given by

$$\begin{aligned} BW &= f_2 + \frac{2\pi}{\tau} - \left(f_1 - \frac{2\pi}{\tau}\right) \\ &= f_2 - f_1 + \frac{2}{\tau} = 2\left(\Delta f + \frac{1}{\tau}\right) \end{aligned} \tag{3.11}$$

where τ is the pulse width.

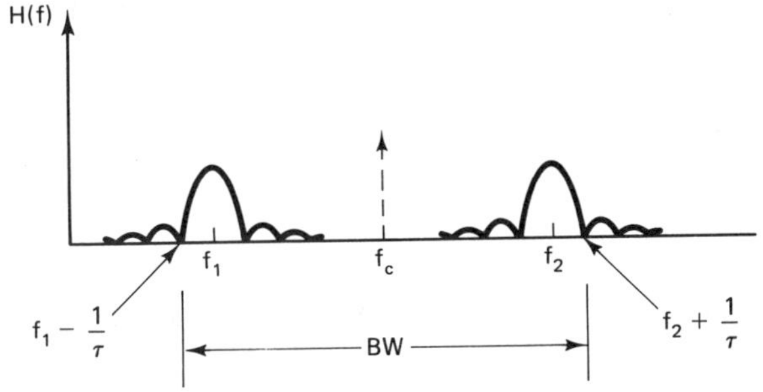

Figure 3.8 FSK frequency spectrum.

EXAMPLE 3.1

Find the bandwidth of an FSK signal that consists of bursts of 49 KHZ and 50 KHZ. The bit rate is 2000 bits/s.

Solution

$$\begin{aligned} BW &= f_2 - f_1 + \frac{2}{\tau} \\ &= 50{,}000 = 49{,}000 + 4000 \\ &= 5{,}000 \text{ Hz} \end{aligned}$$

3.3.1.1 FSK demodulation. The conceptual design of an FSK demodulator (Figure 3.9) is very simple. Referring to the figure, each band-pass filter is centered on one of the two transmitted frequencies. The envelope detectors indicate the total power in each passband and the comparator responds to the largest

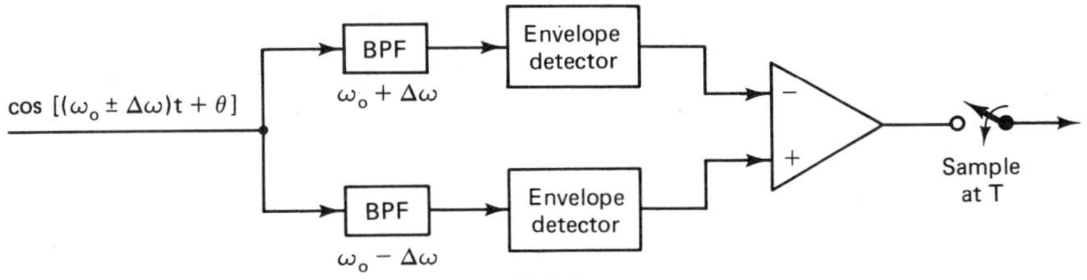

Figure 3.9 FSK detector.

power. This FSK-detection technique is referred to as *noncoherent*. That is, there is no signal involved in the detection process that is frequency- and phase-locked to the incoming FSK signal. The probability of error for noncoherent FSK is given by

$$P_e = \frac{1}{2}\exp\left(-\frac{E}{2N_0}\right) \tag{3.12}$$

where E/N_0 is the energy-to-noise ratio per bit per hertz of bandwidth.

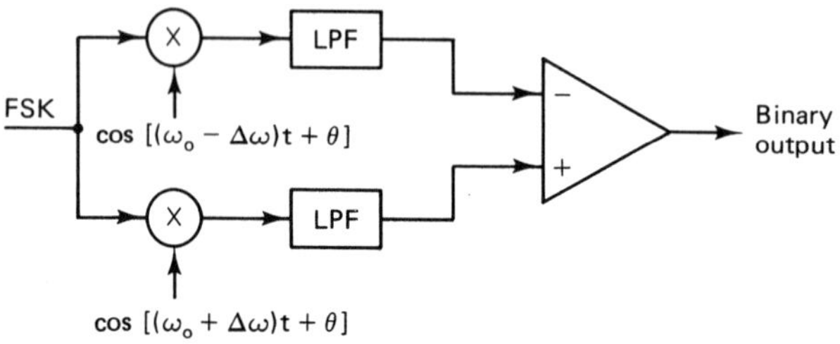

Figure 3.10 Coherent (synchronous) detection.

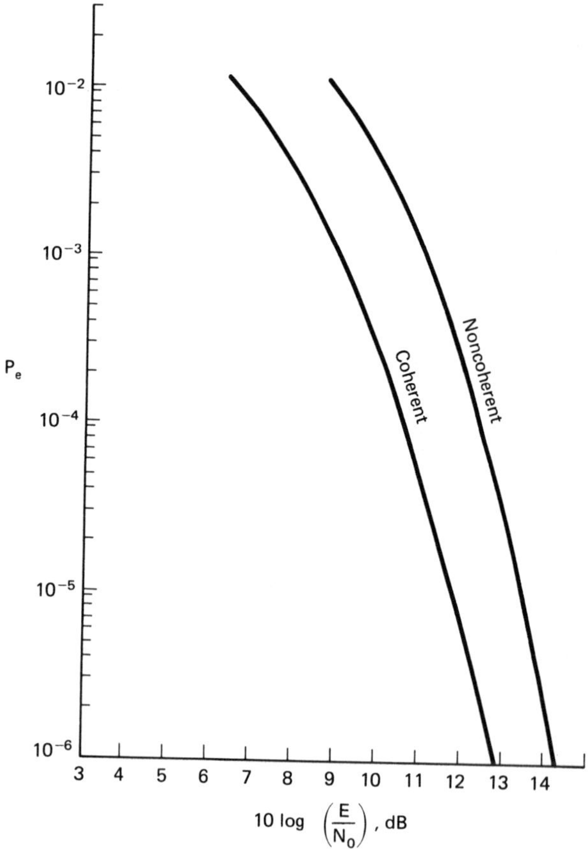

Figure 3.11 FSK performance.

A coherent FSK detector is shown in Figure 3.10. In the figure the incoming FSK is multiplied by stored replicas of the transmitted pulses. The stored replicas (also called the local references) must be exactly in phase (coherent) with the received signal. Since the two transmitted frequencies are not individually continuous, it is usually not practical to derive the local reference. Coherent detection of FSK is generally not implemented in simple FSK systems. The probability of error for synchronous systems of this type is given by

$$P_e = \operatorname{erfc} \sqrt{\frac{E_b}{N_0}} \tag{3.13}$$

Equations (3.12) and (3.13) are plotted in Figure 3.11.

3.3.1.2 FSK detection using a phase-locked loop. A phase-locked loop (PLL) can be routinely applied to FM demodulation. Consequently, it can be applied to FSK decoding. An FSK demodulator is shown in Figure 3.12. Without using inductors, a PLL not only demodulates but also sharply selects frequencies. In general, a PLL is a feedback system consisting of three functional blocks: a

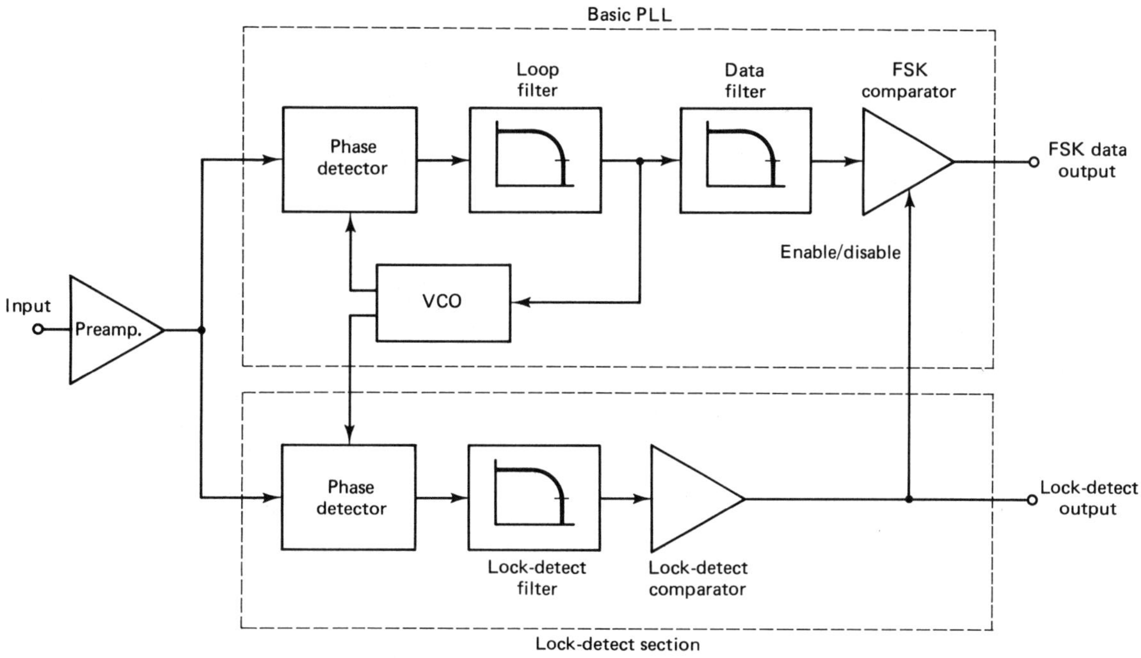

Figure 3.12 FSK demodulation.

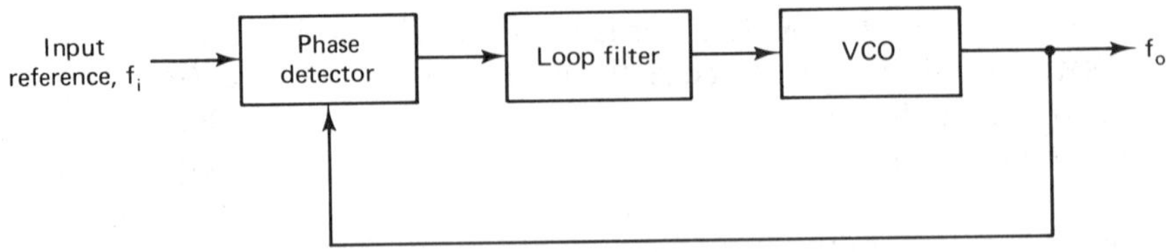

Figure 3.13 Basic phase-locked loop.

phase comparator, a low-pass filter, and a voltage-controlled oscillator (VCO) (see Figure 3.13). The PLL is a closed-loop electronic servo whose output locks onto and tracks an input reference signal. Phase lock is obtained by comparing the phase of the output signal with that of the reference. Any phase difference is converted to an error-correction voltage. This error voltage is applied to the VCO and changes the phase of the VCO's output to make it track the input.

Referring to Figure 3.13, let the input signal be sinusoidal with arbitrary phase. That is,

$$V_i(t) = A_1 \sin(\omega_i t + \theta_i) \tag{3.14}$$

The output of the VCO is

$$V_o(t) = A_2 \cos(\omega_o t + \theta_o) \tag{3.15}$$

The phase comparator may be modeled as a four-quadrant multiplier. Thus at the multiplier output we have

$$V_d = A_1 A_2 K_m \sin(\omega_i t + \theta_i) \cos(\omega_o t + \theta_o) \tag{3.16}$$

where K_m is the multiplier gain. If equation (3.16) is expanded, this results in an equation that contains sums and differences of the input frequency and the oscillator frequency. The sum-frequency term is filtered by the low-pass filter. Consequently, the low-pass filter output may be described by

$$V_f = \frac{A_1 A_2}{4} K_m \sin(\omega_i t - \omega_o t + \theta_i - \theta_o) \tag{3.17}$$

If the PLL is locked to a signal, equation (3.17) reduces to

$$V_f = \frac{A_1 A_2}{4} K_m \sin(\theta_i - \theta_o) \tag{3.18}$$

where $\theta_i - \theta_o$ is the phase error. This equation describes the error voltage that is applied to the VCO to maintain lock (see Figure 3.14). If a disturbance (noise)

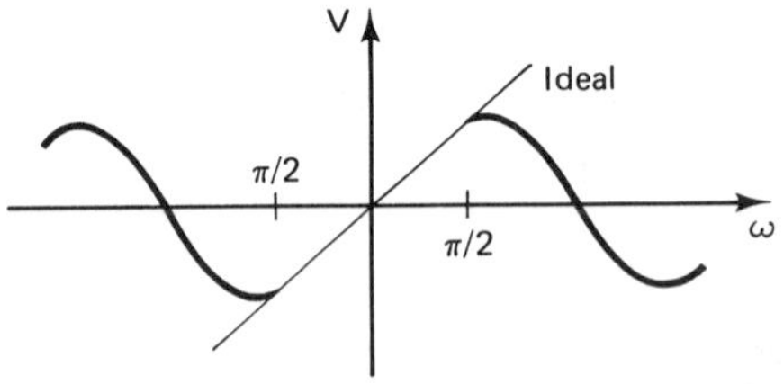

Figure 3.14 Typical phase detector characteristics.

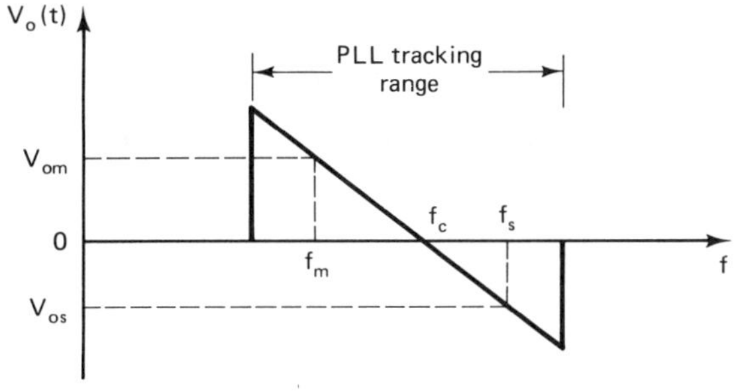

Figure 3.15 PLL output error voltages.

enters the PLL and causes the phase error to exceed $\pm\pi/2$ rad, the PLL will skip one or more cycles.

Now consider the operational aspects of the PLL. If the input signal changes phase or frequency, this results in a change of the error voltage. That is, the PLL demodulates input frequency and phase changes to output voltage changes. As the PLL with an FSK input locks onto either the mark or space frequency, f_m or f_s, the loop-output voltage shifts to V_{om} or V_{os}. This is indicated in Figure 3.15. These voltages are detected and amplified and then converted into a logic level.

EXAMPLE 3.2

Two common loop filter configurations are shown in the accompanying figure. Determine the voltage transfer functions and sketch the magnitude of the transfer function. Discuss the merits of one filter versus the other.

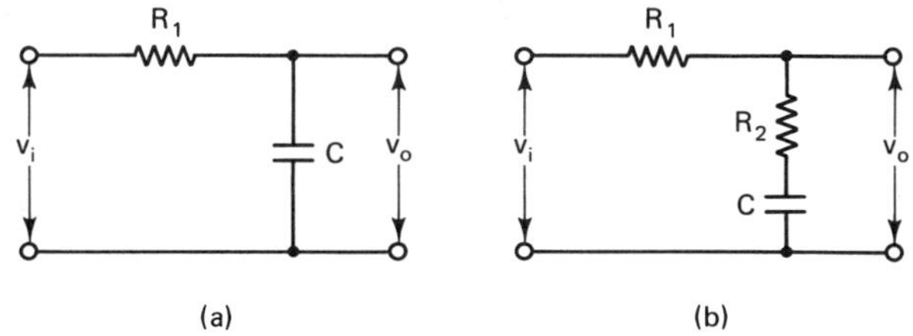

Solution

(a) $$\frac{V_o(t)}{V_i(t)} = H(j\omega) = \frac{1/j\omega C}{R_1 + 1/j\omega C} = \frac{1}{1 + j\omega R_1 C}$$

(b) $$\frac{V_o(t)}{V_i(t)} = H(j\omega) = \frac{R_2 + 1/j\omega C}{R_1 + R_2 + 1/j\omega C} = \frac{1 + j\omega R_2 C}{1 + j\omega C(R_1 + R_2)}$$

Comparing the sketches of the two filters' passbands, we see that the passband of filter (b) trails off with frequency. This filter allows the PLL to have a greater capture range. That is, the difference frequency can be initially greater and the loop will still be able to lock up.

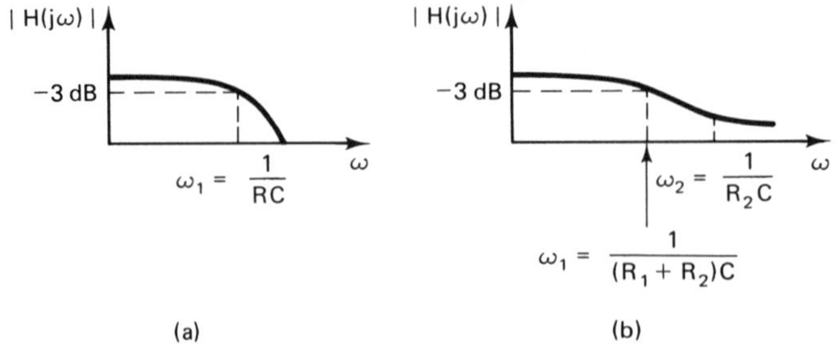

EXAMPLE 3.3

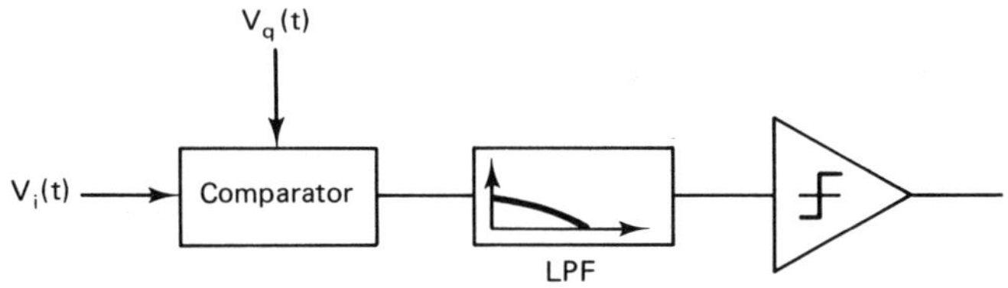

Analyze the operation of the lock-detect circuit in Figure 3.12.
The output of the comparator is

$$V_i(t)V_q(t) = A_1A_2 \sin(\omega_i t + \theta_i) \sin(\omega_o t + \theta_o)$$

$$= \frac{A_1A_2}{2} \cos[(\omega_i - \omega_o)t + (\theta_i - \theta_o)]$$

$$- \frac{A_1A_2}{2} \cos[(\omega_i + \omega_o)t + (\theta_i + \theta_o)]$$

At phase and frequency lock, the output of the filter is

$$V_f = \frac{A_1A_2}{2} \cos(\theta_i - \theta_o)$$

$$\approx \frac{A_1A_2}{2} \text{ V}$$

The level detector detects the voltage (V_f) and supplies a positive level at the lock-detect output.

3.3.2 PSK Modulation

PSK was developed during the early days of the space program. It is now widely used in both military and commercial communication systems. PSK is considered to be an efficient form of data modulation for telemetry applications because it provides the lowest probability of error for a given received signal level when measured over one symbol period. In PSK modulation, the incoming binary pulse code modulation waveform shifts the phase of the output waveform to one of a fixed number of states. We can express the output signal as

$$V_o(t) = E \sin\left[\omega_o t + \frac{2\pi(i - 1)}{M} \right] \tag{3.19}$$

where

$i = 1, 2, \ldots, M$
$M = 2^N$, the number of allowable phase states
$N =$ the number of data bits needed to specify the phase state, M

Three common versions of PSK modulation are binary (BPSK), where $M = 2$; quadrature (QPSK), where $M = 4$ and 8ϕ − PSK, where $M = 8$. The phase states are illustrated in Figure 3.16. Note that increasing the number of phases (signal states) results in an increase in the bit rate while still maintaining the same baud rate. Recall from Chapter 1 that a price is paid for this increase in data rate. That is, an increase in the SNR is necessary to maintain equivalent BERs. Note that all the signal states lie on a circle, so that they may be characterized as constant energy–constant amplitude signals. This characteristic is important in satellite communication links, where AM to PM conversions must be held to a minimum.

From Figure 3.16, we see that if a given amount of information must be transmitted in *a* given time interval, the signaling rate can be reduced in an **M**-ary system by a factor of N. The required channel bandwidth is proportional to the symbol rate. Thus a reduced signaling rate results in a narrower channel bandwidth. As another way of looking at it, if the symbol rate is fixed, the higher-order systems transmit more bits of information through a given bandwidth. We

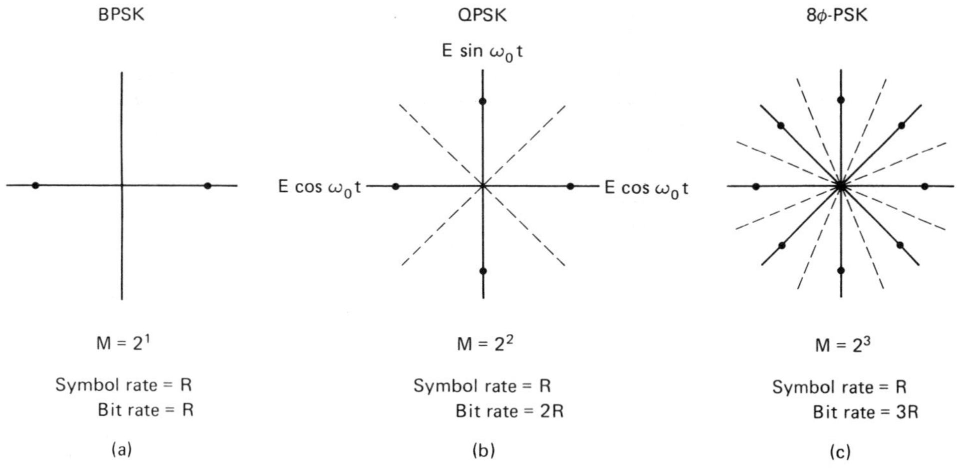

Figure 3.16 PSK phase states.

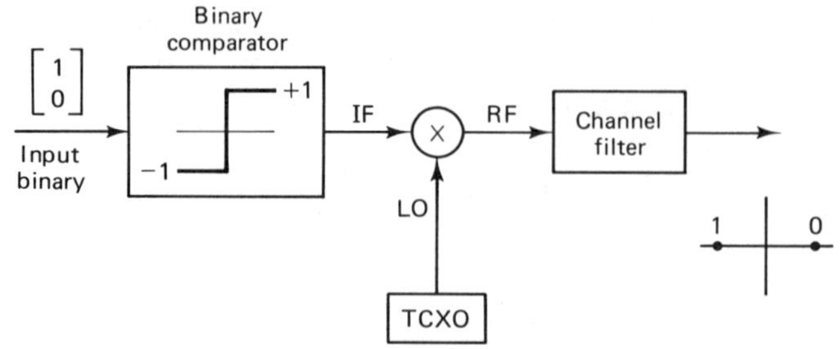

Figure 3.17 BPSK modulator.

say, then, that M-ary systems are *bandwidth efficient*. The price paid is an increase in the system probability of error. The signal space is more crowded for the higher-ordered systems. For the BPSK system, the receiver must distinguish between signals that are 180° out of phase. This reduces to 45° for the 8ϕ system.

A typical BPSK modulator is shown in Figure 3.17. The incoming unipolar waveform is converted to bipolar form, which causes current to be switched either into or out of the IF port of a double balanced mixer. This results in a 0° or 180° phase shift to the local oscillator signal, $\sin \omega_o t$. The mixer output is sometimes filtered to minimize intersymbol interference.

The block diagram for a QPSK modulator is shown in Figure 3.18. In this figure, two carriers in phase quadrature (orthogonal) are each phase shifted and then summed together. The outputs of the modulators in the A and B channels are tabulated in Table 3.3. We can introduce a mathematical description of the QPSK generation process by using the model shown in Figure 3.19. The two orthogonal components of the QPSK generator are often called the I (inphase) and Q (quadrature-phase) channels. This is a carryover of terminology from the inphase and quadrature channels in radar signal processors.

In a QPSK system, the I and Q carriers do not necessarily have the same power level. When the power levels are equal, this is referred to as *balanced QPSK*. In general, balanced QPSK is used when the data rate and performance of the two channels are to be equal. The unbalanced situation is appropriate when the two data rates differ significantly or when the performance specification of the two channels are different. For example, a system with binary data on one

TABLE 3.3 FOUR-PHASE MODULATOR SIGNAL

	Binary value	Voltage	Modulator output
B	0	V	$V \cos \omega_c t$
	1	$-V$	$-V \cos \omega_c t = V \cos (\omega_c t + \pi)$
A	0	V	$V \cos \left[\omega_c t + \frac{\pi}{2}\right]$
	1	$-V$	$-V \cos \left[\omega_c t + \frac{\pi}{2}\right] = V \cos \left[\omega_c t + \frac{\pi}{2}\right]$

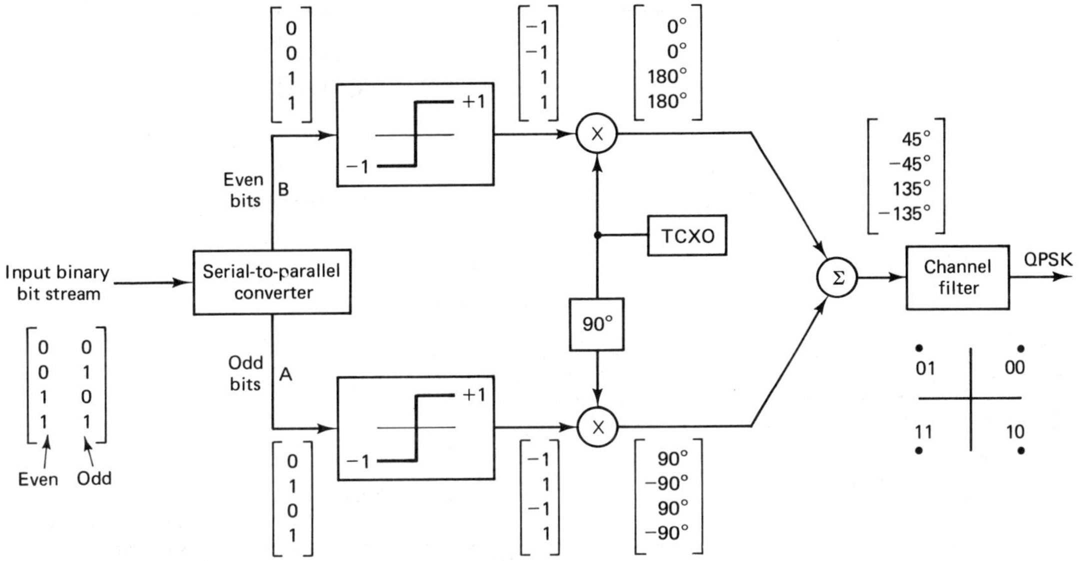

Figure 3.18 QPSK modulator.

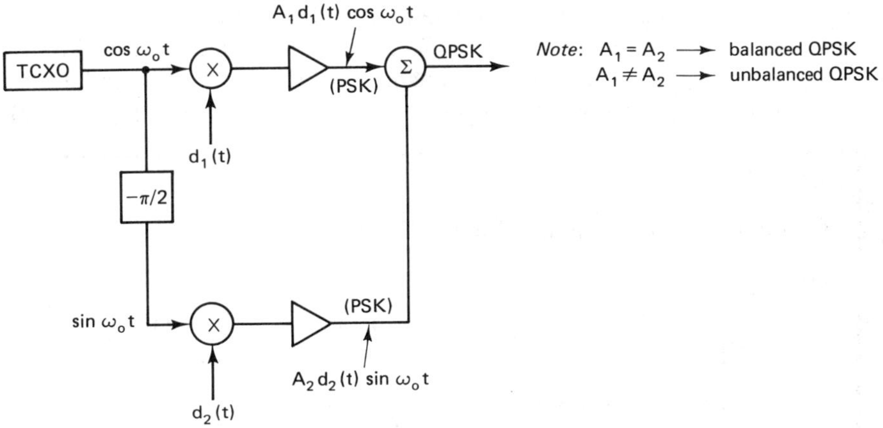

Figure 3.19 QPSK generation.

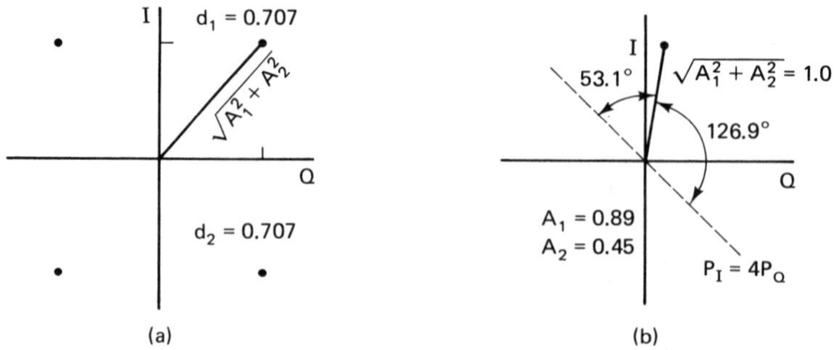

Figure 3.20 (a) Balanced QPSK. (b) Unbalanced QPSK.

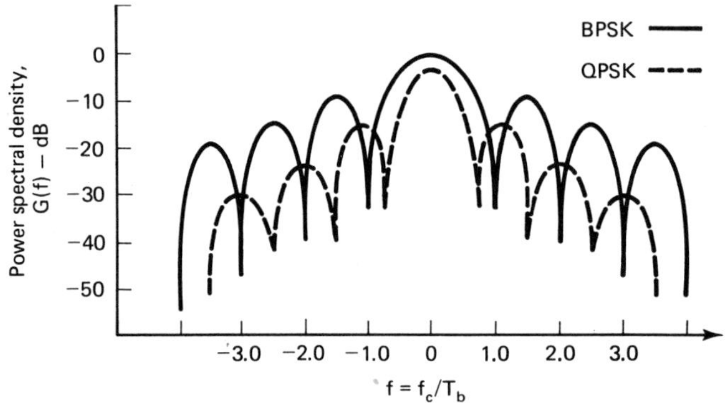

Figure 3.21 The spectrums for BPSK and QPSK are continuous and exhibit nulls at multiples of the symbol rate. The two techniques are compared for equal signal rates.

channel and a ranging signal on the other would represent unbalanced QPSK. The phase states for balanced and unbalanced QPSK are compared in Figure 3.20. In this figure, for unbalanced QPSK the carrier amplitudes are 0.89 and 0.45. This results in phase angles between transmitted phases of 53.6° and 126.4°. A comparison of the spectrums for BPSK and QPSK is shown in Figure 3.21.

EXAMPLE 3.3

Compare the performance for BPSK and QPSK.

Solution

	BPSK	QPSK
Total power	P_C	$P_I + P_Q = P_C$
Total symbol rate	$R = \dfrac{1}{T_B}$	$R_I + R_Q = \dfrac{1}{T_B}$
		with $R_I = \dfrac{1}{2T_B}$
		$R_Q = \dfrac{1}{2T_B}$

Energy/symbol	$E = P_C T_B$	$E = 2P_I T_B$
		$= 2P_Q T_B = P_C T_B$

Conclusion: Equal energy/symbol = equal performance

$$P_e(\text{QPSK}) = P_e(\text{BPSK})$$

3.3.3 PSK Demodulation

PSK demodulation can be divided into three major events (Figure 3.22). These may be itemized as follows:

- *Coherent recovery*. The carrier in the incoming signal is suppressed. A phase-coherent carrier must be derived from the incoming PSK signal.
- *Timing recovery*. Coherent multiplication is used to produce the raw data. Clock synchronization is obtained from the raw data.
- *Channel filtering*. The raw data is passed through a channel filter to shape the pulse train. This minimizes intersymbol distortion effects. The channel filter is often placed at the input of the demodulator where it produces equal results. The shaped pulse is then routed to the data sampler, which outputs the demodulated data.

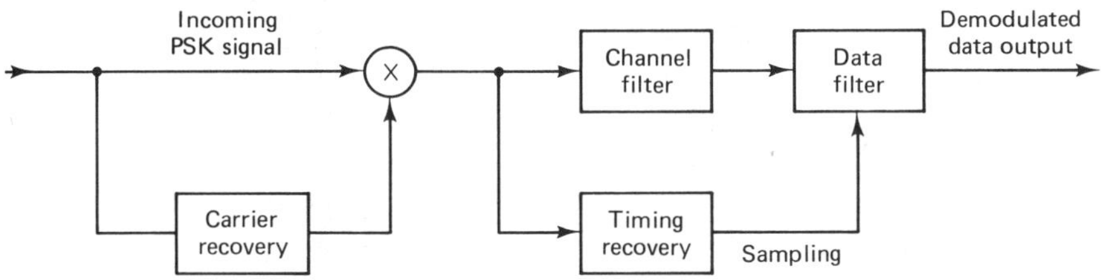

Figure 3.22 PSK demodulation.

PSK (biphase) modulation of an RF (Radio Frequency) carrier is the most efficient method for transmitting digital data over a link. This particular method of binary data encoding corresponds to suppressed carrier square-wave modulation of a CW (Continuous Wave) signal. This modulation scheme eliminates the carrier from the transmitted signal, and consequently recovery of the binary data at the receiver requires reinsertion of a locally generated carrier. Optimum recovery requires that the phase and frequency of the reinserted carrier be identical to the phase and frequency of the incoming signal. Reinsertion of the carrier can be performed by a variation of the standard PLL called a Costas loop (Figure 3.23). Operation of this circuit is considered in the following section.

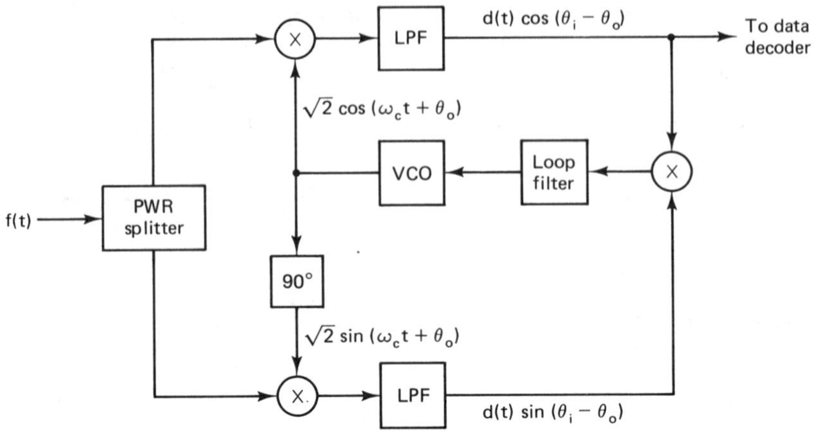

Figure 3.23 Costas carrier recovery loop.

3.3.3.1 Costas loop. Refer to Figure 3.2.3. The data are assumed to be random switching between $\frac{\pm\pi}{2}$ rad each data transition. At the input to the Costas loop, the PSK signal may be expressed as

$$f(t) = \sqrt{2}A \sin\left[\omega_c t + \frac{\pi}{2} d(t)\right] \tag{3.20}$$

where $d(t)$ represents the data and takes on values of ± 1. The rms value of the signal is A. We can write the input signal $f(t)$ as

$$f(t) = \sqrt{2}A d(t) \cos(\omega_c t + \theta_i) \tag{3.21}$$

Operation of the loop is easily analyzed if we assume that it is already in lock. The two low-pass arm filters are designed to eliminate twice-frequency terms. Thus at the filter outputs we have

$$Y_1(t) = \frac{\sqrt{2}A d(t)}{2} \sin(\theta_i - \theta_o) \tag{3.22}$$

and

$$Y_2(t) = \frac{\sqrt{2}A d(t)}{2} \cos(\theta_i - \theta_o) \tag{3.23}$$

where $\theta_i - \theta_o$ is the phase error. The input to the loop low-pass filter is $Y_1 Y_2$, or

$$e_f = \frac{K_m}{4} A^2 d_2^2(t) \sin 2(\theta_i - \theta_o) \tag{3.24}$$

where K_m is the multiplier constant in volts per radians. Since $d(t) = \pm 1$, $d^2(t)$

$= +1$. We can rewrite equation (3.24) as

$$e_f = K_m \frac{A^2}{4} \sin 2(\theta_i - \theta_o) \tag{3.25}$$

Note that this expression is equivalent to equation (3.18) for the standard PLL. The difference is that $\sin 2(\theta_i - \theta_o)$ is zero for 0° and 180°. That is, the loop can lock with the VCO at two different angles relative to the phase of the input signal. In Figure 3.23, the recovered data are taken from the arm where $Y_2(t) = d(t)\cos(\theta_i - \theta_o) \approx d(t)$. If the loop locks on the alternate phase, data appear in the alternate arm. The Costas loop performs both phase coherent reconstruction of the suppressed carrier and synchronous data detection.

A Costas loop can also be used for demodulation of a QPSK signal. From Figure 3.18 the QPSK signal, $f(t)$, is given by

$$f(t) = A_1 d_1(t) \cos \omega_c t + A_2 d_2(t) \sin \omega_c t \tag{3.26}$$

Compare this expression to that for biphase (3.21). Repeating the analysis leading to equations (3.22) and (3.23) will show that $d_1(t)$ is available from the upper arm of the Costas loop (Figure 3.22) and $d_2(t)$ is available from the lower arm. Best results are obtained using unbalanced quadraphase with a power ratio not exceeding 20/30. Analysis will show that at the input to the loop filter, a term exists that contains the product of the two data streams, $d_1(t)$ and $d_2(t)$. This effectively limits the lower bit rate, since the low frequencies are passed by the loop filter. This causes the VCO to lose lock. Proof is left as an exercise for the student (see Problem 3.24).

3.3.3.2 The remodulator. Another popular technique for carrier recovery is remodulation. This technique can be used to extract the data for both BPSK and QPSK. A BPSK implementation is shown in Figure 3.24 (see page 58). The operation of the remodulator, loop can be easily verified by assuming it is already in lock. This is similar to the manner in which we analyzed the Costas loop. The biphase data described by equation (3.21) serves as the PSK input. As in the Costas loop, the remodulator generates a loop error signal proportional to twice the phase error between the incoming phase and the VCO phase.

A remodulation technique for QPSK carrier recovery is shown in Figure 3.25 (see page 58). This loop exhibits a somewhat faster acquisition time compared to the conventional 4ϕ Costas loop demodulator.

3.3.3.3 Multiply-filter-divide circuit. A third method of QPSK carrier recovery used in high-rate burst mode systems is the multiplier-filter-divide circuit (Figure 3.26, see page 59). Referring to the figure, the output, $Y(t)$, of the times-four multiplier may be expressed as

$$Y(t) = [x(t)]^4 \tag{3.27}$$

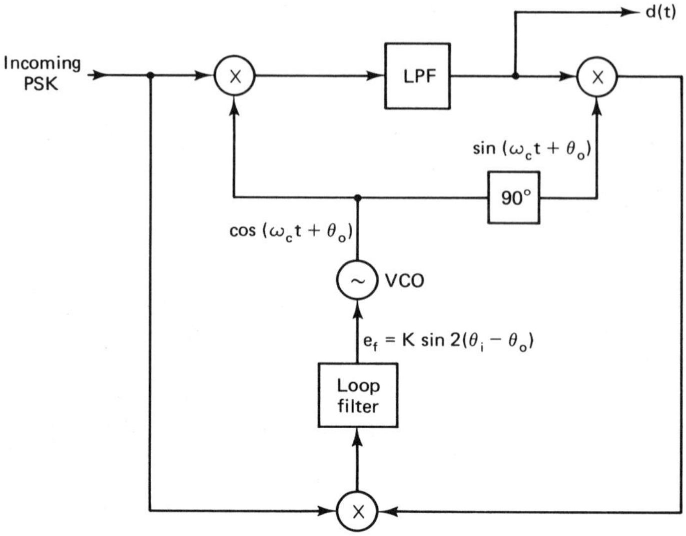

Figure 3.24 Remodulator loop.

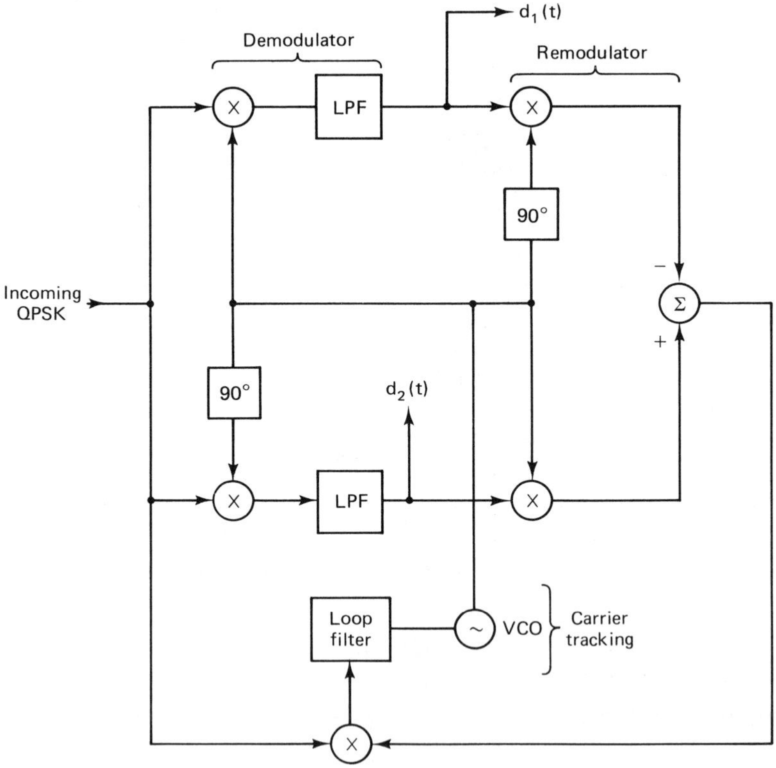

Figure 3.25 QPSK remodulation loop.

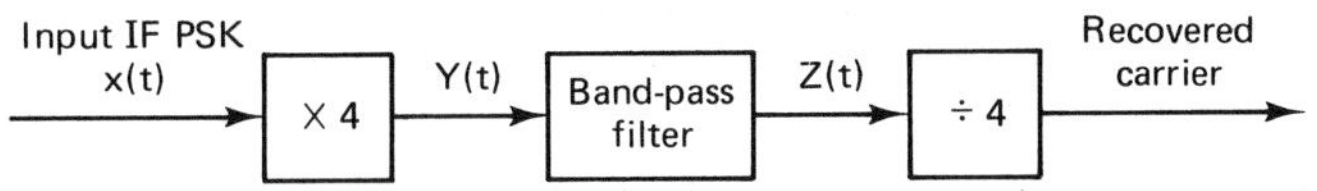

Figure 3.26 Multiply-filter-divide circuit.

where

$$x(t) = \sin\left[\omega_c t + \frac{2\pi(i-1)}{4}\right] \tag{3.28}$$

and

$$i = 1, 2, 3, 4 \text{ QPSK}$$

Substituting equation (3.28) into (3.27) and using trigonometric identities, we have

$$\begin{aligned} Y(t) = {} & \frac{3}{8} - \frac{1}{2}\cos[2\omega_c t + \pi(i-1)] \\ & + \frac{1}{8}\cos[4\omega_c t + 2\pi(i-1)] \end{aligned} \tag{3.29}$$

The output of the band-pass filter, $Z(t)$, contains the desired harmonic at $f = 4f_{IF}$ with phase zero (modulo 2). Finally, frequency division by four yields the desired coherent carrier of frequency ω_c. Once the carrier is recovered, the four-phase PSK can be demodulated synchronously, as shown in Figure 3.27. The results of a four-phase demodulated signal are tabulated in Table 3.4 (see page 60).

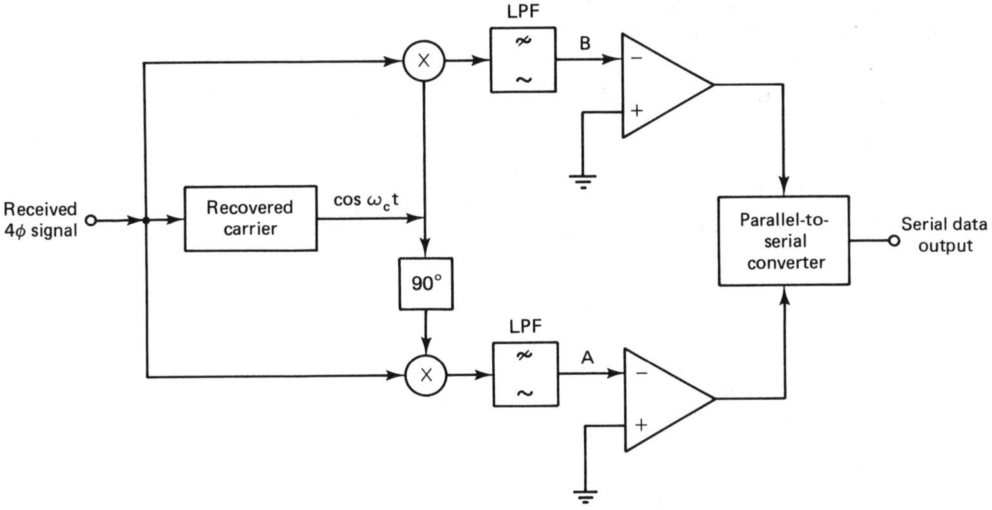

Figure 3.27 Four-phase demodulator.

TABLE 3.4 FOUR-PHASE DEMODULATOR SIGNAL

Received signal phase	*A* output	*B* output	Data output *A*	Data output *B*
45°	$\frac{1}{2}\cos 45° = 0.35$	$\frac{1}{2}\cos 45° = 0.35$	0	0
135°	$\frac{1}{2}\cos 45° = 0.35$	$\frac{1}{2}\cos 135° = -0.35$	0	1
225°	$\frac{1}{2}\cos 135° = -0.35$	$\frac{1}{2}\cos 225° = -0.35$	1	1
315°	$\frac{1}{2}\cos 225° = -0.35$	$\frac{1}{2}\cos 315° = 0.35$	1	0

3.3.4 Differential PSK

Differential PSK represents a form of signaling wherein the coding and signaling schemes are closely intertwined. Consequently, an argument could be made that a discussion of this technique be deferred to Chapter 4, which deals with coding. However, since it is referred to both as differential PSK and differential coding, we discuss it now.

Differential encoding is a scheme whereby the bit transmitted is arrived at by comparing a pair of adjacent bits. For example, in convential PSK a logic 0 may correspond to a 0° phase shift, whereas a logic 1 corresponds to a 180° phase shift of the transmitted signal. Differential encoding of the data can consist of a situation where a logic 1 corresponds to a 0° phase change between successive signals and a logic 0 corresponds to a 180° phase shift between two adjacent signals.

A block diagram for a differential PSK encoder (DPSK) is shown in Figure 3.28. At the output of the exclusive-NOR, a logic 1 is produced if the two inputs are the same and a logic 0 is produced if the inputs differ. Encoding for a typical data bit stream is shown in Table 3.5. To start encoding, a reference logic 1 is

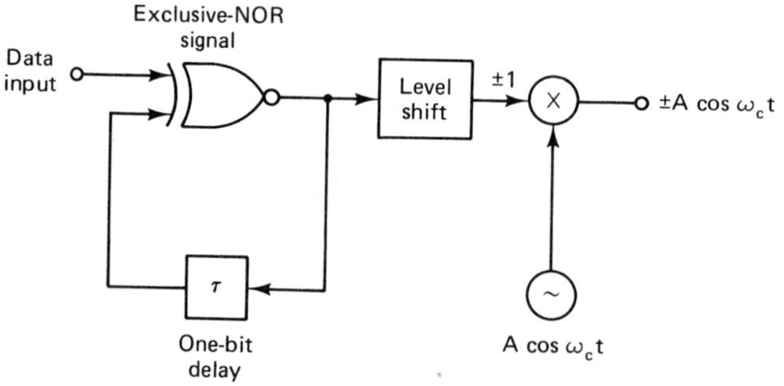

Figure 3.28 DPSK modulator.

TABLE 3.5 ENCODING FOR A TYPICAL DATA BIT STREAM

Data input		1	1	0	1	1	1	0	0	1	1	0
Encoded message	1	1	1	0	0	0	0	1	0	0	0	1
Reference digit	↑											
Transmitted phase	0	0	0	π	π	π	π	0	π	π	π	0

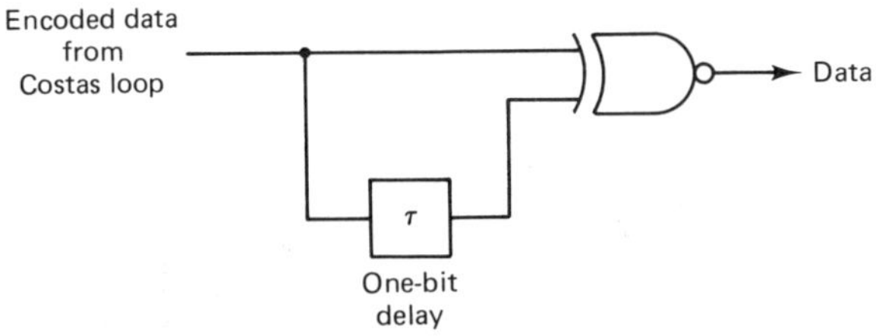

Figure 3.29 Differential data decoder.

TABLE 3.6 DIFFERENTIAL DATA DECODING

Encoded message	1	0	0	1	1	1	1	0	1	1	1	0	
Shifted message		1	0	0	1	1	1	1	0	1	1	1	0
Data		0	1	0	1	1	1	0	0	1	1	0	

chosen. The 1-bit delay allows the present bit to be compared with the previous bit. The encoded data are then level shifted to ± 1 and the output of the multiplier is the final DPSK signal. That is, the carrier is phase shifted 0 or π rad.

The block diagram for a data decoder is shown in Figure 3.29. The decoded data are shown in Table 3.6.

3.3.5 Other Digital Modulation Techniques

At this point we begin to see that a multitude of digital signaling techniques are available to us. To the basic ones we have discussed we can add others. One of these is staggered QPSK, or SQPSK. In a standard QPSK signal, an instantaneous phase shift of 180° occurs whenever both data signals change at the same time. If only one data signal changes state, the phase shift is either plus or minus 90°. Furthermore, if the data rates are equal, the phase shifts occur at intervals that are at least equal to the channel symbol interval.

In SQPSK, we synchronize the two data channels and then delay one by one-half symbol period with respect to the other. With this accomplished, the two data signals cannot change state simultaneously. Transitions in phase occur at two times the rate of QPSK. The instantaneous phase shift of the carrier cannot exceed 90°. The generation of SQPSK is shown in Figure 3.30 (see page 62). An important characteristic of SQPSK is that if we bandlimit and then amplitude limit the signal, the power spectrum remains bandlimited. By way of contrast a band-limited BPSK or QPSK signal tends to return to its original power spectrum when amplitude limited. As a result SQPSK is useful in satellite communications for bandwidth conservation. A traveling wave tube (TWT) amplifier is often operated in the saturated mode in satellites.

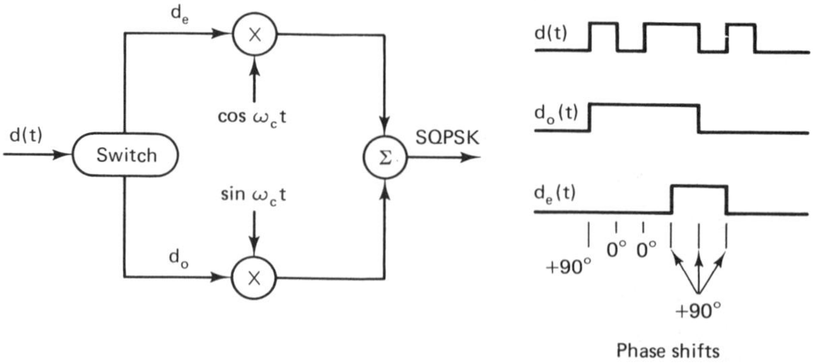

Figure 3.30 SQPSK generation.

M-ary PSK (MPSK) is an extension of the idea of QPSK to a large number of phases. A classical eight-phase modulator is illustrated in Figure 3.31 below. The multiplier splits the incoming f_b data rate into three parallel binary data streams, each with a data rate equal to $f_b/3$. The *a* and *b* outputs are one of four analog levels (see Example 3.4). These levels are multiplied with the output of the local oscillator and both modulator outputs are summed together. Referring to the signal-state-space diagram in Figure 3.31b, we see that all eight signals possess equal amplitudes and are spaced at 45° on a circle. The band-pass filter

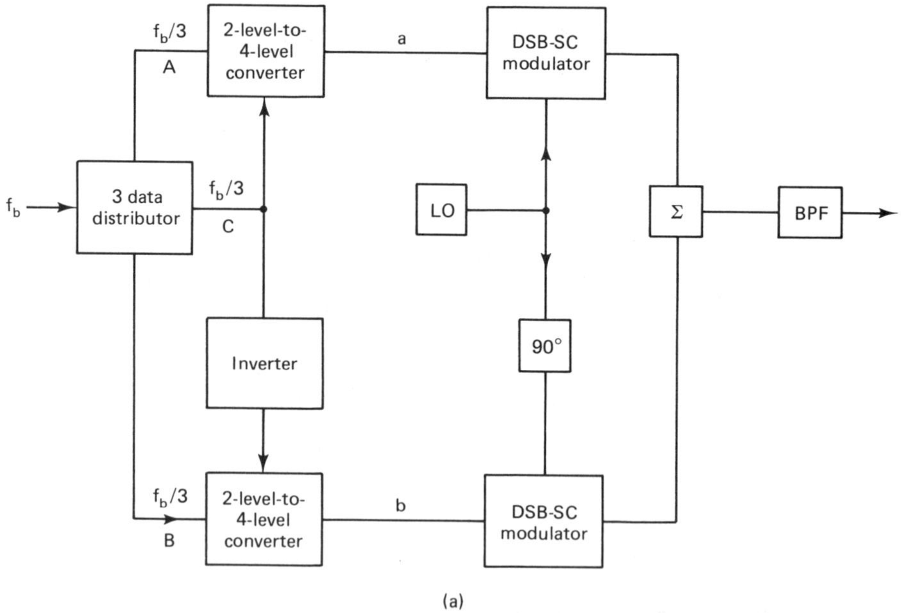

Figure 3.31 Classical eight-phase PSK modulator block and signal-state space diagrams.

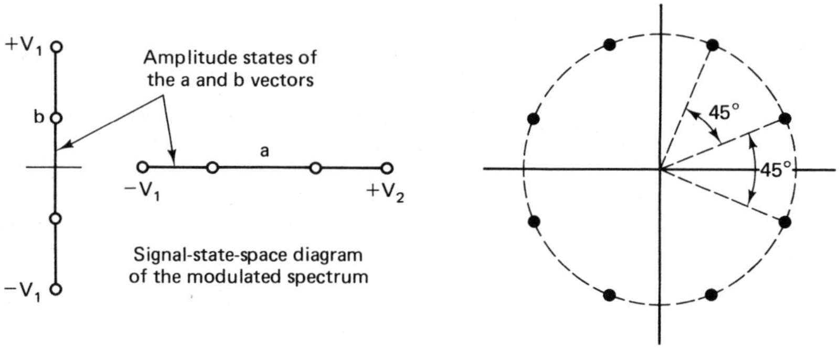

(b)

Figure 3.31 *(Continued)*

output is a bandlimited 8-PSK signal. Sixteen-ary and 64-ary PSK has also received considerable attention in recent years.

EXAMPLE 3.4

Assume that the discrete baseband levels at the output of the 2-level to 4-level converters in Figure 3.31a can have the following values: ± 10 MV and $\pm X$ MV. If the maximum value is 10 MV, find the value of X such that 45° spacing is achieved in the signal-space diagram. Draw the signal-state-space diagram if $X = 2.5$ MV.

Solution

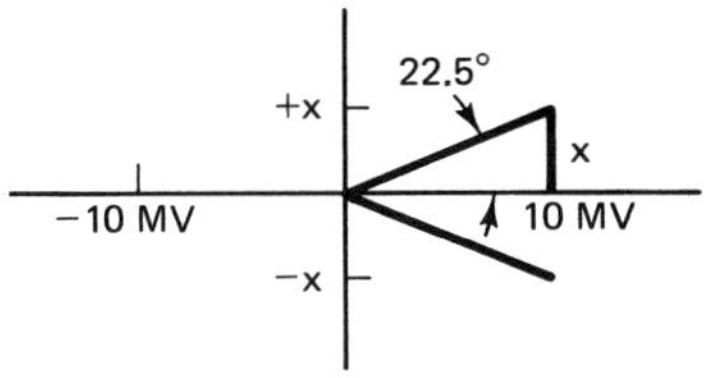

$$\tan 22.5^\circ = \frac{x}{10}$$

$$x = 4.412 \text{ MV}$$

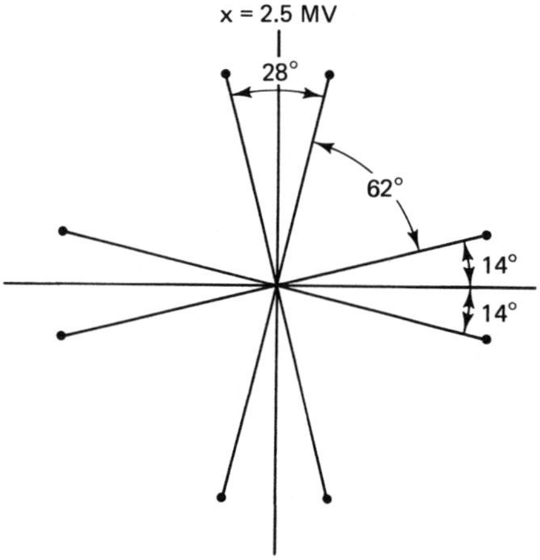

$$\tan\theta = \frac{2.5}{10}$$

$$\theta = 14.04^\circ$$

M-ary FSK is an extension of binary FSK (see Section 3.3.1). Each symbol represents $\log_2 M$ data bits, where M is equal to a power of 2. MFSK, however, occupies a large bandwidth and offers no performance advantage. When the phase continuity of a modulated FSK signal is preserved at the instants of modulating bit stream transitions, a signal form called *continuous-phase FSK* (CP-FSK) is created. It is also often referred to as *minimum shift keying* (MSK). This technique is receiving increasing attention for data communications because it uses less bandwidth than BPSK, is easier to synchronize than QPSK, and performs as well as QPSK. Many signaling techniques are available. In the next section we further compare a number of these techniques to aid the systems designer in choosing one.

3.3.6 Digital Modulation Techniques Compared

There are two theorems central to an understanding and comparison of digital modulation. These are the theorems of Nyquist and Shannon, both of which relate bandwidth to transmission rate. We are already acquainted with Nyquist's theorem (see Chapter 1). Nyquist's theorem relates bandwidth to the symbol rate or modulation rate, R, where a symbol represents 1 or more bits. This theorem states that a maximum of $2W$ symbols per second can be transmitted over a bandwidth of W hertz. The symbol rate R is also known as the baud rate in terms of data transmission terminology.

As an example, a 3-kHz telephone circuit can transmit a maximum symbol rate of 6 kilobaud. This maximum can be achieved only with single-sideband (SSB) modulation. Recall that modulation produces sideband frequencies above and below the carrier, with both sidebands containing the same information. One of the sidebands must be eliminated in order to transmit the maximum amount of information in the given bandwidth. Removal of a sideband in FM or PM would result in a signal with varying amplitude. As a result, the symbol rate usually is limited to W baud. Our telephone circuit, then, is limited to 3000 baud. Actually, better performance is achieved by limiting the rate to 2400 or less. Double-sideband transmission at 1 symbol/Hz can be achieved. However, it requires filters with extremely sharp cutoff at the band edges. Digital modulation produces a $\sin x/x$–type spectrum with a resultant need to confine the power to the given bandwidth. A rate of 0.8 to 0.9 symbol/Hz is easier to achieve.

As stated previously, Nyquist's theorem relates channel bandwidth to bit rate. Shannon's theorem relates channel bandwidth to bit rate, or information rate, and to the channel carrier-to-noise ratio. This result is expressed by equation (1.3) and is repeated here:

$$C = W \log_2 \left[1 + \frac{C}{N} \right] \quad \text{bits/s} \tag{3.30}$$

This equation states the maximum information rate or channel capacity, C, in the presence of random noise. For example, a telephone channel with a 3000-Hz bandwidth has a channel capacity of 39,864 bits/s. Shannon's result states that it is possible to achieve this rate free of error. The Shannon rate has not been reached in practical systems. The application of M-PSK achieves an increase in the information rate but with an increase in the required SNR for a specified BER. The closer we try to approach the predictions of equation (3.30), the more complex the equipment becomes.

Actually, the medium over which we must transmit limits our choices of modulation techniques. For example, microwave relays as well as satellite circuits use repeaters with restricted amplitude linearity. Different amplitude levels can be used, but amplitude level tends to become a problem. Intermodulation results if the repeaters are operated too close to their power limits. Constant-amplitude FSK and PSK techniques are more suitable. PSK is favored because of the reduced bandwidth. Above 20 GHz, however, phase-jitter problems outweigh spectrum economy and FSK is preferred. The choice is FSK in other situations also. On circuits that depend upon radiowave reflections from the ionosphere, phase distortions and rapid amplitude changes often occur because of ionospheric changes. This is characteristic of high-frequency (HF) digital circuits (3 to 30 MHz). Often FSK is also used on cable circuits (1200 bits/s and below) wherever the occupied bandwidth does not press the limits of the voice channel. At higher data rates, however, PSK is the desired choice for spectrum conservation. Spectrum economy is often achieved by a combination of PSK and ASK, also called QAM (see Section 3.3.6.1).

The large bandwidth of fiber optic systems is a continuing interest to users. The linearity of the light sources does present some problems. The most satisfactory mode of operation appears to be the use of two amplitude levels, low and high. This results in only 1 bit/symbol. However, the large bandwidth available helps to compensate for this. Even so, a fiber optic system capable of a 400 megabit/s transmission rate can handle only five digitized channels at 90 megabits/channel. However, good digital video at 30 megabits/s using monomode fibers capable of 600 to 900 megabits between repeaters are improving the prospects for digitized cable TV. A detailed discussion of fiber optics is contained in Chapter 10.

3.3.6.1 Multilevel modulation. Transmission of more than 1 bit/symbol requires modulation techniques that incorporate a number of digital states. Specifically, for n bits/symbol, 2^n states are needed. As previously seen, in PSK systems the digital states consist of equally spaced phases (see Figures 3.16 and 3.31). The 8-PSK system of Figure 3.31 provides 3 bits/symbol. Diagrams of this type are often referred to as signal-state-space diagrams, or constellations. The bit combinations can be assigned in various ways to the different phases. However, adjacent phases are usually made to differ by only 1 bit to minimize the effect of phase error.

One technique that is used is a combination of phase shifting and amplitude keying. A typical signal waveform for four symbol periods using both ASK and

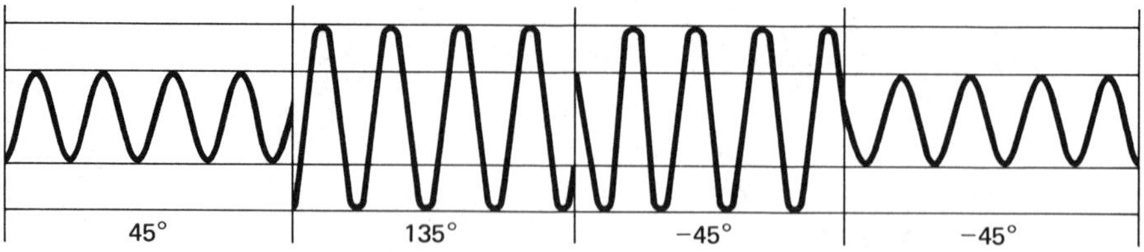

(a)

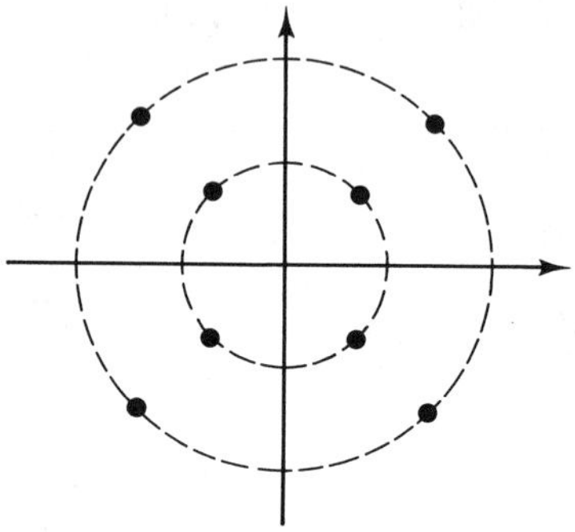

(b)

Figure 3.32 (a) ASK and PSK signals are used in a digital communications system. (b) The two amplitude levels and four phases yield eight distinct combinations and therefore three bits of digital information.

PSK is shown in Figure 3.32. A 3-bit code at the transmitter determines the signal state. A variation on this theme is QAM, often called amplitude phase keying (APK). Generation is accomplished by multilevel amplitude modulating two carriers that are shifted in phase by 90°. Figure 3.33 illustrates the multilevels for two modulation levels. When this modulation is performed simultaneously on both carriers, a total of L^2 states results, where L is the number of amplitude levels on each of the two quadrature carriers. The four amplitude levels of Figure 3.33 can be used to produce the signal-state space shown in Figure 3.34. Work on 64-QAM microwave systems has been done.

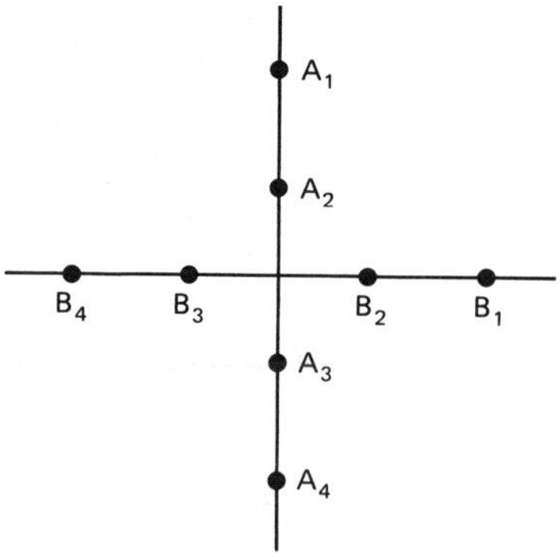

Figure 3.33 QAM multiamplitude levels.

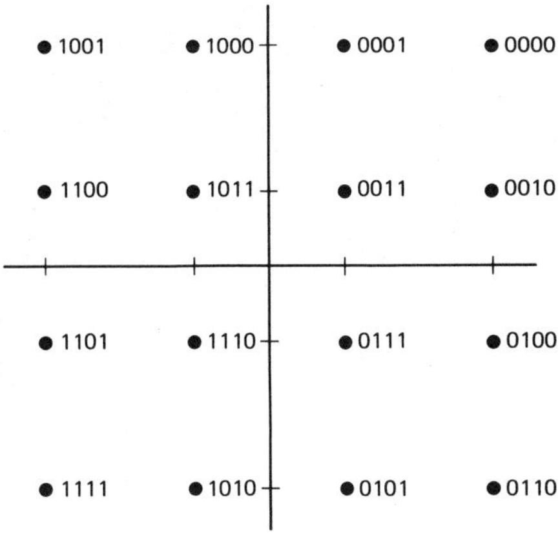

Figure 3.34 Sixteen-QAM signal state space.

3.3.6.2 Correlative level encoding. Another way of transmitting a data stream involves modifying the stream before it is placed on a carrier. Methods such as this may be referred to as *baseband techniques*. Correlative level encoding is a baseband technique in which a predetermined amount of ISI is intentionally introduced at the transmitter and then removed at the receiver. The ISI can be added either digitally (Figure 3.35) or by analog filtering (Figure 3.36). This cor-

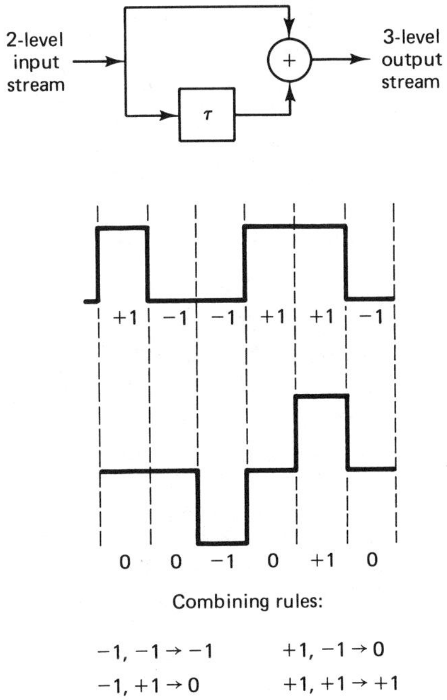

Figure 3.35 Combining adjacent bits in a data stream provides ISI that produces bandwidth.

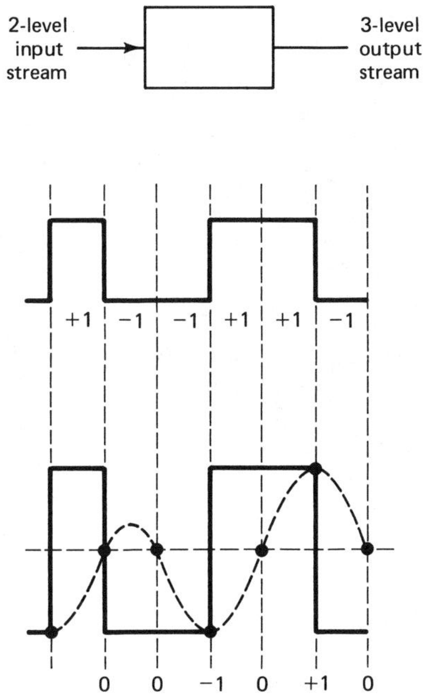

Figure 3.36 Overfiltering the data stream is another method of introducing intersymbol interference.

relative technique is sometimes called *partial response signaling* (PRS). If only adjacent bits or symbols are used, the technique is called *duobinary*. If bits or symbols separated by one keying interval are combined, the technique is called *modified duobinary*. Separation by two or more keying intervals results in a technique called *polybinary*. Longer correlation spans result in more energy being concentrated at lower frequencies in the baseband, thus conserving spectrum. However, this also requires larger memories and greater transmitter power.

EXAMPLE 3.5

Discuss a decoding technique for the duobinary example shown in Figure 3.35.

Solution Examining the combining rules, the inverse rules can be stated regarding subtraction of adjacencies at the three-level output.

- +1, decide 1
- −1, decide 0
- 0, decide opposite of previous decision

The required circuit is shown in the accompanying figure.

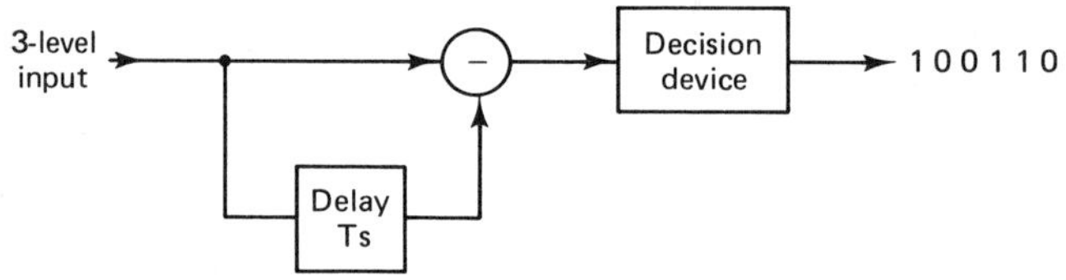

3.3.6.3 Noise effects. At the receiver designed to demodulate MPSK, the detector must be able to distinguish differing phases in the presence of noise. In Figure 3.37 it is clear that a noise vector that results in a phase shift of more than ±22.5° will cause a symbol error. It is obvious that as the number of phases increases, the system becomes more susceptible to channel noise. The distance between points is related by the following:

$$\sin \theta = \sin \frac{360°}{2n} = \frac{d/2}{D}$$

or

$$\frac{d}{D} = 2 \sin \frac{\pi}{n} \tag{3.31}$$

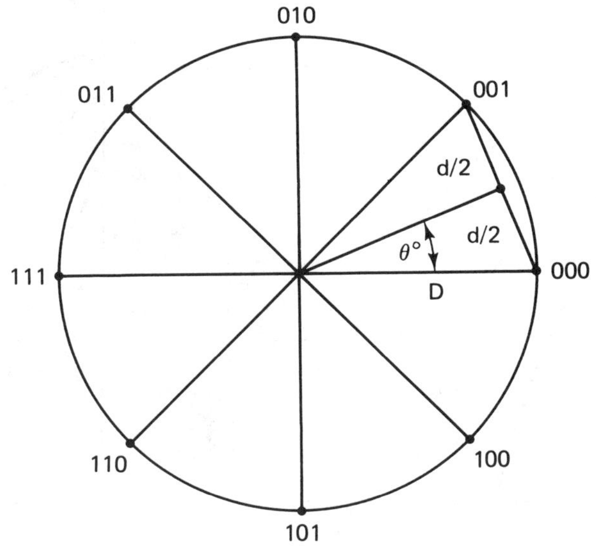

Figure 3.37 The eight states of a 3-bit/symbol PSK modulation system can be diagrammed as shown.

It can be shown that the BER is given by

$$\text{BER} = (\log_2 n)^{-1} \operatorname{erf}(z) \tag{3.32}$$

where

$$z = \sqrt{\sin \frac{\pi}{n} (\log_2 n) \frac{E_b}{N_0}}$$

and the error function erf(z) is given by

$$\operatorname{erf}(z) = \frac{2}{\sqrt{\pi} \int \exp(-t^2)\, dt}$$

In this expression E_b/N_0 is the ratio of the energy per bit to the noise power spectral density. This equation is plotted in Figure 3.38 for different values of n.

In multilevel APK (MAPK) systems, the detector in the receiver must be able to distinguish between different amplitudes' phase points in the presence of noise. The more points, the more susceptible the system is to noise. Since amplitude as well as phase is used, the system is less susceptible to noise. For an MAPK using L levels, the distance between points is related to the peak amplitude

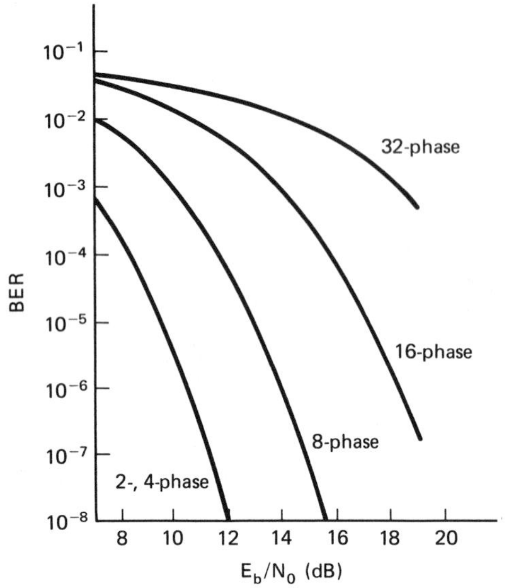

Figure 3.38 The more phases in an MPSK system, the more susceptible it is to bit errors caused by noise.

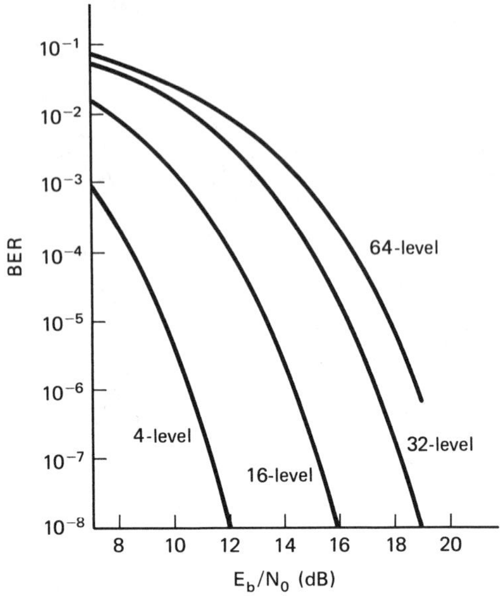

Figure 3.39 An MAPSK is less susceptible to noise error than is an MPSK system at higher levels.

by the following expression:

$$\frac{d}{D} = \frac{\sqrt{2}}{L - 1} \tag{3.33}$$

A plot of the BER versus E_b/N_0 is shown in Figure 3.39. Comparing the BERs for a BER $= 10^{-6}$, note that 16-PSK requires $E_b/N_0 = 18$ dB versus 14.5 dB for a 16-level MAPK system.

A technique known as *forward error correction* can be used to improve the error rates of Figure (3.38) and (3.39). This technique is discussed in the next chapter. Basically, the technique involves the use of redundant bits to detect and correct errors in real time at the receiver without retransmission. The addition of redundancies, however, results in a slower bit rate and the circuitry becomes more complex. Developments in integrated circuits, however, are making circuit complexity less of a factor in system design.

Figure 3.40 compares a number of modulation techniques. Numerals on the plots are the number of signaling states. For comparison, the Shannon limit for a bandlimited channel is shown. The curves are based on a 10^{-4} BER. The curves can be interpreted as follows: the point marked BPSK is at the coordinates (8.4 dB, 1 bit/s/Hz). This means that using BPSK, 1 bit/s/Hz of bandwidth can be transmitted with a probability of error $P_e = 10^{-4}$ if $E_b/N_0 = 8.4$ dB. As another example, the point marked QPSK indicates that the data rate can be doubled to

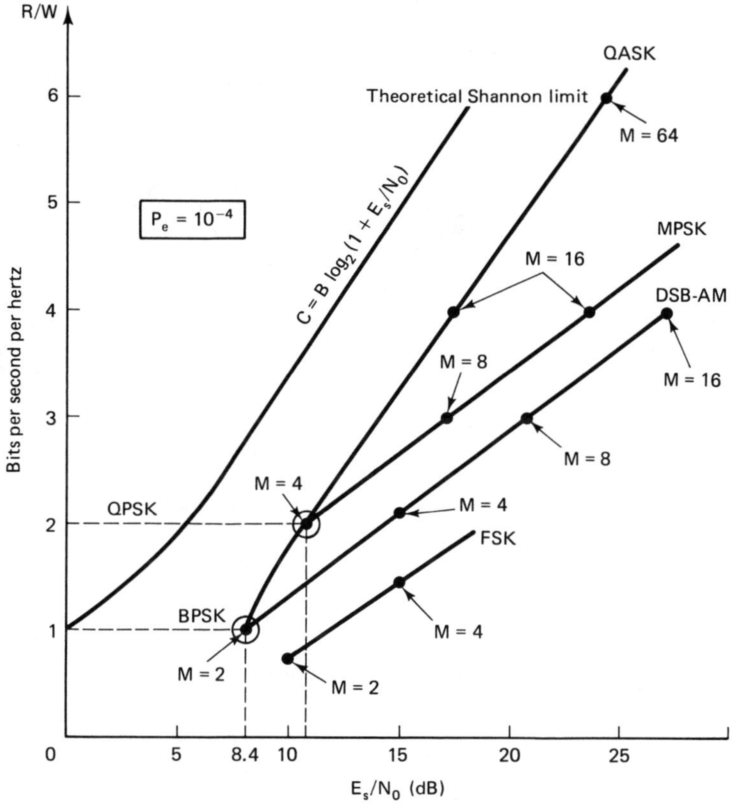

Figure 3.40 Comparison of modulation techniques.

2 bit/s/Hz if E_b/N_0 is increased to 11.4 dB (a 3-dB increase over BPSK). That is, to maintain equal performance, the power must be doubled if the information rate is doubled. Similar curves can be plotted for other BER values.

3.4 BANDWIDTH EFFICIENCY PLANE

The bandwidth efficiency of different modulation techniques can be illustrated on a bandwidth efficiency plane (Figure 3.41, page 72), in which the abscissa is energy per bit per unit of N_0 instead of carrier power versus N_0. We can rewrite equation (3.30) as

$$\begin{aligned} C &= W \log_2\left(1 + \frac{S}{N}\right) \\ &= W \log_2\left[1 + \frac{E_b R}{N_0 W}\right] \end{aligned} \tag{3.34}$$

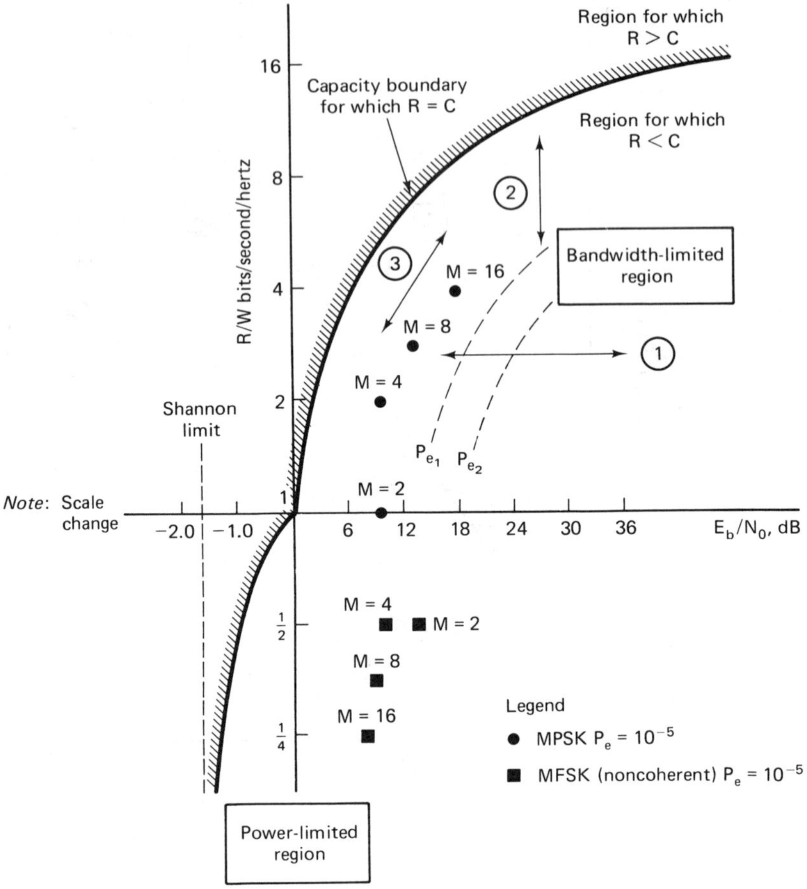

Figure 3.41 Bandwidth efficiency plane.

where

E_b = energy/bit
R = bit repetition rate
N_0 = noise spectral density in watts per hertz

If we take the antilog of equation (3.34), we can rewrite this equation as

$$\frac{E_b}{N_0} = \frac{W}{R}\,[2^{C/W} - 1] \tag{3.35}$$

This equation is plotted on the R/W versus E_b/N_0 plane in Figure 3.41. This plane is called the *bandwidth efficiency plane*. Since the ordinate R/W is a measure of how much data can be transmitted in a specified bandwidth in a given time, it reflects how efficiently the bandwidth resource is used. The abscissa, E_b/N_0, is in decibels. A bit rate $R > C$ cannot be supported by the communication system. For $R < C$, potentially error-free communication is possible.

Operating points for MPSK and FSK modulation are plotted on the plane, each at $P_e = 10^{-5}$. Note that for MPSK modulation, R/W increases with increasing M. On the other hand, for MFSK modulation, R/W decreases with increasing M. For the same E_b/N_0, QPSK ($M = 4$) has a bandwidth efficiency of 2 bits/s/Hz compared to 1 bit/s/Hz for BPSK ($M = 2$). This feature originates from the fact that QPSK is a composite of two BPSK signals transmitted on waveforms orthogonal to one another, both having the same spectral occupancy. Also plotted on the bandwidth efficiency plane are operating points for noncoherent MFSK modulation at BER = 10^{-5}. Note that binary FSK ($M = 2$) and quaternary FSK ($M = 4$) have the same bandwidth efficiency even though BFSK requires a greater E_b/N_0 for the same error rate. Bandwidth efficiency for noncoherent FSK varies with the modulation index.

Assume that an equal increment of bandwidth is required for each MFSK tone that the system must support; then it can be seen that for $M = 2$, $R/W =$ 1 bit/s/(2 Hz) $= \frac{1}{2}$ and for $M = 4$, $R/W =$ 2 bits/s/(4 Hz) $= \frac{1}{2}$. System trade-offs are indicated on the figure. Movement of the operating point along line 1 is equivalent to a trade-off between the BER versus E_b/N_0 performance, with R/W fixed. Movement along line 2 represents a trade-off of BER versus the bandwidth, W, with E_b/N_0 fixed. Finally, movement along line 3 illustrates trading bandwidth versus E_b/N_0 performance, with BER fixed. We can effect movement along line 1 simply by changing E_b/N_0. Movement along lines 2 or 3 can be effected by appropriate changes to the system modulation or coding scheme.

The power-limited region represents a system in which system bandwidth is available but power is limited. This could be, for example, a space-to-earth link. The following trade-offs can be made:

- Improved BER can be achieved by expanding bandwidth (for a given E_b/N_0).
- The required E_b/N_0 can be reduced by expanding bandwidth (for a given BER).

We see that the bandwidth efficiency plane provides a ready reference for system design trade-offs. Using this plane, we can verify whether or not a candidate encoding or modulation method offers improvement in the required E_b/N_0 (coding gain) for a specified BER or whether the method offers improvement in the BER for a given E_b/N_0. The subject of coding gain is treated in Chapter 4.

3.5 BANDWIDTH COMPRESSION

Digital encoding of an analog signal expands the bandwidth that is required. Generally speaking, the signal is sampled at a rate somewhat greater than the Nyquist rate. Each sample is quantized to a given number of levels, which determines the number of bits in the digital word that represents the sample. The basic bandwidth expansion over the analog signal bandwidth is the sampling rate multiplied by the number of levels, which is then divided by the analog bandwidth.

This increase in baseband requirements imposed by elementary coding schemes particularly impacts satellite communication services. The fixed bandwidth allocations for satellites require bandwidth compression techniques. At the same time, signal-quality standards must be met. Regardless of the bandwidth-compression technique chosen, we must allow for signal degradation imposed by the bandlimiting by channel filters. This means that a lower BER at the receiver output must be chosen.

One technique for substituting digital processing for transmitted signal power is called *forward error-correction* coding. Using this technique, redundancy is introduced into the basic bit stream, which is then encoded into blocks. Coding algorithms establish block lengths and contents so that particular bits in certain block locations can be checked as indications of error. This technique is discussed in Chapter 4.

3.5.1 Bandwidth Compression: How It Works

The exponential variation of the required SNR with bandwidth results in a slope of 6 dB for every doubling of the bandwidth (or bit rate). This means that twice as many pulses can be used for every doubling and the number of amplitude levels is squared in the coding scheme. A quantizer having equal quantization steps will generate varying SNRs as the amplitude of the analog signal varies. We can reduce the bit rate significantly by quantizing with step sizes that increase as the signal amplitude increases. This technique is called *companding* (see Section 2.1.2); it maintains the desired SNR and increases the dynamic range with a consequent reduction in bandwidth.

Delta modulation was discussed in Chapter 2. Using this technique, only differences in sizes between successive samples are encoded. If we choose to use adaptive delta modulation, the quantization steps vary with amplitude, similar to companding. For example, adaptive delta modulation applied to digitally encoded speech can reduce the bit rate required for quality voice transmission from 64 to 16 kilobits/s. Comparable techniques applied to television results in compressions of 4:1 in bandwidth.

Refer to Figure 3.42. Bandwidth compression can be achieved by the use of a phase-amplitude signal space to represent the bit in a word with complex voltage vectors. We discussed these modulation techniques in previous sections. For example, 16-level quadrature ASK, in which 4 phases and 4 amplitude levels are combined to offer the equivalent of 4 bits/symbol (Figure 3.34) via the 16 available state vectors, has been found to exhibit a 6-dB advantage over 16-phase PSK in the SNR required for BER $= 10^{-6}$, with both having the same information content per unit bandwidth. In general M-ary QASK exhibits advantages over M-PSK for more than 4 states.

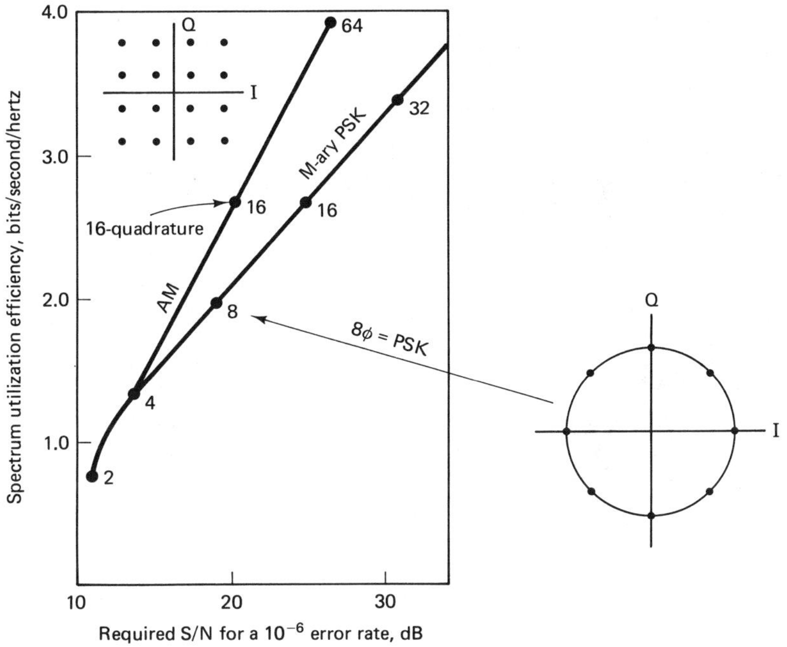

Figure 3.42 Relative modulation efficiencies of multiple-level phase and amplitude modulation.

Convolutional coding, in which the coding and convolution of blocks in the coded transmission follow particular algorithms, can be used to provide powerful compression with BERs at least as good as forward error–correcting coding. Automatic repeat query is used with this technique. Convolution coding is discussed in Chapter 4. If automatic repeat query is used with forward error–correcting coding, bit error rates as low as 10^{-9} can be achieved. We can see from this discussion that coding and bandwidth compression techniques are intimately related and will continue to be an area of active research. Coding techniques are discussed in the next chapter.

PROBLEMS

3.1. Digital communication from a satellite in geostationary orbit (36,000 km from the earth) to an earth station is to be investigated. The transmission power is 10 W, the frequency is 250 MHz, and the antenna gains of the satellite and Earth station are 10 and 30 dB, respectively. The additive white Gaussian noise (AWGN) has a spectral density $N_0 = 10^{-20}$ W/Hz, and an error probability of 10^{-5} is desired.

(a) Assuming ideal PSK modulation, what is the maximum data rate R this system allows?

(b) The binary data uses M-ary orthogonal modulation. Find the maximum binary rate for $M = 32$.

(c) The error probability is desired to be vanishingly small. If an arbitrarily complex encoding scheme and an arbitrarily large bandwidth are allowed, what is the maximum transmission rate possible? Compare with parts (a) and (b).

3.2. If the main objective is improved BER performance, which of the following two modulation schemes would you select? Show computations.

$$\text{Binary coherent FSK with } \frac{E_b}{N_0} = 12 \text{ dB}$$

$$\text{Binary noncoherent FSK with } \frac{E_b}{N_0} = 14 \text{ dB}$$

3.3. If your main objective is improved BER performance, which of the following two modulation schemes would you select? Show computations.

$$\text{Binary noncoherent FSK with } \frac{E_b}{N_0} = 13 \text{ dB}$$

$$\text{Binary coherent PSK with } \frac{E_b}{N_0} = 8 \text{ dB}$$

3.4. A digital signal is received from a satellite with an SNR of $P/N_0 = 20$ dB.

(a) What is the maximum data transmission rate possible with an FSK modulation if the BER is required to be 10^{-5}?

(b) Compare your result with the capacity of this system if a bandwidth of 100 Hz is available. What is the required E_b/N_0 at capacity? What is the bandwidth efficiency of this system?

(c) If an arbitrarily large bandwidth and an arbitrarily complex system are available, what is the maximum transmission rate possible? Compare with part (a).

3.5. A binary PSK demodulator operates at the rate of 1000 symbols/s. During any 24-h period it makes 100 bit errors, on the average. The system noise density $N_0 = 10^{-10}$ W/Hz.

(a) What is the system demodulation rate in bits per second?

(b) What is the bit-error probability?

(c) If the value of received signal power per bit is adjusted to be 10^{-6} W, will such received power be adequate to maintain the error rate found in part (b)?

3.6. Consider a binary transmission system whereby each bit received in error is automatically retransmitted. If p is the bit-error probability on the forward channel, what is the maximum average rate (in bits per channel use) that can be used over this channel with no errors delivered to the user?

3.7. **(a)** If the symbol error rate is $P_E = 10^{-5}$, what is the BER for a 16-ary PSK system?

(b) Repeat part (a) for a 16-ary FSK system.

3.8. Why is a 111111 pattern used for clock recovery in differentially encoded PSK?

3.9. The output of the phase detector of a PLL is given by $[\sin \phi(t)]$, where $\phi(t)$ is the phase-error signal. Why is a sine function used? Can the phase-detector output be $\cos[\phi(t)]$? Explain.

3.10. A QPSK demodulator produces quadrature data outputs. Why is the overall error rate the same as that for either of the quadrature data paths? Assume that the data are random and equally likely to be ± 1 at the input to the QPSK modulator.

3.11. We wish to transmit a 100-character alphanumeric message within 2 s using 7-bit ASCII coding followed by an 8-bit parity check per character. The modulation scheme is 32-ary PSK. Assume there are no special start, stop, or synchronization symbols.

(a) Calculate the bit rate, the symbol rate, and the number of baud. Provide an analytical expression for the set of digital waveforms used in the transmission process.

(b) Repeat (a) for 16-ary PSK.

(c) Repeat (a) for 8-ary PSK.

(d) Repeat (a) for 4-ary PSK.

(e) Repeat (a) for binary PSK.

(f) Repeat (a) for M-ary PSK.

3.12. Explain why QPSK has the same ideal error rate as BPSK.

3.13. Show that the rules for accomplishing differential encoding are

$$\text{BPSK:}\quad C_n = C_{n-1} \oplus D_n$$

where C_n represents the encoded data, D_n represents the input data, and $\oplus$ denotes the logical exclusive-or operation, and

$$\text{QPSK:}\quad C_n = C_{n-1}\overline{I}_n\overline{Q}_n + \overline{C}_{n-1}I_nQ_n + D_{n-1}I_n\overline{Q}_n + \overline{D}_{n-1}\overline{I}_nQ_n$$

$$D_n = C_{n-1}\overline{I}_nQ_n + \overline{C}_{n-1}I_n\overline{Q}_n + D_{n-1}\overline{I}_n\overline{Q}_n + \overline{D}_{n-1}I_nQ_n$$

where C_n and D_n are the quadrature differential encoder outputs and I_n and Q_n are the quadrature inputs to the encoder.

What are the corresponding decoding rules?

3.14. Find the average number of bit errors made in a day by an optimum coherent demodulator operating continually with binary PSK waveforms. The data rate is 5000 bits/s. The input digital waveforms are antipodal signals and can be expressed as:

$$s_1(t) = A \cos \omega_0 t$$

$$s_0(t) = -A \cos \omega_0 t$$

where $A = 1$ mV, and noise density $N_0 = 10^{-11}$ W/Hz. Assume that signal power and energy per bit are normalized relative to an impedance level of 1 Ω.

3.15. An integrate-and-dump circuit is used as a detector in various demodulators. The circuit integrates the received baseband data for a prescribed period of time, at which point the integrator output is sampled. Once the output has been sampled, the integrator is returned to a zero state, so that energy from one bit does not affect subsequent bit decisions. Suggest a circuit that would perform the function.

3.16. We want to transmit 800 characters per second, using 7-bit ASCII coding followed by an eighth parity check per character. The modulation is 16-ary FSK. There are no special start, stop, or synchronization symbols.

(a) What is the bit rate?

(b) What is the symbol rate?

(c) What is the number of baud?
(d) Provide an analytical expression for the set of digital waveforms used for modulation.

3.17. Explain why FSK with modulation index $0.5 \leq h \leq 1$ would be of primary interest in many communication systems.

3.18. An M-ary choice that is not generally used in an actual design is $M = 3$. Explain why not.

3.19. (a) Set $C = R$ in equation 3.35 and compute several points to verify Figure 3.41.
(b) A communication system is desired that will transmit 60 kilobits/s of data over a 10-kHz channel. The channel noise, N_0, is 10^{-19} W/Hz. If the received signal power is specified to be 10^{-12} W, is this a feasible design?

3.20. Quadrature QPSK is to be used to transmit two data streams, which differ in bit rate by a factor of 10. What must be the ratio of their amplitudes if the SNR of the demodulated data streams is to be the same? Draw a phasor diagram as in Figure 3.20.

3.21. A channel is bandlimited to 25 kHz and has an average power limitation of 10 W. The one-sided spectral density of the noise to the channel is 10^{-5} W/Hz.
(a) Find the maximum rate at which information can be sent over the channel without error.
(b) Find the limiting capacity of the channel if the bandwidth is not limited.

3.22. A message bandlimited to 4 kHz is transmitted over a channel that is bandlimited to 100 kHz. What is the smallest channel SNR in the message bandwidth that will yield an output SNR of 30 dB?

3.23. A message bandlimited to 4 kHz is sampled at the Nyquist rate and quantized to L levels, where $L = 2^q$. The message is transmitted over a binary channel with $E_b/N_0 = 25$. What is the value of q that will provide approximately equal contributions to the SNR from quantizing and channel noise and what is the resulting SNR? What is the channel bit-error probability? What is the bit rate for this channel?

3.24. Refer to Figure 3.23. Assuming a QPSK input to the Costas loop, derive an expression for e_f, the input to the loop low-pass filter.

3.25. Evaluate the spectrum for OOK (On-Off Keying) signaling that is keyed by a test pattern consisting of alternating 1s and 0s. Assume rectangular binary pulses occurring at a rate of 5000 bits/s and a carrier frequency of 100 kHz. Indicate the value of the first zero crossing of the spectrum.

3.26. Repeat Problem 3.25 for BPSK signaling.

3.27. A multilevel PCM system is to be used to transmit an analog signal. The number of levels is eight. The absolute bandwidth is 2.7 kHz and a full scale accuracy of $\pm 1\%$ is desired. The analog voltage varies between $\pm$ 1 V.
(a) Determine the required received SNR.
(b) Determine the minimum bit rate and baud rate of the PCM signal.
(c) Determine the channel bandwidth required.

3.28. Given the following PSK signal:

$$e_{PSK}(t) = A \sin[\omega_c t + \phi + d(t)\theta], \qquad 0 < t \leq T$$

where the binary data $d(t) = \pm 1$.

(a) For $\theta = \pi/2$, show that the PSK signal is a BPSK signal.

(b) For $0 \leq \theta \leq \pi/2$, show that in addition to a BPSK signal, a discrete carrier term also exists.

4

ERROR-CONTROL CODING

4.0 INTRODUCTION

In Chapter 3 we described signaling and encoding schemes for transmitting digital data. We observed that the probability of error for a particular digital modulation format was a function both of the SNR at the receiver and of the data rate. In a practical system, the available signal power and the bandwidth are, more often than not, limited. This relates directly to the probability of error. With these restrictions in place, the only practical alternative available for reducing the probability of error is to make use of *error-control coding*. This technique is also referred to as *channel coding*.

Specifically, error-control coding is the deliberate use of redundancy. The channel encoder adds digits to the transmitted message according to predetermined rules. The additional digits do not add any information themselves but do make it possible for the channel decoder to detect and correct errors in the information digits. Through the device of detecting and correcting errors we lower the overall probability of error. In this chapter we wish to become familiar with the characteristics of different coding schemes and means of implementation. We also wish to consider combined modulation (signaling) and coding approaches that are applicable to satellite and terrestrial digital transmission.

4.1 THEORETICAL BASIS FOR CHANNEL CODING

In the receiver, in order ultimately to recover the information, symbol detection must be performed. Even in the absence of noise, the symbols suffer from ISI. The tails of one pulse spill over into the adjacent symbol intervals and interfere with correct detection (see Figure 1.2). As discussed in Chapter 1, Nyquist showed that the theoretical minimum bandwidth needed to transmit X symbols per second without ISI is $X/2$ hertz. In practice a binary system typically requires X Hz of bandwidth for the transmission of X symbols/s. Consider an M-ary system. For a fixed bandwidth, as M increases, the bandwidth efficiency, R/W, measured in bits per second per hertz increases. To illustrate this, Figure 3.38 is repeated as Figure 4.1. Movement along the horizontal line in Figure 4.1 from point a to point b represents trading E_b/N_0 for a reduced bandwidth requirement. With the same system bandwidth, we can transmit at an increased data rate and thus at an increased R/W.

Shannon demonstrated that the system capacity C for channels perturbed by additive white Gaussian noise (AWGN) is a function of the average received signal power, S, the average noise power, N, and the bandwidth, W. This capacity relationship, known as the Shannon-Hartley theorem, is given by equation 1.3 and is repeated here:

$$C = W \log_2 \left(\frac{S}{N} + 1 \right) \tag{4.1}$$

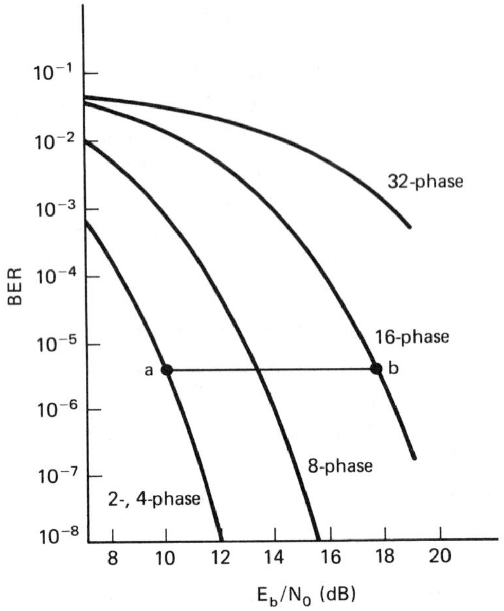

Figure 4.1 The more phases in an MPSK system, the more susceptible it is to bit errors caused by noise.

Shannon's theorem for noisy channels states that it is possible to transmit information over such a channel at a rate R, where $R \le C$, with an arbitrarily small error rate by using a sufficiently complicated coding scheme. If $R > C$, it is impossible to find a code that can achieve an arbitrarily small error rate. Specifically, Shannon's work demonstrated that values for S, N, and W set a limit on the transmission rate, not on accuracy.

Using equation (4.1) it can be shown that the required E_b/N_0 approaches the required Shannon limit of -1.6 dB as the bandwidth, W, increases without bound. That is,

$$\begin{aligned} C &= W \log_2\left(\frac{S}{N} + 1\right) = W \log_2\left(\frac{S}{N_0 W} + 1\right) \\ &= \frac{S}{N_0} \log_2\left(\frac{S}{N_0 W} + 1\right)^{N_0 W/S} \end{aligned} \tag{4.2}$$

Next, apply the relation

$$\lim_{X \to 0} (1 + X)^{1/X} = e$$

to equation (4.2). This results in

$$\begin{aligned} \lim_{W \to \infty} C &= \lim_{W \to \infty} \left[\frac{S}{N_0} \log_2\left(\frac{S}{N_0} + 1\right)^{N_0 W/S}\right] \\ &= \frac{S}{N_0} \log_2 e \end{aligned} \tag{4.3}$$

The signal power, S, may be expressed as

$$S = \frac{E_b}{T_b} = E_b \cdot R \tag{4.4}$$

where

E_b = energy/bit
T_b = time duration/bit
R = bit-repetition rate

Letting $R = C$ and substituting equation (4.4) into equation (4.3), we have

$$\mathcal{C} = \frac{E_b \mathcal{C}}{N_0} \log_2 e$$

or

$$\frac{E_b}{N_0} = \frac{1}{\log_2 e} = \frac{1}{1.44} = -1.6 \text{ dB} \tag{4.5}$$

We can relate this to the probability of error, as shown in Figure 4.2. Note that at the Shannon limit, the probability-of-error curve is discontinuous, going from

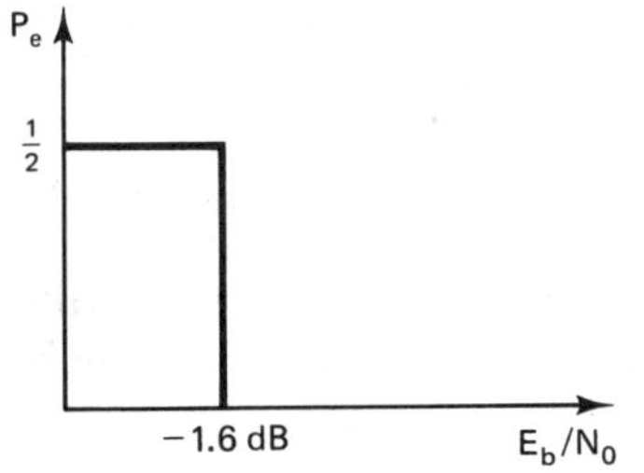

Figure 4.2 Probability of error versus E_b/N_0 at the Shannon limit.

$P_e = \frac{1}{2}$ to $P_e = 0$. It is not possible to reach the Shannon limit (Equation 4.5), because as M increases without bound, the bandwidth requirement and delay become infinite and the complexity of implementing the system increases without bound. Referring to Figure 4.1, for a BER of 10^{-5}, an E_b/N_0 of 9.6 dB is needed for BPSK. Shannon's work, then, promises a theoretical performance improvement of 11.2 dB through the use of coding techniques. Approximately 7 dB of that promised improvement is available today with the coding techniques presently being implemented.

4.1.1 Bandwidth of Digital Data

The theorems of Nyquist and Shannon are based strictly on the assumption of bandlimited channels. That is, no signal power whatsoever is allowed outside the defined bandwidth. In designing practical systems, then, we are faced with a dilemma. Strictly bandlimited signals are not realizable, since this implies infinite transmission time delay. On the other hand, nonbandlimited signals containing energy at arbitrarily high frequencies (wide bandwidth) appear just as unreasonable. Considered in this context, it is easy to see why there is no single universal definition of bandwidth.

Some of the most common definitions of bandwidth are illustrated in Figure 4.3. Other definitions such as −35 or −50 dB down imply even larger bandwidths than those shown. The power spectral density, $S(f)$, for a single rectangular pulse of duration τ has the analytical form

$$S(f) = \tau \left[\frac{\sin \pi (f - f_c)\tau}{\pi (f - f_c)\tau} \right]^2 \tag{4.6}$$

where

$$f_c = \text{carrier frequency in hertz}$$

and

$$\tau = \text{symbol duration in seconds}$$

This spectral density has the same general sin x/x appearance as Figure 4.3. It is characteristic of a sequence of random digital data, where we assume that the spectral density averaging time is long relative to the symbol duration, τ. Note that the spectral density plot has a main lobe along with smaller symmetrical side lobes. The general shape of this plot is characteristic of that for most digital formats. There are, however, some digital formats that do not have well-defined side lobes.

The bandwidth criteria indicated in Figure 4.3 are as follows:

- BW_1—*half-power bandwidth*: This is the traditional −3-dB bandwidth, that is, the interval between frequencies at which $S(f)$ has decreased to half power.

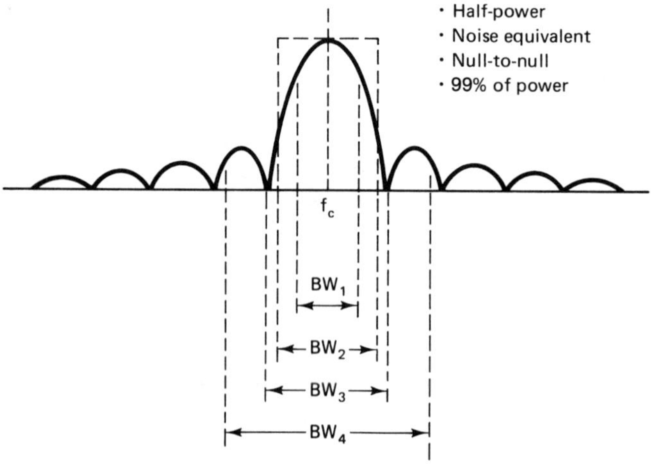

Figure 4.3 Various bandwidth criteria.

- BW_2—*equivalent noise bandwidth* (ENBW): This is an equivalent rectangular bandwidth that would contain the total signal power, P, over all frequencies.

  That is,
  $$P = W_N\, S(f_c) \tag{4.7}$$
  where
  P = total signal power in watts
  W_N = equivalent noise bandwidth in hertz
  $S(f_c)$ = the value of $S(f)$ at the center frequency f_c (assumed to be the maximum value over all frequencies)

- BW_3—*null-to-null bandwidth*: This is the bandwidth of the main lobe. This is a popular measure of bandwidth. Complete generality of this criterion is not possible, however, since some digital formats lack well-defined lobes.
- BW_4—*fractional power containment bandwidth*: This bandwidth criterion is specified in Section 2.202 of the FCC rules and regulations. Specifically, this definition states that the occupied bandwidth leaves exactly 0.5% of the signal power above the upper bandlimit and exactly 0.5% of the signal power below the lower bandlimit. Thus, 99 percent of the signal power is contained inside.

4.2 CHANNEL ENCODING

Look back at Figure 3.1. Channel encoding refers to the data transformation performed after source encoding but prior to modulation that transforms the source bits into channel bits. Channel coding is traditionally partitioned into two groups: waveform coding and structured sequences (codes). We discussed both waveform (or signal-design coding) along with baseband codes such as NRZ in Chapter 3. In the following material, we wish to extend our discussion of structured sequences (codes) and, finally, discuss examples wherein the two partitioned groups are merged.

We can state the objectives of error-control coding as follows:

1. Error detection
   - Determine whether a given segment of the received data stream contains errors. Notify the data source or destination of the error.
   - Minimize the probability of undetected errors.
2. Error correction
   - Obtain a reduction in the probability of error for a given value of E_b/N_0.
   - For a given probability of error, reduce the value of E_b/N_0. The amount of reduction is called the *coding gain* for that probability of error. The concern here is generally with the average probability of error.

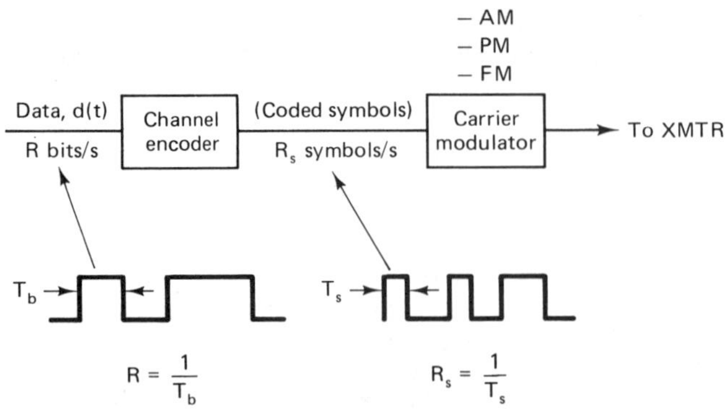

Figure 4.4 Functional block diagram of a digital transmitter employing coding.

Error-detection and error-correction requirements are determined primarily by the type of information being transmitted. For subjective data, such as voice and video, the concern is usually with error correction and average probability of error. The concern with other data types, digital command codes and computer data words, is with error detection and the probability of undetected error. The system throughout is also a matter of concern. The functional block diagram of a transmitter employing coding is shown in Figure 4.4. Referring to this figure, we can make the following observations:

- The symbol rate, R_s, is always greater than the bit rate, R.
- The code rate may be defined as the number of information bits per coded symbol. That is,

$$\text{Code rate} = \frac{R}{R_s} \text{ bits/symbol}$$

4.3 CODING PRINCIPLES

With the foregoing discussion in mind, it is well to examine some of the basic principles involved in implementing codes. The basic problem can be stated as follows: We wish to convert M possible messages into M possible code words. The code words must embody such objectives as error detection, error correction, and security. As an example, consider a message consisting of 3 bits. With 3 bits, there are $2^3 = 8$ possible messages. We wish to substitute a code word for each of the possible messages. As noted previously, to achieve our objectives we employ redundancy. The code words, then, are longer than the corresponding message words from which they are generated.

The error-detection and error-correction capabilities of various redundant codes can be analyzed by first defining the concept of *distance* between two code words. The distance between any two equal-length binary words is defined to be

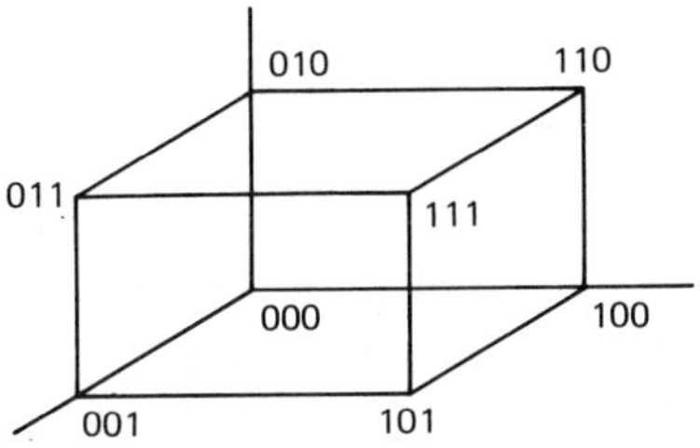

Figure 4.5 Representation of 3-bit words in three-dimensional space.

equal to the number of bit positions by which the two words differ. We can think of distance in terms of multidimensional space. For example, our 3-bit words can be plotted in three-dimensional space (Figure 4.5).

Suppose that we have an elementary message consisting of a 0 or a 1. When the message is 0, we transmit 000. When it is a 1, we transmit 111. Referring to the cube, the distance between the two valid code words representing the two messages is equal to three because it is necessary to traverse along three of the lines (edges of the cube) to get from one to the other. Now suppose that the message has 3 bits and that all combinations are used. Referring again to Figure 4.5, all the corners of the cube are now used. An error in any one of the 3 bits simply moves us to an adjacent corner and an error is made. The distance between the transmitted and received word is 1, and an error is made since the received word corresponds to one of the possible messages. We improve the situation by increasing the distance. For example, the words

0000, 0011, 0101, 0110, 1001, 1010, 1100, 1111

all have a minimum distance of 2. For these 4-bit code words, we need an abstract figure with 16 points (corners) to represent the four-dimensional code words. It is convenient, then, to think in terms of n-dimensional space.

Using our 4-bit example, suppose that an error is made in one place and that we receive the word 1000. This could have resulted from the words 0000 or 1010. An error is detected but we cannot correct it. Next consider code words consisting of 01111, 01000, and 10011. The minimum distance is 3. Now suppose 01000 was transmitted and 01001 was received. The distance to the first code word is 3, the second is 1, and the third is 4. An intelligent receiver can easily be made to decode the received word as 01000. Next suppose that the same message is received as 01011. That is, two errors are made. The received message is now equidistant from 01000 and 10011. An error is detected but cannot be corrected. We say, then, that this code is double-error detecting but only single-error correcting. If the minimum distance between code words is D_{min}, errors involving up to $D_{min} - 1$ can be detected. If D_{min} is an even number, errors up to $(D_{min}/2) - 1$ can be corrected. For D_{min} odd, errors up to $\frac{1}{2}(D_{min} - 1)$ can be corrected.

The previous ideas are quantitative but do provide valuable insight into the coding process. We can extend our understanding by relating the coding process to the probability of error, as shown in Example 4.1.

EXAMPLE 4.1

A 3-bit message is transmitted via PSK and the signal to noise ratio at the receiver is 10 dB.

(a) Compute the probability of 2 bits being in error.

(b) The message is next encoded so that the word length is increased to 5 bits. What is the probability of 2 bits being in error? Assume equal transmitted power in both cases.

Solution

(a) From Figure 3.38, the BER is equal to 10^{-5}. If we assume that individual errors are independent of each other, the probability of two particular bits being in error is $(10^{-5})^2 = 10^{-10}$. For 3 bits taken 2 at a time, there are three possible combinations, the first and second, first and third, and the second and third bits. The probability of two bit errors is given by

$$\begin{aligned} P_e &= \binom{3}{2} (P_B)^2(1 - P_B)^{3-2} \\ &= \frac{3\cdot 2\cdot 1}{2}(10^{-5})^2(1 - 10^{-5}) \\ &\approx 3 \times 10^{-10} \end{aligned}$$

where P_B is the probability of a bit error.

(b) *Without coding*:

$$\frac{E_b}{N_0} = \frac{S}{RN_0} = 10 \text{ dB} \quad (10)$$

With coding: The coded bit rate, R_c, is $\frac{5}{3}$ times the data bit rate. Then

$$\frac{E_b}{N_0} = \frac{S}{R_c N_0} = 10 \times \frac{3}{5} = 6 \quad (7.78 \text{ dB})$$

From Figure 3.38, the probability of a bit error is $P_B = 5 \times 10^{-4}$. The probability of two bit errors is

$$\begin{aligned} P_e &= \binom{5}{2} (P_B)^2(1 - P_B)^{5-2} \\ &= \frac{5\cdot 4\cdot 3\cdot 2\cdot 1}{2\cdot 1}(5 \times 10^{-4})^2(1 - 5 \times 10^{-4})^3 \\ &= 1.497 \times 10^{-5} \end{aligned}$$

Note from this example that the BER actually increased with coding. That is, more bits must be detected during the same time interval with the same available power. The performance improvement due to coding is not yet apparent. This is illustrated in Example 4.2.

EXAMPLE 4.2

Calculate the probability of a message error for Example 4.1.

Solution

(a) Without coding the probability of a message error, P_M, is

$$P_M = 1 - (1 - P_B)^3 = 1 - (1 - 10^{-5})^{11} = 3 \times 10^{-4}$$

(b) the receiver can correct the received word if an error is made in one bit position. A message error is made, then, if 2, 3, 4, or 5 bits are in error.

$$P_M = \sum_{k=2}^{5} \binom{5}{k} (P_B)^k (1 - P_k)^{5-k}$$

A good approximation is made if we use only the first term of the summation:

$$P_M \approx \binom{5}{2} (5 \times 10^{-4})^2 (1 - 5 \times 10^{-4})^3$$

$$= 1.497 \times 10^{-5}$$

We see that the probability of message error has improved by a factor of $(3 \times 10^{-4})/(1.497 \times 10^{-5}) = 28.5$ with the use of coding.

4.4 TYPES OF CODES

We can divide error-control codes into two broad categories: block codes and convolution codes. A third category—concatenated codes—is also mentioned in the literature. This category is actually a combination of block and convolution codes. In block codes, a block of k information bits is followed by a group of r check bits. This results in a block of n code symbols (Figure 4.6). The n-bit block of output bits is called a *code word*. Coding schemes in which the message appears at the beginning of the code word are called *systematic codes*. The ratio of data

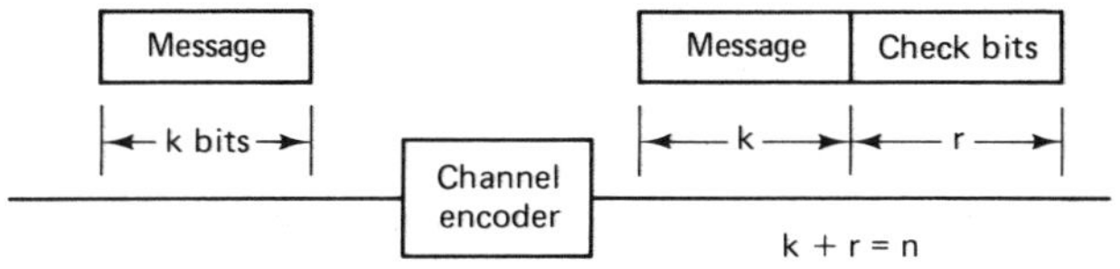

Figure 4.6 Example of a block encoder.

bits to total bits within a block, k/n, is called the *code rate*. The code itself is referred to as an (n, k) code.

The channel encoder of Figure 4.6 attempts to correct possible errors in the received word, that is, the check bits are used to detect and possibly correct errors. Consequently, this scheme is referred to as a *forward-acting error-correction* method. If a reverse channel is available such that the receiver can notify the transmitter that errors have occurred and that the data must be retransmitted, this method of error control is termed *error detection*. This results in a lower probability of bit error. This is achieved at the expense of an overall lower data rate. The performance improvement possible with block coding can be illustrated with the following example of a single-error-correcting code.

EXAMPLE 4.3

BPSK modulation is used to transmit data at a rate of 4800 bits/s. The received SNR is 9.6 dB.

(a) What is the probability of bit error and message error if the data is uncoded? Assume a message is 11 bits.

(b) Repeat (a) for a (15, 11) block code.

Solution

(a)

$$\frac{E_B}{N_0} = \frac{S}{RN_0} = 9.6 \text{ dB } (9.12)$$

From Figure 3.38, $P_B = 1.02 \times 10^{-5}$ and $P_M = 1 - (1 - P_B)^{11} = 1.12 \times 10^{-4}$.

(b) The code rate R_c is

$$R_c = 4800 \times \tfrac{15}{11} = 6545 \text{ bits/s}$$

$$\frac{E_b}{N_0} = \frac{S}{R_c N_0} = 9.12 \times \tfrac{11}{15} = 6.688 \quad (8.25 \text{ dB})$$

The probability of bit error is $P_B = 1.36 \times 10^{-4}$, so

$$P_M = \sum_{k=2}^{15} \binom{15}{k} (P_B)^k (1 - P_B)^{15-k}$$

$$\approx 1.94 \times 10^{-6}$$

Improvement factor

$$= \frac{1.12 \times 10^{-4}}{1.94 \times 10^{-6}} = 58$$

The summation is started with $k = 2$, since the code corrects all single errors within a block of $n = 15$ bits.

This type of coding can be generalized. We defined systematic codes as being those illustrated in Figure 4.6. The simplest possible code word results if $r = 1$. This is known as a single-parity bit-check code. The parity bit is generated from the message bits. In general, if a_i is a message bit and C_i is a parity-check bit, each code word is of the form

$$a_1 a_2 a_3 \cdots a_k \; c_1 c_2 c_3 \cdots c_r$$

where $k + r = n$, the total number of bits in the code word. For k message bits, we have 2^k distinct message words. There are 2^{k+r} possible code words, of which only 2^k are used—one for each message word. We choose each check bit to yield even parity when combined with particular message bits. For example, c_1 can be chosen to yield an even number of 1s when combined with a_1, a_2, a_5 and a_6, and so on.

For even parity the modulo 2 sum of the code word is equal to zero. For the record, modulo-2 addition is the same as the operation of an *exclusive-or* circuit. Carries are ignored. Then, we can express the code words as a matrix equation. For example

$$HT = 0$$

where H is the generator matrix and T is the transmitted codeword given by

$$\begin{bmatrix} h_{11} & h_{12} & h_{13} & \cdots & h_{1k} & 1 & 0 & 0 & \cdots & 0 \\ h_{21} & h_{22} & h_{23} & \cdots & h_{2k} & 0 & 1 & 0 & \cdots & 0 \\ h_{31} & h_{32} & h_{33} & \cdots & h_{3k} & 0 & 0 & 1 & \cdots & 0 \\ \vdots & \vdots & \vdots & & \vdots & \vdots & \vdots & \vdots & & \vdots \\ hr_1 & hr_2 & hr_3 & \cdots & hr_k & 0 & 0 & 0 & \cdots & 1 \end{bmatrix} \times \begin{bmatrix} a_1 \\ a_2 \\ a_3 \\ \vdots \\ a_k \\ c_1 \\ c_2 \\ \vdots \\ c_r \end{bmatrix} = 0 \qquad (4.8)$$

To see that this operation defines code words, as illustrated in Figure 4.6, we can perform the matrix multiplication. The first two rows in the matrix, which is the product of H and T, are

$$\begin{aligned} h_{11}a_1 + h_{12}a_2 + h_{13}a_3 + \cdots + h_{1k}a_k + c_1 &= 0 \\ h_{21}a_1 + h_{22}a_2 + h_{23}a_3 + \cdots + h_{2k}a_k + c_2 &= 0 \end{aligned} \qquad (4.9)$$

Some examples illustrate the use of this method.

EXAMPLE 4.4

The generator matrix for a (6, 3) block code is shown. Find all the code words of this code.

$$H = \begin{bmatrix} 0 & 1 & 1 & 1 & 0 & 0 \\ 1 & 0 & 1 & 0 & 1 & 0 \\ 1 & 1 & 0 & 0 & 0 & 1 \end{bmatrix}$$

Solution The message length k is 3 and the length of the code word is $n = 6$. For $k = 3$, there are eight possible message blocks: 000, 001, 010, 011, 100, 101, 110, 111. The code word computation for message block (111) is

$$\begin{bmatrix} 0 & 1 & 1 & 1 & 0 & 0 \\ 1 & 0 & 1 & 0 & 1 & 0 \\ 1 & 1 & 0 & 0 & 0 & 1 \end{bmatrix} \times \begin{bmatrix} 1 \\ 1 \\ 1 \\ c_1 \\ c_2 \\ c_3 \end{bmatrix} = 0$$

Performing the multiplication,

$$0 + 1 + 1 + c_1 = 0 \Rightarrow c_1 = 0$$

$$1 + 0 + 1 + c_2 = 0 \Rightarrow c_2 = 0$$

$$1 + 1 + c_3 \qquad = 0 \Rightarrow c_3 = 0$$

The code word for message block (111) is then 111000. A tabulation of the code words for all eight messages is as follows:

Message	Code word
0 0 0	0 0 0 0 0 0
0 0 1	0 0 1 1 1 0
0 1 0	0 1 0 1 0 1
0 1 1	0 1 1 0 1 1
1 0 0	1 0 0 0 1 1
1 0 1	1 0 1 1 0 1
1 1 1	1 1 1 0 0 0

Example 4.4 illustrates how the encoder generates the code words. The encoder must store the matrix H and then perform the binary arithmetic to generate the code word. Now suppose that the code word T is transmitted. For the code to be useful to us, it must permit such things as error detection and correction. If an error occurs, the received codeword R is given by

$$R = T \oplus E$$

where once again the addition is performed modulo 2. To see how error detection is accomplished, we first multiply the received code word, R, by the matrix H. That is

$$HR = H(T + E) = HT + HE = HE \tag{4.10}$$

where equation (4.8) has been used. We define the product HE as S, the *syndrome*. An example illustrates how the syndrome permits error detection.

EXAMPLE 4.5

For Example 4.4, message block (111) was transmitted, and 110 was received. An error occurs in the third position. Find the syndrome.

Solution

$$HR = \begin{bmatrix} 0 & 1 & 1 & 1 & 0 & 0 \\ 1 & 0 & 1 & 0 & 1 & 0 \\ 1 & 1 & 0 & 0 & 0 & 1 \end{bmatrix} \begin{bmatrix} 1 \\ 1 \\ 0 \\ 0 \\ 0 \\ 0 \end{bmatrix} = \begin{bmatrix} 1 \\ 1 \\ 0 \end{bmatrix}$$

Note that the syndrome is equal to the third column of H, thus identifying the error in the third bit position. It is clear that in order to correct an error, H must not contain duplicate columns. Also, a column that is all 0s must not occur in H, since we could not be sure if the syndrome meant no errors or there was an error in the bit position corresponding to the all-zero column of H. Suppose now that errors occur in two bit positions. In this case the syndrome is equal to the sum of the corresponding columns of H. For example, if errors occur in the second and third bit positions, the syndrome is

$$S = \begin{bmatrix} 0 \\ 1 \\ 1 \end{bmatrix}$$

This would be incorrectly interpreted as a single-bit error in the first column. We can detect but not correct double-bit errors using three check bits. Correction of more than one error involves finding code words that have minimum distances greater than three. That is,

$$\frac{D_{\min}}{2} - \frac{1}{2} = \frac{3}{2} - \frac{1}{2} = 1$$

4.4.1 Binary Cyclic Codes

Binary cyclic codes are a subclass of the linear block codes discussed previously. These codes are useful for two reasons:

- Encoding and syndrome calculations can easily be implemented with shift registers with feedback.
- A mathematical basis exists that makes it possible to design codes with useful error-correcting properties

In the preceding section we noted that block codes in general are described most succinctly using a matrix representation. However, cyclic block codes can be more usefully represented as polynomials. A cyclic code word consists of a group of message bits and a block check code (BCC). Using the nomenclature of Figure 4.6, let n equal the total number of bits in the code word and let k be the number of message bits: Then $n - k$ is the number of bits in the BCC, that is an (n, k) cyclic code. The code word is derived from two polynomials, the generator polynomial $P(X)$ and the message polynomial $G(X)$. These polynomials are algebraic representations of the respective binary words. We represent the polynomials by a string of powers of X such as $X^n + \cdots + X^2 + X^1 + X^0$. In the binary representation, a 1 is placed in each position that contains a term. The absence of a term is represented by a 0. For example, the binary representation of $X^4 + X^3 + X$ is 1 1 0 1 0. In generating cyclic codes we make use of the following theorem:

Theorem. If $P(X)$ is a polynomial of degree $(n - k)$ and is a factor of $X^n + 1$, then $P(X)$ generates an (n, k) cyclic code in which the code polynomial $C(X)$ for a message polynomial $G(X)$ is generated by

$$C(X) = P(X)G(X) \tag{4.11}$$

An example illustrates the technique.

EXAMPLE 4.6

The generator polynomial of a (7, 4) cyclic code is $X^3 + X + 1$ and the message polynomial is $X^2 + 1$. Find the code polynomial $C(X)$.

Solution

$$n - k = 3 \quad \text{and} \quad k = 4 \Rightarrow n = 7$$

$$\begin{aligned} C(X) &= P(X)G(X) \\ &= (X^2 + 1)(X^3 + X + 1) \\ &= X^5 + X^3 + X^3 + X^2 + X + 1 \end{aligned}$$

Since

$$X^3 + X^3 = (1 + 1)X^3 = 0 \leftarrow \text{modulo-2 sum}$$

then

$$C(X) = X^5 + X^2 + X + 1$$

The code word is 0 1 0 0 1 1 1.

The code word can also be put in the systematic form as follows:

1. Multiply the message $G(X)$ by X^{n-k}, where $n - k$ is the number of bits in the BCC (this adds $(n - k)$ 0 bits onto the end of the binary equivalent of $G(x)$.
2. Divide the resulting product $X^{n-k}[G(X)]$ by the generator polynomial $P(X)$.
3. Disregard the quotient and add the remainder $R(X)$ to the product to yield the code message $C(X)$, which is represented as $X^{n-k}[G(X)] + R(X)$.

An example illustrates the technique.

EXAMPLE 4.7

The generator polynomial of a (7, 4) cyclic code is $X^2 + X + 1$, and the message is 1 0 1 0. Find the code polynomial $C(X)$.

Solution

$X^{n-k}G(X) = X^3[X^3 + X] = X^6 + X^4 \quad (0X^7 + 1X^6 + 0X^5 + 1X^4 + 0X^3 + 0X^2 + 0X^1 + 0X^0)$ The binary equivalent of this product contains 7 bits and is 1 0 1 0 0 0 0. This product is divided by $P(X)$.

```
                       1 1 1 0 1  ← Quotient
P(X) → 1 1 1   )1 0 1 0 0 0 0  ← X^(n-1)G(X)
                1 1 1
                ─────
                  1 0 0
                  1 1 1
                  ─────
                    1 1 0
                    1 1 1
                    ─────
                        1 0 0
                        1 1 1
                        ─────
                          1 1  ← Remainder R(X)
```

$$C(X) = X^{n-k}[G(X)] + R(X)$$

$$= \frac{1\,0\,1\,0}{1} \quad \frac{0\,1\,1}{1}$$

P(X) Message   BCC

A useful result of the code $C(X)$ is the following. The code word is transmitted and the receiving station divides it by the same generator polynomial $P(X)$. If there is no error, a remainder of zero is produced. The message is presumed to be correct. The modulo-2 binary division is as follows:

```
          1 1 1 0 1
      ______________
1 1 1  )1 0 1 0 0 1 1
        1 1 1
        -----
          1 0 0
          1 1 1
          -----
            1 1 0
            1 1 1
            -----
              1 1 1
              1 1 1
              -----
              0 0 0
```

In Example 4.7, note that the first 4 bits of the code word are the message and the remaining 3 bits are the block BCC. A shift register can be used to compute the BCC. The number of stages in the register is equal to $n - k$, that is, the degree of the generating polynomial. The configuration is shown in Figure 4.7.

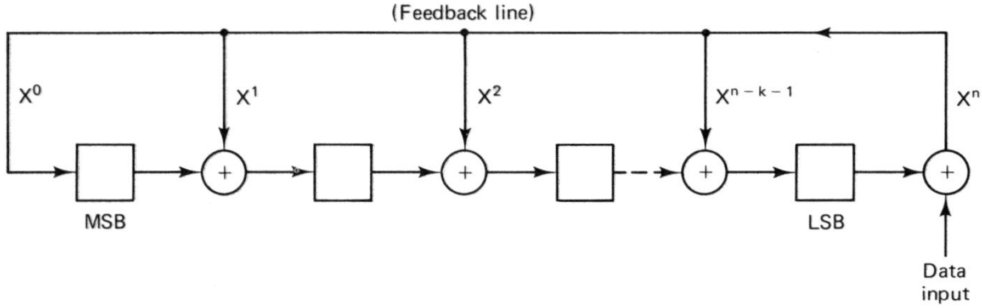

Figure 4.7 An $(n - k)$-stage encode for an (n, k) binary cyclic code.

EXAMPLE 4.8

A (7, 4) binary cyclic code is generated by $P(X) = X^3 + X + 1$. Design the encoder and verify its operation with message (0 1 0 1).

Solution

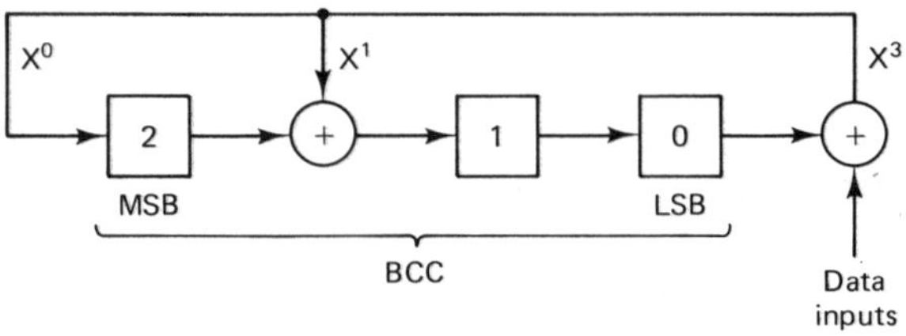

Shift Number	MSB (Most Significant Bit)		LSB (Least Significant Bit)		
Start	0	0	0	—	1
1	1	1	0	1	0
2	0	1	1	0	0
3	0	0	1	1	1
4	1	1	0	0	0

At the conclusion of the fourth shift operation the register contains the block check code 0 1 1. Note that the LSB of the register contains the MSB of the BCC code.

At the receiver, the message 0 1 0 1 is applied to a register identical to the register at the transmitting end. If the message is correct, this again produces the BCC code 0 1 1. The BCC code is transmitted following the message and is also shifted into the register on the receiving end. If the BCC codes are identical, the contents of the register at the receiving end are all 0s, thus indicating correct message reception. We see that one of the main advantages of cyclic codes over other linear block codes is the ease of implementation using shift registers. Error detection is accomplished by the same easy process.

4.4.2 Pseudonoise Code

Pseudonoise (PN) codes have proven to be very useful cyclic codes. These codes also can be easily generated with shift registers. The spectrum is similar to that of white noise, hence the name. Figure 4.8 illustrates the generation of a 7-bit PN code. With 3 bits, there are $2^3 = 8$ possible bit combinations. We can initialize the register with any one of the combinations with the exception of all 0s. Suppose that we start with 0 0 1. The immediate output is a logic 1 and this is summed in

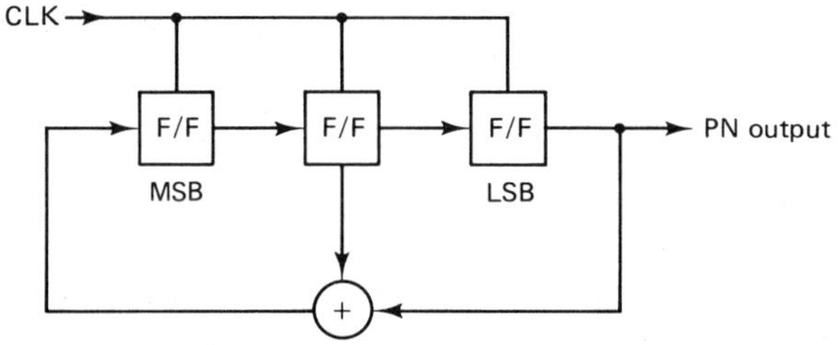

Figure 4.8 Seven-bit PN generator.

an exclusive-or circuit with a zero from the center flip/flop (F/F). The output is a logic 1 to the MSB F/F. When the register is clocked, the contents become 1 0 0. Continuing this operation we see that the outputs are

1 0 0 1 0 1 1 | 1 0 0 1 0 1 1

Note that the sequence begins repeating itself after the first 7 bits. The output PN code is periodic with a period of 7 bits.

If we examine the first 9 bits of the preceding sequence in groups of three adjacent bits, note that every possible 3-bit combination occurs. That is,

1 0 0   0 0 1   0 1 0   1 0 1   0 1 1   1 1 1   1 1 0

Furthermore, taking 7 bits at a time, each 7-bit code is a cyclic shift of every other sequence. The seven possible PN code words are:

1 0 0 1 0 1 1	0 0 1 0 1 1 1	0 1 0 1 1 1 0
1 0 1 1 1 0 0	0 1 1 1 0 0 1	1 1 1 0 0 1 0
1 1 0 0 1 0 1		

and the distance between any pair of code words is four units. In general, we can state that for a register of n bits, the code word length will be $2^n - 1$. Also, the distance between code words will always be 2^{n-1}.

One of the favorite uses of PN codes lies in *spread-spectrum* communications. We use a higher bit rate PN code to spread (widen) the spectrum of the digital data before transmitting. This can be accomplished by imputing the digital data and the higher-symbol-rate PN code to an exclusive-or gate. The spectrum of the signal at the output of the gate occupies a wider bandwidth than the original digital data. At the receiver, the data is recovered by imputing the spread spectrum data and the PN code into another exclusive-or gate. Of course this means that the receiver must generate the code and it must be synchronized with the code generator at the transmitter. This technique allows multiple signals occupying the same RF bandwidth to be simultaneously transmitted without interfering with one another. The security inherent in this method of coding is apparent: Knowledge of the code can be controlled. Spread-spectrum communications are of such importance that we will devote chapters five and nine to this topic.

4.5 CONVOLUTIONAL CODES

Convolutional codes represent an important class of codes. In a convolutional code, a block of n code bits generated by the encoder depends not only on the block of k message bits but also on the preceeding $(n - 1)$ blocks of message bits. This contrasts with the block code, where the block of n bits generated by the encoder depends only upon the block of k message bits for which a code word is being formed.

The technique for generating a convolutional code is to *convolve* a sequence of source bits U_k with an encoding function according to the following convolution equation:

$$t_n = \sum_{k=-\infty}^{n} U_k h_{n-k} \tag{4.12}$$

The h's in equation (4.12) are either 1 or 0. This equation can be implemented with a shift register and modulo-2 adders. An example is shown in Figure 4.9.

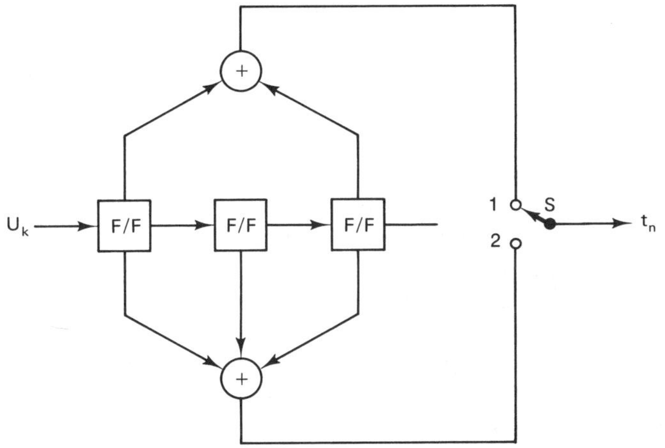

Figure 4.9 Convolution encoder (rate − 1/2).

The encoder is mechanized with a K-stage shift register, where K is said to be the *constraint* length. In this particular example $K = 3$. Operation of the encoder is as follows: At the kth unit of time, data bit U_k is shifted into the first stage of the register. At the same time, all bits in the register are shifted one stage to the right and the outputs of the modulo-2 summers are sampled by switch S and transmitted. Since there are 2 code bits per data bit, the code rate is $\frac{1}{2}$ for this example. We refer to this as a rate $-\frac{1}{2}$ encoder. For the general case of n code bits per data bit, the code rate is $1/n$. An example illustrates the operation of the encoder.

EXAMPLE 4.9

For the rate $-\frac{1}{2}$ convolutional encoder of Figure 4.9, the first set of h's is given by 1 0 1 and the second set by 1 1 1. If the input sequence is 1 1 0 0 1 0 0 1 1 1, find the output sequence. The most recent bit is on the right.

Solution With the switch in position 1, we find the first, third, fifth, . . . , bits. For switch position 2, we get the second, fourth, sixth, . . . , bits. The output is

1 1 1 0 1 0 1 1 1 0 1 1 1 1 1 0 0 0 1 0 1 1

In arriving at this output sequence, we assumed that the register is initialized with all 0s prior to receiving the first bit of the input sequence and that the input sequence is followed by a string of 0s.

The encoding operation of Figure 4.9 can also be represented in a graphical fashion. One method is a tree (Figure 4.10), which represents the output for every possible input. It would appear that for long input sequences, the tree would continue to grow. Actually it is limited by the number of storage elements (the constraint length) in the register. In this example, after three input elements, the tree begins to repeat itself. Referring to Example 4.9, the first input bits are 1 1 0. This corresponds to path *ACGJ* or an encoder output of 1 1 1 0 1 0, which agrees with Example 4.9.

Note that after three levels of splitting, the top half of the tree is identical to the bottom half. After the tree settles down, there are only four possible states. It is common practice to specify the points in the tree as a trellis (Figure 4.11, see page 104). The four states are obvious from this figure. This reflects the fact that the two previous input bits contain all the past history that affects a particular input bit. We can deduce from this line of reasoning that a register with four storage elements would result in a trellis that would settle down to eight states. The four states are labeled as 1, 2, 3, and 4.

Thus state 2 is equivalent to points *C*, *E*, *I*, and *M*. At each state there are two possible input bits. For each of these inputs, there are two corresponding output bits. For example, at state 2, an input of 1 causes a transition to state 4 with output 10. An input of 0 causes a transition to state 3 with output 01. An input of 0 always results in the upper path, whereas an input of 1 results in the lower path.

4.5.1 Decoding Convolutional Codes

The structure of the tree and the trellis suggest a possible means for decoding the received sequence. For example, by tracing through the trellis we can compute all the possible transmitted sequences. From these sequences we than choose the one whose distance is nearest to that of the received sequence. The number of paths for an L-bit sequence is 2^L. This approach becomes impractical as L increases. A simple decoding technique known as the *Viterbi algorithm* simplifies this procedure. We shall see that this algorithm reduces the number of computations to a linear rather than an exponential function of L.

Referring to the trellis of Figure 4.11, suppose that the received sequence

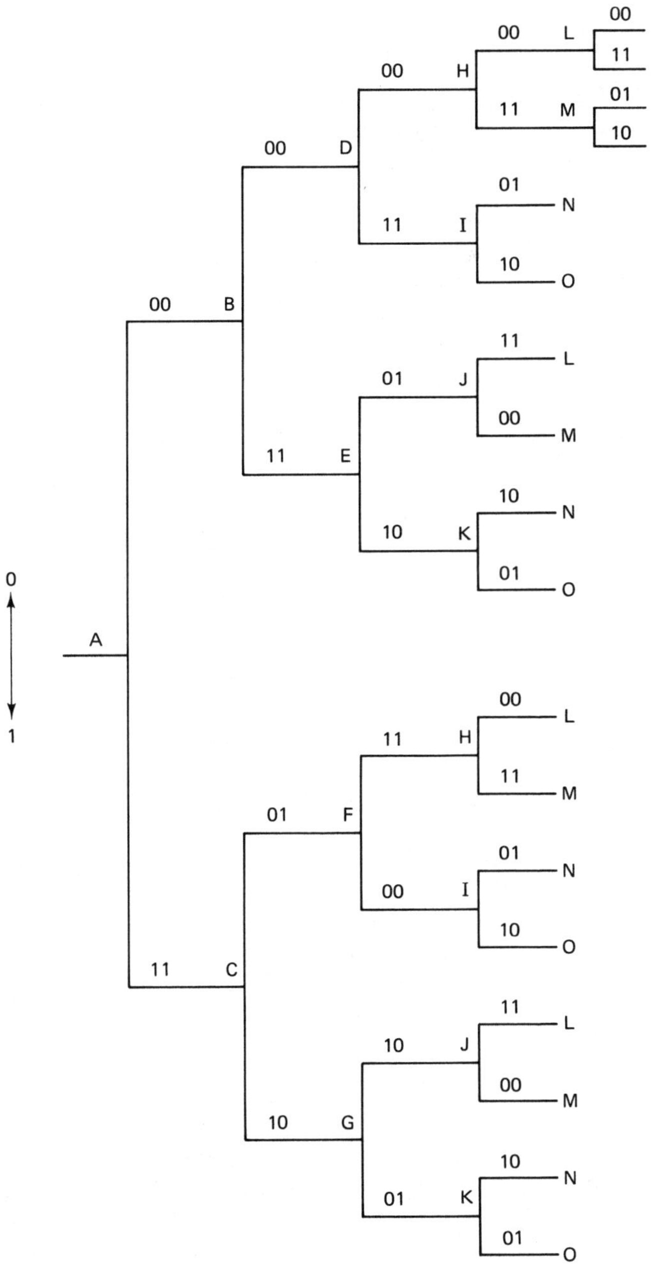

Figure 4.10 Tree for Example 4.8.

is 1 1 1 0 0 1 1 1. To use the algorithm to decode this, we redraw the trellis, indicating the distance between each pair of received bits and the corresponding bits of the trellis. The 0 on path *AC* results from comparing the received bits 1 1

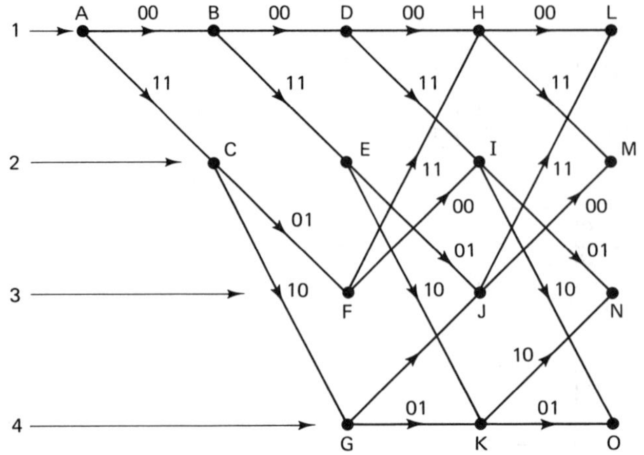

Figure 4.11 Trellis for Example 4.8.

to 1 1 as indicated on Figure 4.11. The cumulative distances to different points in the trellis are indicated in brackets. To reach point F we go through a distance of 2. At the third level there are two possible paths. For example, at point F the

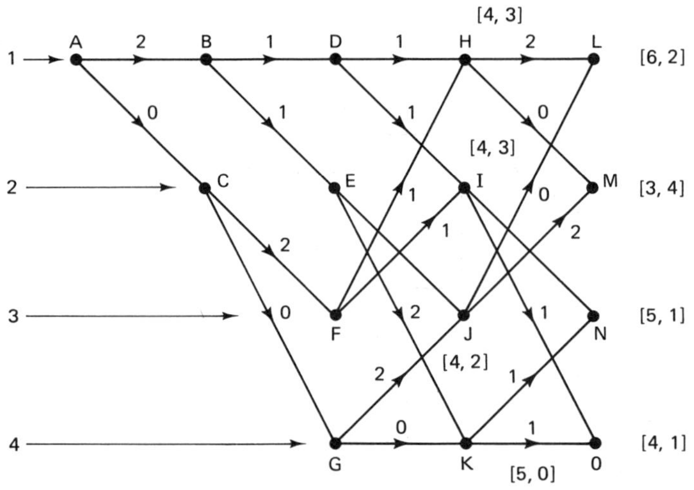

path distances are 4 and 3. The first entry in the brackets corresponds to the upper path. At this point the path with the largest distance is discarded. If two paths have equal distance, either one can be chosen. The probability of decoding error will be the same. Rejecting one of the paths is the essence of the Viterbi algorithm. This makes the number of path combinations a linear function of L. At the output we find a path $ACGKO$. This path is generated by an input of 1 1 1 1. An input of 1 1 1 1 is coded into 1 1 1 0 0 1 0 1, which has distance 1 from the received word.

4.6 CHANNEL CODING WITH MULTIPHASE SIGNALS

The previous material has considered channel coding of the algebraic type. We considered both block and convolutional coding schemes. The ability to detect and/or correct errors is provided for by the transmission of redundant bits, thus lowering the effective information rate per transmission bandwidth. There exists also the possibility of concatenation. That is, we assign short block code words to state transitions in a trellis structure. This represents a merging of block coding and convolutional coding.

Recent work has considered a combined modulation and coding approach that is appropriate for satellite or terrestrial digital transmission. For example, if rate $-\frac{1}{2}$ codes are mapped (modulated) onto 4-PSK modulation, the bandwidth efficiency is that of binary PSK. However, substantial coding gain results. We enter a situation where the minimum distance between wave forms enters into the probability of error rather than the Hamming distance between code words. This results in a trellis wherein the branches are labeled with phase states rather than transmitted code bits. For example, with 8-PSK, each phase represents 3 bits. The penalty paid is an increase in complexity of the receiver. This is an area of active research. The reader is referred to the literature for current trends in this area.

PROBLEMS

4.1. Consider transmitting sequences of $k = 4$ of information through an AWGN channel with a noise spectral density N_0 watts per hertz.

(a) Assume an ideal PSK modulation scheme; what is the required value of E_b/N_0 if the BER rate is $p = 10^{-5}$? What is the uncoded message-error probability for this system?

(b) A (7, 4) single-error-correcting block coding scheme is now utilized. Assuming that the E_b/N_0 value of part (a) is unchanged, find the coded word error probability. Compare with part (a).

(c) Repeat parts (a) and (b) if $p = 10^{-4}$ and $p = 10^{-3}$. What can you conclude about coding systems in general as the noise becomes increasingly severe?

4.2. (a) Plot the capacity versus the bandwidth of a satellite channel with additive white Gaussian noise of spectral density N_0 and average signal power P. What is the asymptotic value of the capacity?

(b) Let $S/N_0 = 20$ dB. Find the maximum information transmissions rate R in bits per second if ideal PSK modulation is used and an error probability of 10^{-5} is required. Repeat for an error probability of 10^{-4}.

(c) Repeat part (b) if QPSK modulation is used.

(d) Let $S/N_0 = 20$ dB and assume a channel bandwidth of 1000 Hz. What is the maximum transmission rate R in bits per second if an arbitrarily complicated system may be used and an arbitrarily small error probability is required? Compare with parts (b) and (c).

(e) Repeat part (d) if the bandwidth may be as large as desired. Compare your results with those of part (b).

4.3. Discuss the main differences between block and convolutional coding.

4.4. Show that addition and subtraction are equivalent in modulo-2 arithmetic.

4.5. In Example 4.2, why does the BER decrease after coding? Explain how a coding gain is realized.

4.6. Use the rate $-\frac{1}{2}$ convolutional coder shown.

(a) Determine the code tree.

(b) Determine the Viterbi decoding trellis diagram.

(c) What is the constraint length for this coder?

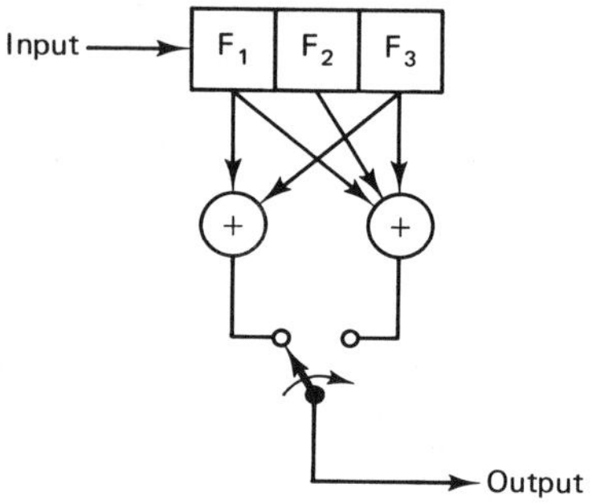

4.7. Calculate the probability of message error, P_M, for a 12-bit data word encoded with a (24, 12) linear block code. Assume the code corrects all 1-bit and 2-bit error patterns. The probability of a bit error is $P_B = 10^{-3}$.

4.8. Consider a (6, 3) linear block code capable of single-error corrections. Assume that the modulation is coherent BPSK and that the uncoded $E_b/N_0 = 10$ dB.

(a) Does the code provide any message-error-rate performance improvement? If it does, how much? If it does not, explain why not.

(b) Repeat part (a) with $E_b/N_0 = 6$ dB.

4.9. Calculate the probability of message error, P_M, for a 12-bit data word encoded with a (24, 12) linear block code. Assume the code corrects all 1-bit and 2-bit error patterns. Also, assume that the probability of error for a coded bit is $P_B = 10^{-3}$.

4.10. Calculate the message-error-rate performance improvement of a (24, 12) double-error-correcting linear block code. Assume that the modulation is coherent BPSK and that the uncoded $E_b/N_0 = 10$ dB.

4.11. Consider a (127, 91) linear block code capable of triple error corrections.

(a) What is the message error rate for an uncoded block of 92 bits if the uncoded BER equals 10^{-3}?

(b) What is the coded message error rate using the (127, 92) block code if the coded bit error rate equals 10^{-2}?

(c) Does the code provide any message-error-rate improvement? If it does, how much? If it does not, explain why not.

4.12. Find the required predetection signal-to-noise density ratio (P_r/N_0), in decibels, in order to receive output data at the decoded data rate of 1 megabit/s with a BER of 10^{-5}. Assume binary noncoherent FSK modulation. Also, assume convolution coding with the following decoder relationship

$$P_b = 2000\ (P_B)^4$$

where P_B and P_b are the BERs into and out of the decoder, respectively.

4.13. Suppose you were trying to find the quickest way to get from London to Vienna by boat and train. The following diagram was constructed from various schedules:

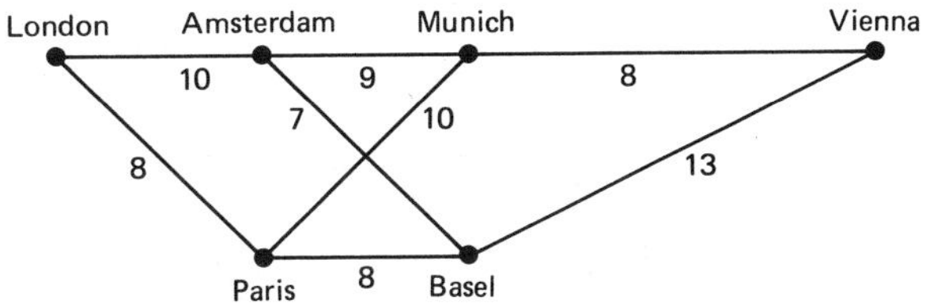

The labels on each path are travel times. Using the Viterbi algorithm, find the fastest route from London to Vienna. In a general sense, explain how the algorithm works, what calculations must be made, and what information must be retained in the memory used by the algorithm.

4.14. Consider a (7, 4) code whose generator matrix is:

$$H = \begin{bmatrix} 1 & 1 & 1 & 1 & 0 & 0 & 0 \\ 1 & 0 & 1 & 0 & 1 & 0 & 0 \\ 0 & 1 & 1 & 0 & 0 & 1 & 0 \\ 1 & 1 & 0 & 0 & 0 & 0 & 1 \end{bmatrix}$$

(a) Find all the code vectors of this code.
(b) Find C, the parity check matrix of this code.
(c) Compute the syndrome for the received vector 1 1 0 1 1 0 1. Is this an allowable code vector?

4.15. If the channel BER is given by P_e, show that the error rate at the output of a differential decoder for both PBSK and QPSK is given by

$$P_b = 2P_e(1 - P_e)$$

For small P_e, what is the effect of differential decoding? What is the E_b/N_0 degradation if $P_e = 10^{-4}$ for PBSK and QPSK?

4.16. The following code word is received:

1 1 0 1 0 1 1 0 0 1

Using the accompanying Viterbi decoding diagram, determine the original data sequence.

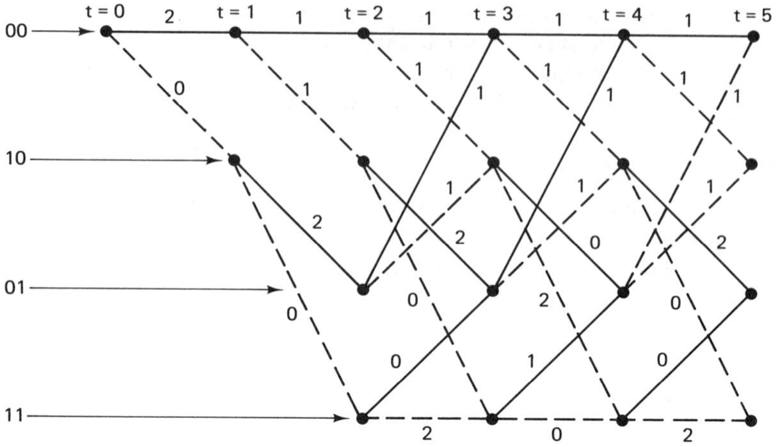

4.17. Use the convolutional encoder shown.

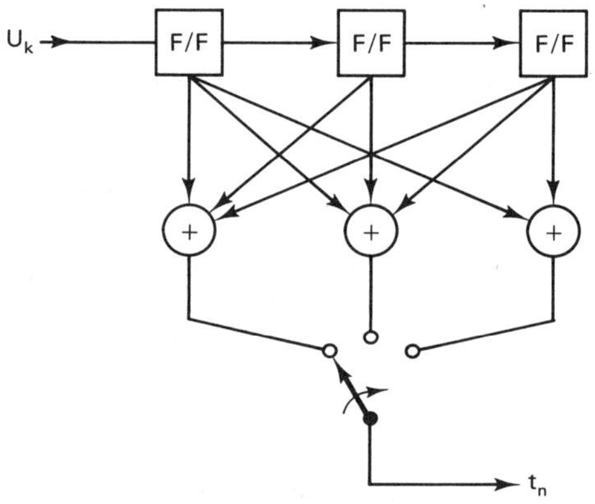

(a) Draw the tree.
(b) Draw the trellis diagram.
(c) Let the information sequence be 1 0 1 1 0 0 0 → t. Determine the transmitted code word.
(d) Indicate on the trellis diagram the path corresponding to the information sequence.
(e) Assume that the received code word sequence is

110, 110, 001, 101, 011, 110, 001

Using Viterbi decoding, determine the decoded information bits. Compare with the encoded data.

4.18. A (7, 3) binary cyclic code is generated by $P(x) = x^4 + x^2 + x + 1$.
(a) Draw the encoding circuit.
(b) If the message to be encoded is 1 0 1, determine the register contents as the message is shifted into the register.

4.19. A (15, 5) cyclic code is generated by

$$P(x) = x^{10} + x^8 + x^5 + x^4 + x^2 + x + 1$$

(a) Find the code polynomial and code word for the message

$$G(x) = x^4 + x + 1$$

(b) Find the code polynomial and code word in the systematic form.

(c) The polynomial $r(x) = x^{14} + x^{13} + x^9 + x^7 + X^6 + 1$ is received. Does this represent a valid code word?

4.20. Determine the encoded output message for the (3, 3) convolutional encoder shown.

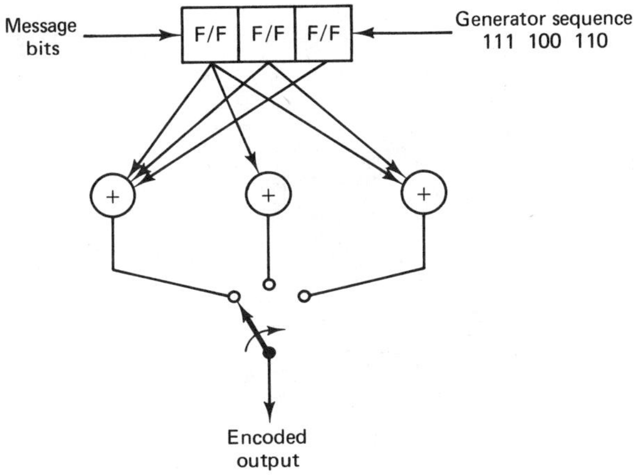

4.21. Determine the code tree for a (3, 3) convolutional encoder with generator sequence 111 101 100.

4.22. Draw the trellis for the encoder of Problem 4.21.

5

SPREAD SPECTRUM TECHNIQUES

5.0 INTRODUCTION

The body of knowledge known as spread spectrum represents one of the most exciting innovations in communications technology. This technique is a direct application of Shannon's original work in information theory. We define spread spectrum as a technique in which a transmitted signal occupies a bandwidth that is much larger than that required by the baseband information signal. This technique trades transmission bandwidth for enhanced detectability and interference rejection. This derives from the fact that one of the most common techniques for making a signal different from all other signals (including noise) is to increase its bandwidth W, so that it is wide compared to the information bandwidth. In fact, we shall see in Chapter 9 that the ratio of the spread bandwidth to the information bandwidth is a measure of the performance of the system.

5.1 PRINCIPLES OF SPREAD SPECTRUM

The basic concept of a spread spectrum is illustrated in Figure 5.1. Clearly, the idea is to take the energy in bandwidth B_D and spread it over the wider bandwidth B_{RF}. Basically, three techniques can be used to accomplish this:

- Direct sequence pseudonoise (DSPN)
- Frequency hopping (FH)
- Time hopping (TH)

A combination of these techniques is also available for use. Pseudorandom (PN) codes are used to achieve spreading of the spectrum. This coding technique was introduced in Chapter 4. Prior to considering methods of generating these codes, it is instructive first to consider the techniques for widening the spectrum of a signal.

5.1.1 Direct-Sequence Pseudonoise Spreading

One of the most straightforward ways to widen the spectrum of a signal is simply to multiply (modulate) it by a wideband signal. We shall see that the spreading

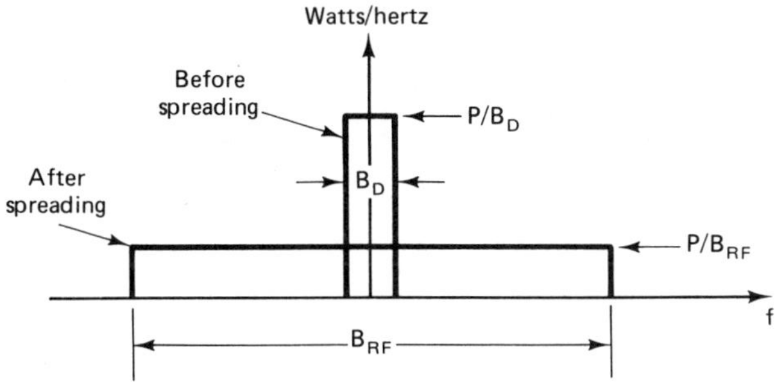

Figure 5.1 Illustration of spread spectrum.

signal must have properties that aid in acquiring and tracking the signal. That is, these properties are necessary to enable the receiver to despread (demodulate) the received signal. The binary PN signal possesses these properties. The most common spread-spectrum technique is direct-sequence PN spreading in conjunction with PSK data modulation. This system is illustrated in Figure 5.2. In this system, $d(t)$ is the message waveform representing the message sequence. $C(t)$ represents the wideband code waveform. These signals have bandwidths B_M and B_W, respectively. Suppose now that the message waveform is equal to $\sqrt{2Ps}\,d(t)$ where Ps represents the signal power and $d(t)$ is a binary waveform

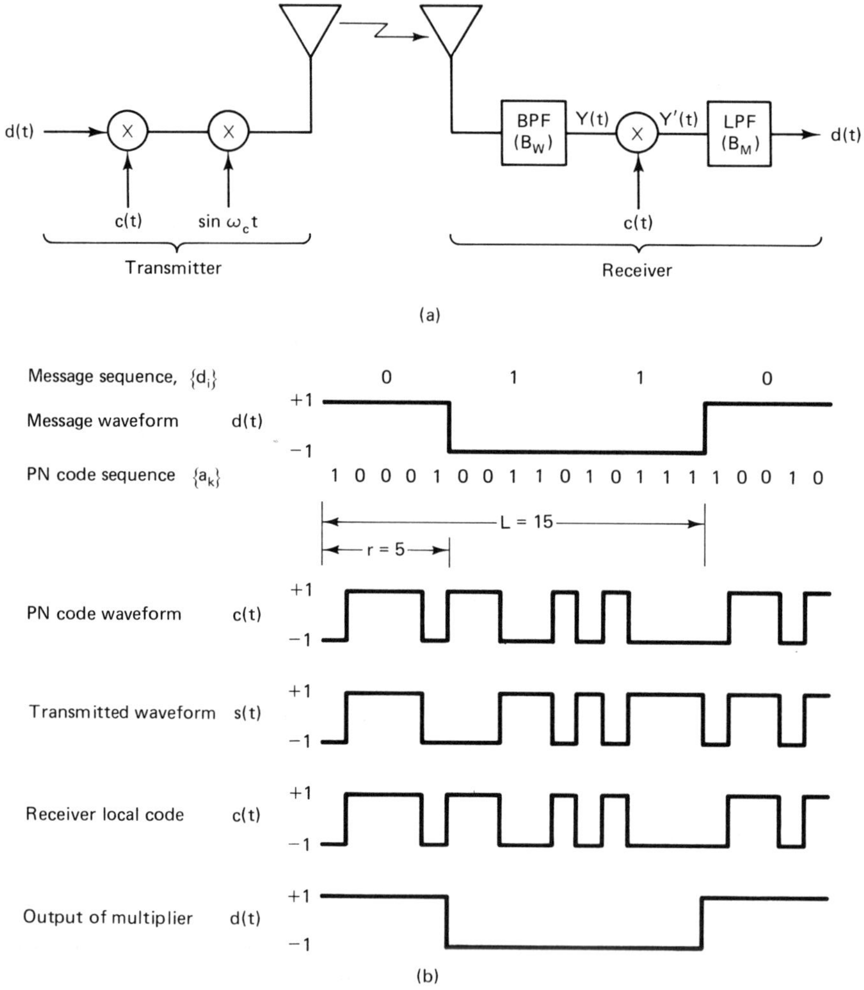

Figure 5.2 (a) Biphase PN communication system. (b) DSPN waveforms.

that takes on values of ± 1. Also, let $C(t)$ be a pseudo-noise PN sequence with values ± 1 whose bit rate (referred to as *chip rate*) is much higher than the bit rate of the message signal. The received signal $Y(t)$ may be expressed as

$$Y(t) = \sqrt{2Ps}\, d(t)C(t) \sin(2\pi f_c t) \tag{5.1}$$

where f_c is the frequency of the RF carrier. Recovery of the message waveform is accomplished by mixing the received signal, $Y(t)$, with a replica of $C(t)$. This results in

$$Y'(t) = \sqrt{2Ps}\, d(t) \sin 2\pi f_c \tag{5.2}$$

where $C^2(t) = (\pm 1)^2 = 1$ is used. Referring to equation (5.2), $d(t)\sin(2\pi f_c t)$ simply represents a PSK signal that can be demodulated by conventional means. In actual practice, an IF stage is generally utilized in the receiver. A general discussion of spread-spectrum communication systems is delayed until Chapter 9.

5.1.2 Frequency Hop Coding

The second method of spreading the message signal is accomplished by transmitting each bit or subinteger bit (chip) on a different carrier selected from a wide range of frequencies. Refer to Figure 5.3. The selection of frequencies is controlled by the PN sequence. There are 2^n frequency slots and for each successive message bit or fraction of a bit, two slots are selected to denote a 1 or 0. The signal is dehopped using a replica of the PN code by the receiver shown in Figure 5.4 (see page 116). We see that, as in the DSPN case, acquisition and tracking of the PN sequence must be accomplished before the data can be retrieved.

5.1.3 Time Hop Spreading

Time hop spreading represents the third technique for widening the spectrum. TH channelizes via time slots within a given time frame. This contrast with FH, where channels are established via frequency slots. The basic TH modulator is shown

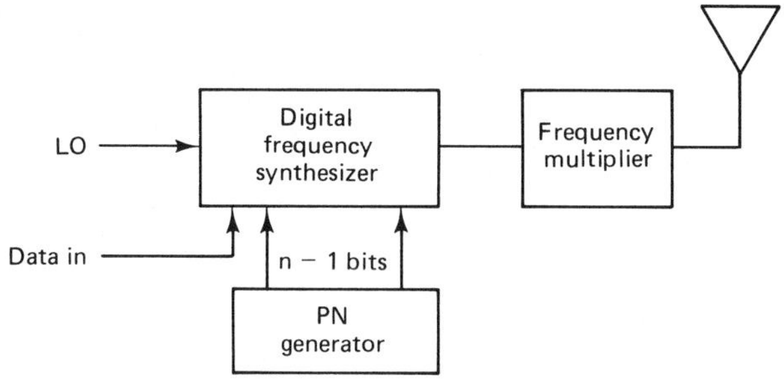

Figure 5.3 Frequency-hopping modulator.

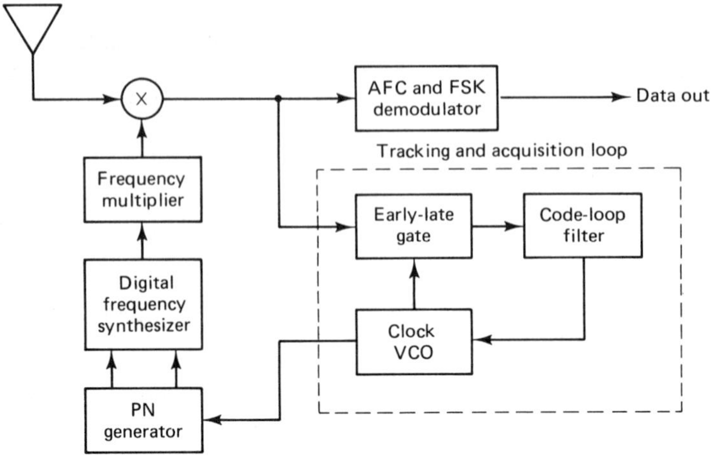

Figure 5.4 Frequency-hopping receiver.

in Figure 5.5. Burst transmission is employed within the slot by a PN code. The data bits within a frame are stored for transmission at high speed during the burst. A time hop receiver is shown in Figure 5.6. The time slot selector selects the desired portion of the frame, and demodulation is then performed. A conventional bit synchronizer (see Chapter 7) uses data transitions to permit code tracking. Code tracking is performed by making the data clock and the PN clock synchronous.

From this discussion, we see that TH and FH systems are analogous in time and frequency. A combination of the three basic spread-spectrum modulation techniques may be appropriate for different applications. Each technique possesses special advantages, which can be combined for a particular system design.

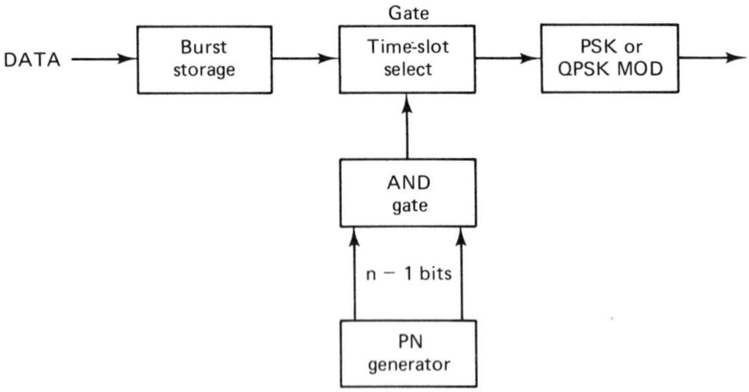

Figure 5.5 TH modulator.

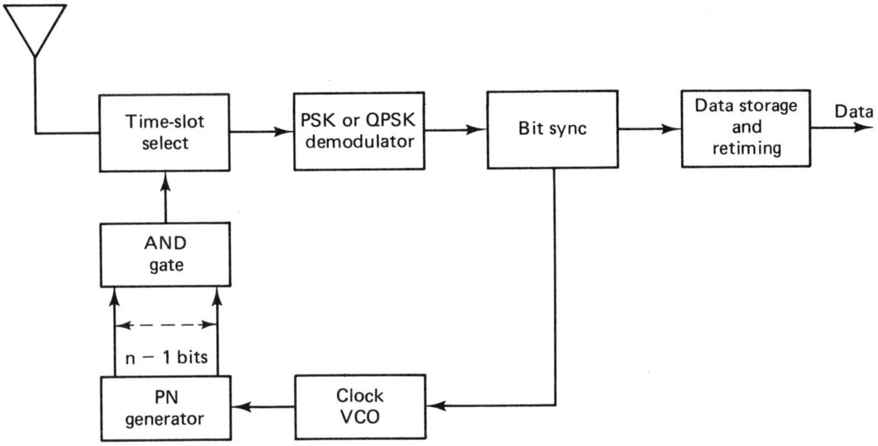

Figure 5.6 TH receiver.

A general discussion is deferred to Chapter 9. In the following, we discuss the generation of spread spectrum codes and their properties.

5.2 GENERATION OF PSEUDORANDOM SEQUENCES

Now that we have acquired preliminary information relating to spread-spectrum techniques, we wish to examine the nature of pseudorandom (codes) sequences and become familiar with their properties. We define a random sequence as a series of numbers, which when observed for a sufficient time interval, exhibits independence of any previous or future time interval. A considerable body of knowledge exists relating to the study of random sequences. For example, the mathematical tools used to describe a random sequence include the use of probability density functions, time-correlation functions, and spectral densities. The sequences generated by shift registers are not truly random, being finite in length. The length of the sequence varies with the length of the register. Stated another way, the sequence repeats itself periodically; hence the name pseudorandom. Even so, there are certain properties associated with PN codes that agree with the concept of randomness. The sequences generated by shift registers as in Figure 4.8 are a series of binary bits (1s and 0s). For a sufficiently long segment of the series, the number of 1s is approximately equal to the number of 0s. For the same long segment we may observe that about one-half the runs, either 0 or 1, are 1 bit long, one-fourth are 2 bits long, one-eighth are 3 bits long, and so on. If the sequence is compared to a time-shifted version of itself, the correlation coefficient is maximum only when the time shift is equal to zero (see Figure 5.7). In the next section we shall see that this is a measure of the *correlation* of the sequence.

(a) Sequence $1\ 0\ 1$

Sequence repeated; $\tau = 0$ $1\ 0\ 1 \rightarrow \tau$

$$R = 3 - 0 = 3$$

$$\rho = \frac{3 - 0}{3} = 1$$

(b) Sequence $1\ 0\ 1$

Sequence repeated; $\tau = +1$ $0\ 1\ 0 \rightarrow \tau$

$$R = 0 - 3 = -3$$

$$\rho = \frac{0 - 3}{3} = -1$$

Correlation function

τ	R	ρ
0	3	1
1	−3	−1
2	1	$\frac{1}{3}$

Figure 5.7 Example of correlation computation.

5.2.1 Correlation Properties

In addition to the properties discussed in Section 5.2, an additional property that must be imposed on the random sequence is in terms of the autocorrelation and cross-correlation. We define the unnormalized correlation between two sequences as

$$R = \text{number of agreements} - \text{number of disagreements} \tag{5.1}$$

If the sequence is correlated with itself, the result is referred to as *autocorrelation*. If a sequence is correlated with another sequence, the result is called *cross-correlation*. It is often desirable to use normalized correlation in our computations. We define this to be

$$\rho = \frac{\text{number of agreements} - \text{number of disagreements}}{\text{number of agreements} + \text{number of disagreements}} \tag{5.2}$$

An example of autocorrelation computation using a 3-bit sequence is shown in Figure 5.7. Note the mechanics of computing the correlation function is to fix the original sequence in time and then slide a replica of the sequence past in steps equal to the bit time, τ, computing the difference between number of agreements and disagreements at each time interval. Only one-half of the computation is shown, that is, $\tau = 0, +1$, and $+2$. The complete range is $\tau = -2, -1, 0, +1, +2$. Both sides of the correlation function are symmetrical. The reader is invited to verify this. We may further note that 100 percent correlation results in a positive peak at $\tau = 0$ with $\rho = 1$. Thus autocorrelation measures the amount of similarity between the sequence and its time shifted version.

The operation illustrated in Figure 5.7 may be expressed mathematically. Let $\{a_n\}$ represent a random sequence given by $\{a_n\} = \{a_0, a_1, a_2, \ldots\}$, where a_i is either $+1$ or -1. If a_n is periodic with period P, then the autocorrelation function may be defined as

$$R_a(\tau) = \sum_{i=1}^{P} a_i(a_{i+\tau}) \tag{5.3}$$

where τ is a time shift of the sequence $\{a_n\}$. Cross-correlation is defined between two different sequences $\{a_n\}$ and $\{b_n\}$, both of period P as

$$R(\tau) = \sum_{i=1}^{P} a_i(b_{i+\tau}) \tag{5.4}$$

The normalized autocorrelation function of a PN sequence is shown in Figure 5.8.

Pseudorandom sequences are structured to have a strong correlation at the zero-shift point and weak correlation elsewhere. Two completely different pseudorandom sequences would have an ideal cross-correlation value of zero. An ideal pseudorandom sequence exhibits the statistics of random noise; thus any search for correlation would be in vain. Use of the auto- and cross-correlation concepts permits another property and figure of merit for a PN sequence to be defined. We say that two sequences are *orthogonal* if the degree of correlation approaches zero over the entire sequence. It has been found that the greater the number of prime factors in the code length ($P = 2^n - 1$), the higher the cross-correlation and the lower the orthogonality. For example, a sequence with 2047 states has two factors: 23 and 89. Another sequence with a length $P = 4095$ states, has four prime factors: 3, 5, 7, and 13. The first sequence exhibits less long-term correlation than the second. Therefore, the first sequence is better than the second.

We can generalize this to mean that a sequence length with the least number of prime factors should be chosen. The orthogonality is best if the length of the sequence $2^n - 1$ is a prime itself. In this case the number n is referred to as a

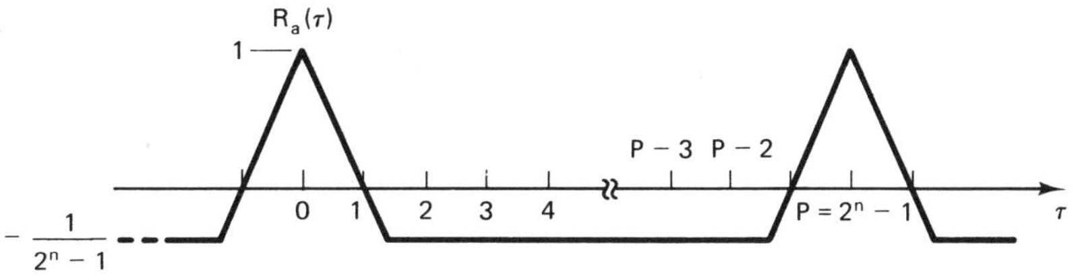

Figure 5.8 Normalized autocorrelation function of a PN sequence.

TABLE 5.1 MERSENNE PRIME MAXIMAL LENGTH SEQUENCES UP TO $n = 17$

Mersenne primes (n)	Sequence length ($2^n - 1$)
2	3
3	7
5	31
7	127
13	8,191
17	131,071

primitive, or *Mersenne prime* (see example 5.1). We usually choose Mersenne prime–based sequences to produce pseudorandom sequences for communications. Between $n = 2$ and $n = 100$, very few Mersenne prime numbers exist. The sequence length for the first six Mersennes are shown in Table 5.1.

Note that this table is labeled in terms of maximal-length sequences. Maximal-length sequences have periods given by $2^n - 1$ and, therefore, the shift register takes on all possible states during one cycle of the PN code sequence. This is the longest sequence that can be generated by a digital shift register of length N stages. There are 2^{n-1} odd binary numbers and $(2^{n-1}) - 1$/nonzero even numbers. This implies an output sequence of 2^{n-1} ones and $2^{n-1} - 1$ zeros. We see that there exist an approximately equal number of 1s and 0s. This corresponds to the first desirable property of random sequences.

EXAMPLE 5.1

Show that the polynomial $x^5 + x + 1$ is reducible, that is, it is not primitive and thus can be factored.

Solution A number of methods are available for factoring polynomials. One procedure is to choose a trial devisor and divide this into the polynomial. The first trial devisor is chosen as the two lowest-ordered terms of the polynomial.

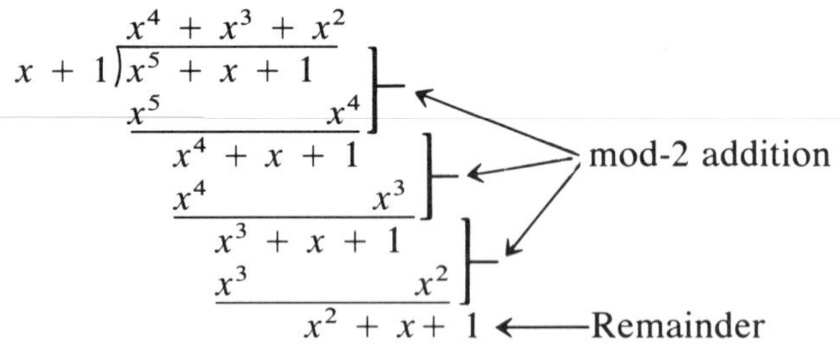

A remainder of $x^2 + x + 1$ is left. Divide the remainder into the original polynomial.

$$
\begin{array}{r l}
 & x^3 + x^2 + 1 \\
x^2 + x + 1 \,\big) & x^5 + x + 1 \\
 & x^5 \qquad\quad x^4 + x^3 \\
\hline
 & x^4 + x^3 + x + 1 \\
 & x^4 + x^3 \qquad\quad + x^2 \\
\hline
 & x^2 + x + 1 \\
 & x^2 + x + 1 \\
\hline
 & 0 \quad 0 \quad 0 \leftarrow \text{Remainder}
\end{array}
$$

(The subtraction steps are marked "mod-2 addition".)

Thus

$$x^5 + x + 1 = (x^3 + x^2 + 1)(x^2 + x + 1)$$

The actual generation of PN sequences was introduced in Chapter 4. Feedback tap points for a maximal length sequence are represented by the presence or absence of the coefficients in the characteristic (generating) polynomial. An example of maximal-length sequence generation is illustrated in Figure 5.9. Maximal-length register tap points for generators consisting of 6 to 10 stages are shown in Table 5.2 on page 122.

The use of PN codes permits a multiplexing technique called code-division multiple access (CDMA) to be used. The use of CDMA allows many users to occupy the same channel without interference. The zero cross-correlation (orthogonality) of different codes makes the technique viable. It is similar to frequency-division or time-division multiplying with the exception that each user is assigned a unique orthogonal code. Maximal-length sequences based on Mersenne primes are ideal for CDMA channels. The problem is that the sequences have a limited number of codes available for use (see Table 5.2). This limitation is overcome by a composite sequence, which is generated by the combination of linear maximal-length sequences. These sequences, called Gold codes, possess certain properties that make their use attractive. These codes, discussed in Section 5.2.2, allow for the construction of large families of independent PN codes.

Since PN codes are periodic, the frequency spectrum can be determined from the fourier series. In the frequency domain, the spectral density of a binary

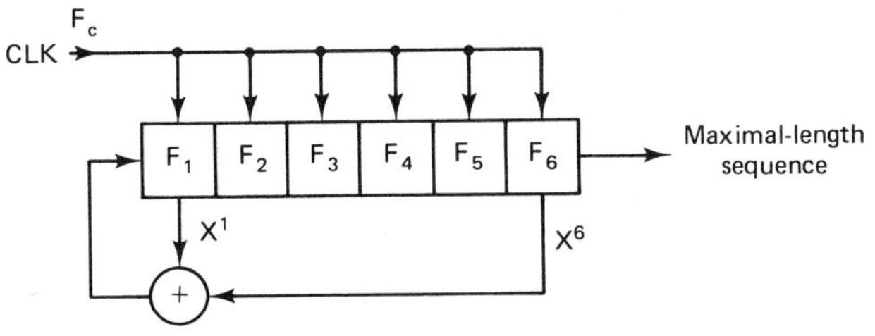

Figure 5.9 Maximal length sequence generator (n = 6).

TABLE 5.2 FEEDBACK CONNECTIONS FOR MAXIMAL LENGTH SEQUENCES FOR N = 6 TO 10

Number of stages	Code length	Maximal-length sequences tap points
6	63	[6, 1] [6, 5, 2, 1] [6, 5, 3, 2]
7	127	[7, 1] [7, 3] [7, 3, 2, 1] [7, 2, 3, 2] [7, 6, 4, 2] [7, 6, 3, 1] [7, 6, 5, 2] [7, 6, 5, 4, 2, 1] [7, 5, 4, 3, 2, 1]
8	225	[8, 4, 3, 2] [8, 6, 5, 3] [8, 6, 5, 2] [8, 5, 3, 1] [8, 6, 5, 1] [8, 7, 6, 1] [8, 7, 6, 5, 2, 1] [8, 6, 4, 3, 2, 1]
9	511	[9, 4] [9, 6, 4, 3] [9, 8, 5, 4] [9, 8, 4, 1] [9, 5, 3, 2] [9, 8, 6, 5] [9, 8, 7, 2] [9, 6, 5, 4, 2, 1] [9, 7, 6, 4, 3, 1] [9, 8, 7, 6, 5, 3]
10	1,023	[10, 3] [10, 8, 3, 2] [10, 4, 3, 1] [10, 8, 5, 1] [10, 8, 5, 4] [10, 9, 4, 1] [10, 8, 4, 3] [10, 5, 3, 2] [10, 5, 2, 1] [10, 9, 4, 2]

bit sequence waveform can be shown to have a standard sin x/x distribution of energy versus frequency. Specifically, the power spectrum of a PN wave form is given by

$$S(\omega) = \frac{P+1}{P^2}\left(\frac{\sin \omega t_0/2}{\omega t_0/2}\right)^2 \sum_{\substack{n=-\infty \\ n\neq 0}}^{\infty} \delta\left(\omega - \frac{2\pi n}{P t_0}\right) + \frac{1}{P^2}\delta(\omega) \tag{5.5}$$

where $\delta(\omega)$ is the Dirac delta function, P is the sequence length, and t_0 is the bit period (or signaling period) of the PN waveform. Note that this equation defines a line spectrum with frequencies at multiples of the fundamental frequency $f_0 = 1/t_0$. Equation (5.5) is illustrated in Figure 5.10. Nulls in the PSD envelope

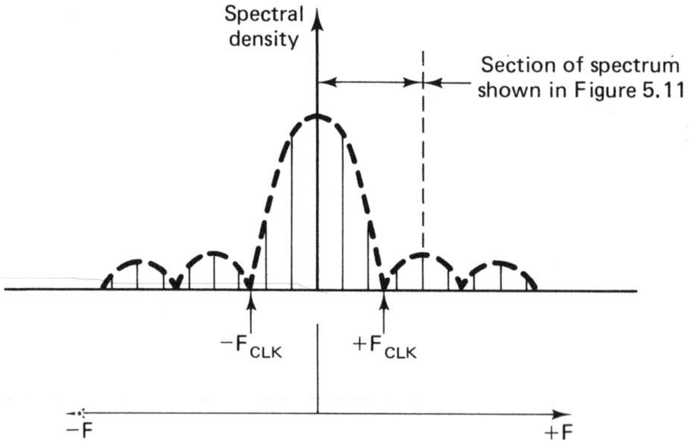

Figure 5.10 Line spectrum of pseudorandom sequence showing code repetition lines and clock frequency nulls.

function occur at multiples of the shift register clock frequency, F_c. The spectral lines occur at

$$\omega = \frac{2\pi n}{Pt_0} = \frac{n\omega_0}{P}$$

where P is the sequence length. For long sequences, the spectral lines degenerate to (a continuous filled) in power spectral density envelope.

This in fact, is one of the figures of merit of a PN sequence. That is, there are no discernible spectral lines. Reasonable spectral properties such as this are not realized until the code length is on the order of 1000 states or greater. Referring to Figure 5.9, this would be a 10-bit generator. The ideal PN sequence spectrum has both the main and side lobes filled in completely. In addition, the nulls between the lobes are deep and clean.

5.2.2 Gold Codes

Gold codes allow us to construct large families of codes from pairs of maximal-length shift registers. This is particularly useful for CDMA systems. The following are properties of Gold codes:

- Constructed from maximal-length sequences but are not maximal length themselves.
- Large number of codes available.
- Codes have well-defined correlation characteristics.
- Simple feedback logic.

A Gold code is produced by modulo-2 addition of two maximal sequences to produce a new code, which is nonmaximal in length. The length is given by

$$L_{GC} = 2^n - 1 \tag{5.6}$$

An example illustrates the procedure.

EXAMPLE 5.2

For $n = 6$, obtain the generating functions and design a Gold code generator.

Solution Refer to Table 5.2. Using the first two maximal-length sequence tap points, the generating polynomials are:

$$F_1 = X^6 + X + 1$$

$$G_1 = X^6 + X^5 + X^2 + X + 1$$

The Gold code generator is shown in the accompanying figure.

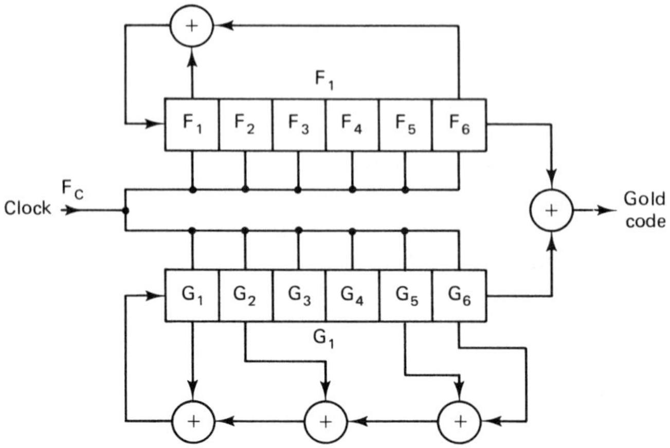

Large families of Gold codes can be constructed from a single pair of maximal-length shift registers. For example, if register F_1 is preloaded with one of its possible codes, this represents a time shift of the maximal length sequence from register F_1. The actual sequence from F_1 remains the same. However, this shift relative to register G_1 allows a new Gold code to be produced. For register F_1, 63 time shifts are possible, thus producing a like number of new Gold codes.

Experimentally derived pictures of the correlation and spectral properties of the PN generator of Example 5.2 are shown in Figure 5.11. Note in the figure that the maximal-length sequence shows a very strong response only once for every repetition of the code. The Gold code (Figure 5.11(b)) shows a strong response at each repetition plus several minor correlation peaks per code cycle. Clearly, in terms of application, the maximal-length sequence is superior to Gold code. Cross-correlation is shown for each of the two sequences in Figure 5.11(c) and (d). Ideal PN codes have zero cross-correlation. The cross-correlation properties between different maximal-length sequences and different Gold code sequences are about the same. Figure 5.11(e) and (f) illustrate the power spectral densities for these sequences. The photographs appear to indicate a continuous spectrum. Actually, this is due to the resolution of the spectrum analyzer (300 Hz). From equation (5.5), the frequency separation between spectral lines is given by $1/Pt_0$. For a clock rate of 3200 PPS, this is $(3200)/(63)(2) = 25.396$ Hz. The number of spectral lines in the main lobe is $3200/25.396 = 126$. Increasing the resolution of the spectrum analyzer results in the display of the line spectrum in the first lobe being as shown in Figure 5.12 on page 126.

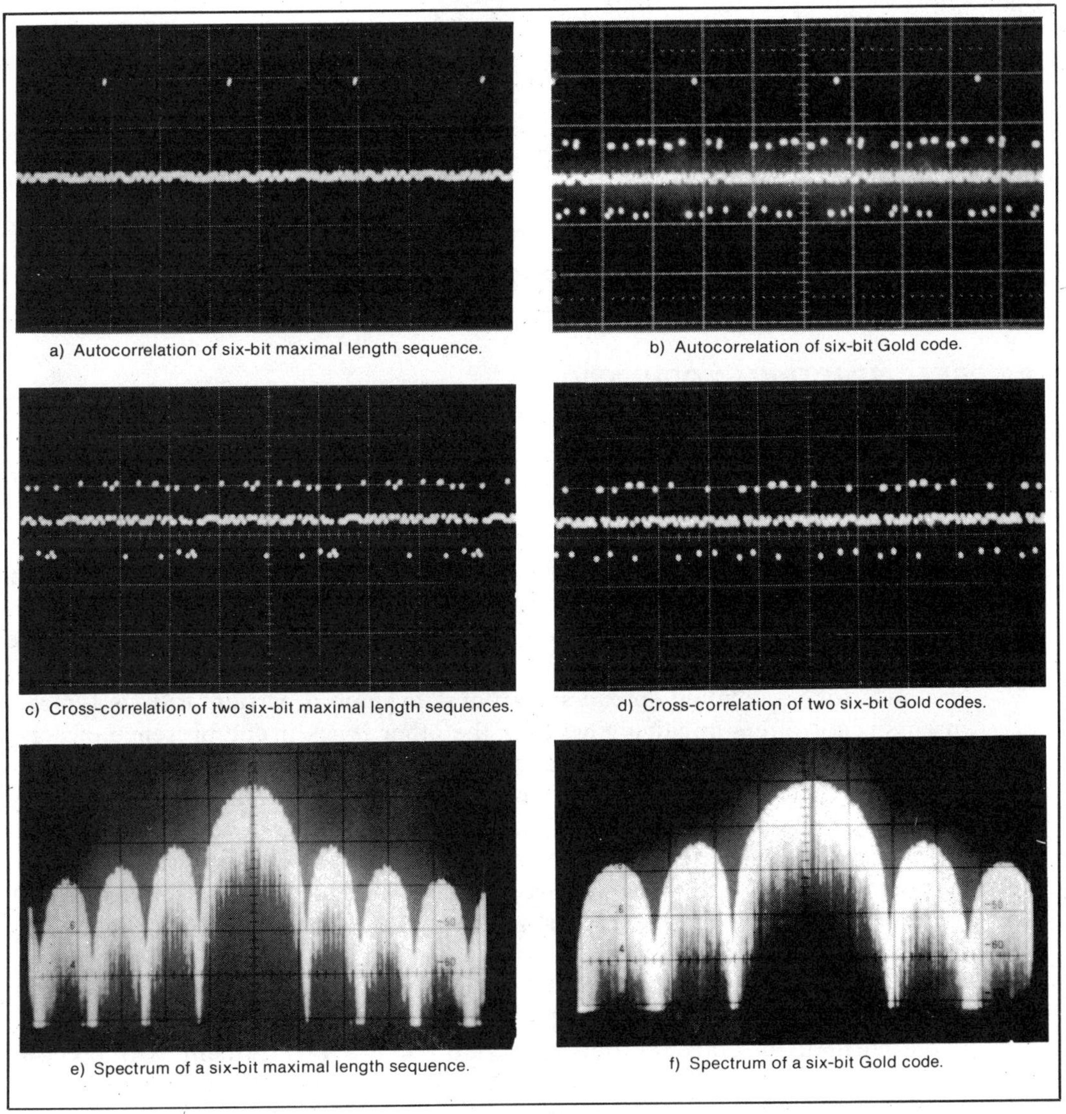

a) Autocorrelation of six-bit maximal length sequence.

b) Autocorrelation of six-bit Gold code.

c) Cross-correlation of two six-bit maximal length sequences.

d) Cross-correlation of two six-bit Gold codes.

e) Spectrum of a six-bit maximal length sequence.

f) Spectrum of a six-bit Gold code.

Figure 5.11 Comparison of the correlation and spectral properties of maximal-length sequences and Gold codes.

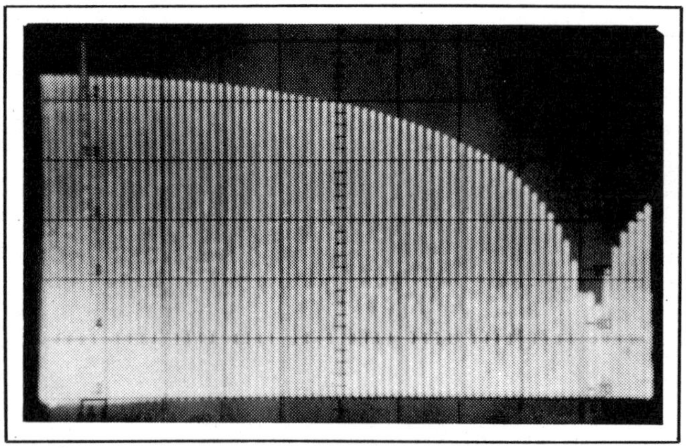

Figure 5.12 Line spectra resulting from narrow-band measurement of pseudorandom sequence.

5.3 SPREAD-SPECTRUM MODULATION

The discrete modulation of the carrier frequency is usually performed either as a multiple level (M-ary) PSK or frequency-shift keyed (FSK) signal. For example, if $M = 2$, the carrier phase is shifted 0° or 180° at a rate determined by the clock of the PN sequence generator. For $M = 256$, an FSK modulator would shift the carrier frequency to one of 256 orthogonal frequencies, all centered around the rest carrier frequency. The PSK modulated frequency spectrum is referred to as a *direct-sequence spread spectrum*, whereas the FSK-modulated carrier frequency is referred to as a *frequency hopped spread spectrum*. Of the numerous methods for modulating spread-spectrum data upon a carrier, PSK is typically used since it offers an advantage over the other types in the presence of noise. Recall from Chapter 3 that PSK requires less signal power to produce an equivalent SNR and to deliver data at an equivalent level of probable error. Specifically, for a specified error probability, a phase-coherent PSK system requires 3 dB less signal power than an FSK system and 6 dB less power than an on-off keyed system. Considering the various PSK schemes, continuous phase-shift modulation (CPSM) transmits with the smallest power loss into the spectral side lobes, thus increasing the efficiency.

The modulation technique chosen, FSK or PSK, depends to a large extent upon the specific application of the spread spectrum system. For example, for fixed positions of both the transmitter and receiver or a satellite with precisely known orbital parameters, time to synchronize is less important than system considerations such as weight and power. In situations where large amounts of data are transmitted for long periods of time, a synchronization time of several seconds is inconsequential. On the other hand, in dynamic situations such as a voice link between high-speed aircraft, synchronization becomes the dominant determining design parameter. A design goal for synchronization time is on the order of 50 ms or less. In terms of bandwidth, the null-to-null spectral width of the main lobe for PSK modulation is equal to twice the clock frequency of the PN generator. The bandwidth of FSK is equal to the number of frequency slots occupied multiplied by the bandwidth of an individual frequency. The final choice between FSK and PSK may depend on the geometry of the application, the subsequent ease of synchronization, and the performance of uncontrollable parameters such as multipath propagation.

Based on efficiency, the following schemes show the advantages of coherent PSK modulation:

- PSK or Manchester coding
- 4-PSK
- CPSM

The circuitry for conventional PSK is simple and economical. In addition, we have already noted the power advantage of PSK over FSK. The $(\sin x/x)^2$ power spectral density distribution results in relatively high power in the sidebands. This makes PSK conducive to interference from adjacent channels. QPSK was discussed in Chapter 3. It handles data twice as fast as biphase modulation with the same bandwidth and power. It does suffer from phase discontinuities. Staggered QPSK (SQPSK) may be used for frequency conservation. This topic is discussed in Section 3.3.5 and is illustrated in Figure 5.13. Orthogonality for this scheme is provided by the phase orthogonality of the sine and cosine carriers. The duration of the baseband keying, T, must be longer than the reciprocal of the carrier frequency, f_c. Staggering the two data streams reduces the maximum possible phase changes to one-half that of conventional QPSK. To reduce the amount of power

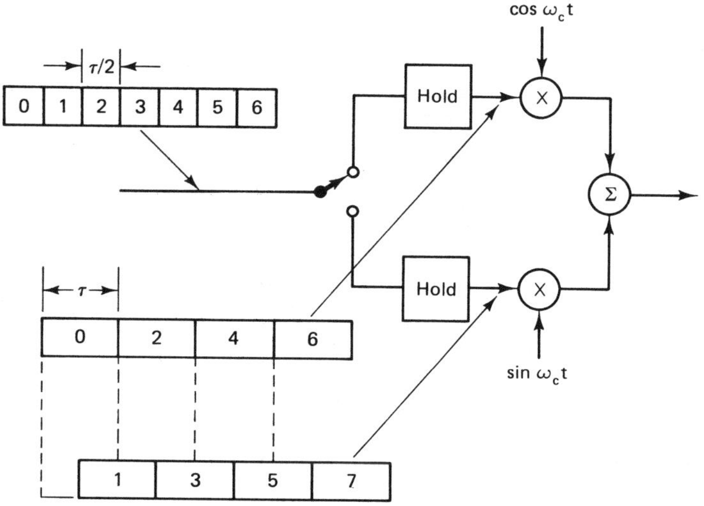

Figure 5.13 SQPSK modulator.

lost into the sidebands, pseudo 4-PSK is used. The biphase modulator input signal is shaped (rounded) when a phase transition is imminent. This causes the modulator output to change gradually. Shaping of the input signal is accomplished by multiplying the keying bit stream by a half-sine wave with the same polarity as the keying bit. This effectively converts each keying element into a half-sine wave with the same polarity as the keying bit. The side lobe characteristics of the output spectrum are improved over that of the SQPSK spectrum.

5.3.1 Continuous-Phase-Shift Modulation

A disadvantage associated with 4-PSK is a reduction in efficiency of the transmitter power amplifier. The half-sine-wave shape of the transmitted carriers reduces the average power. If the half-sine waves are staggered so that the peak value of one coincides with the zero crossing of the other, a constant-envelope RF signal is produced, that is, CPSM. A scheme for producing CPSM is shown in Figure 5.14. The sample and hold-circuit functions are the same as in Figure 5.13 for 4-PSK. The input-keying scheme is staggered while simultaneously doubling the pulse widths. Next, the staggered keying stream is mixed with the quadrature carrier components. The quadrature carrier components are described by

$$M_c(t) \cos \omega_c t$$

and

$$M_s(t) \sin \omega_c t$$

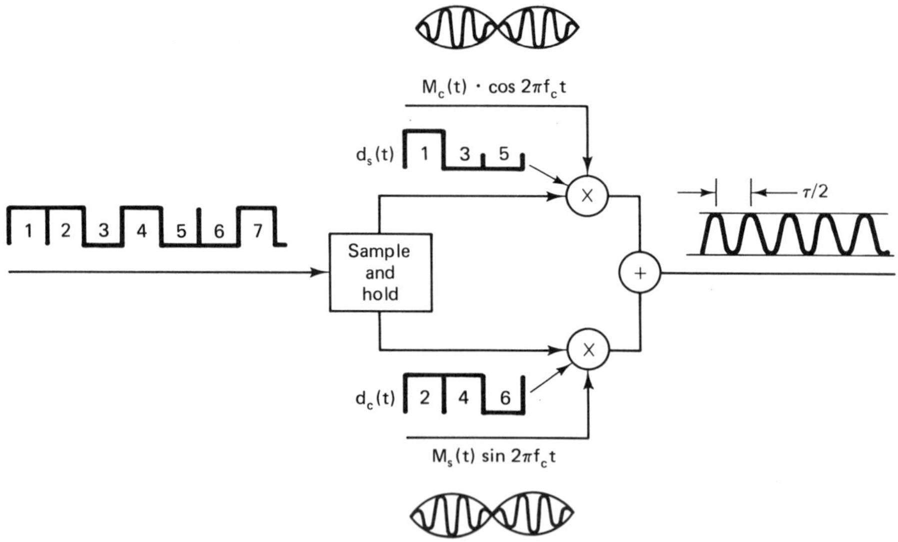

Figure 5.14 CPSM modulators generate constant-envelope outputs in spread-spectrum systems by staggering the input-keying stream in a sample-and-hold circuit. At the same time, the pulses in the stream double in width.

where $M_c(t)$ and $M_s(t)$ describe the envelopes of the quadrature carrier. In order that a constant envelope signal can be derived, the envelopes are adjusted so that

$$M_c(t) = \left|\sin \frac{\pi t}{T}\right| \tag{5.6}$$

and

$$M_s(t) = \left|\cos \frac{\pi t}{T}\right| \tag{5.7}$$

where T is the duration of the doubled pulse width. Since the carrier components are in quadrature, it is clear from equations (5.6) and (5.7) that adding the two together produces a constant envelope signal.

Referring to Figure 5.14, each carrier component is mixed with the processed keying stream, $d_s(t)$ and $d_c(t)$. Each of the mixed components is phase modulated at one-half the PN sequence (chip) rate. Thus the composite CPSM signal is modulated at the chip rate. With $T/2$ as the duration of each element of the incoming keying stream, CPSM is equivalent to continuous-phase PSK with a modulation index M_p equal to 0.5. During the $T/2$ input keying element, the phase term changes by $\pm\pi/2$. Thus in CPSM, each input element changes the phase linearly by $\pm 90°$. The sign of the phase change is determined by the signs of two adjacent input bits, $d_s(t)$ and $d_c(t)$.

The phase characteristics of CPSM are illustrated by the state diagram of Figure 5.15. This diagram correlates the phase of the CPSM signal to the binary

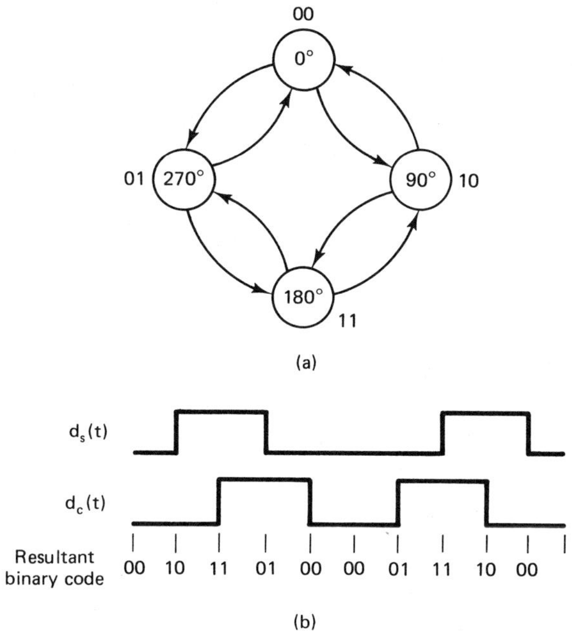

Figure 5.15 Phase shifts of 90° in either direction characterize the CPSM-output signal. (a) The CPSM state diagram correlates the binary-input values to (b) the output phase.

input signal. By limiting the phase changes to 90°, we have retained the feature of reduced side-lobe power, and the constant envelope increases the efficiency of the transmitter power amplifier. Further discussion along these lines leads to the development of practical spread-spectrum communication systems—a subject we wish to defer until Chapter 9. In the next two chapters we develop and discuss techniques common to all digital communication systems: synchronization techniques.

PROBLEMS

5.1. A truly random waveform has no DC term. Why is there a DC term in the power density of the PN code?

5.2. **(a)** What are the sequences generated by the polynomials $x^5 + x^4 + x^3 + x^2 + 1$ and $x^5 + x^4 + x^2 + x + 1$? Assume an initial condition of (11111) in both cases.
(b) Compute and plot the autocorrelation function for each sequence in part (a).
(c) Compute and plot the cross-correlation function for the two sequences.
(d) Draw the circuit diagram for generating each of the sequences.

5.3. A PN sequence is $2^{15} - 1$ in length. How many runs of four 1s would be expected?

5.4. Plot the detailed power spectrum and the autocorrelation function for the maximal length sequence generated by $P(x) = x^4 + x + 1$. The shift register clock rate, F_c, is 2 kHz.

5.5. The code in Problem 5.4 biphase modulates a 2.0-MHz carrier. Plot the power spectrum of the modulated carrier.

5.6. The following sequence of symbols is received from a spread spectrum transmitter: 1 0 1 1 1 1 1 0 1 1 0 0 1. Spectral analysis of the received signal indicates that the power spectrum consists of discrete lines that are spaced at 322.6 HKz and that the spreading code rate is 10 MHz. What code generator is being used in the transmitter?

5.7. Prove that the generating function $X^3 + X + 1$ is primitive.

5.8. Consider the generating function $X^6 + X + 1$. Is this function primitive? Demonstrate that approximately an equal number of 1s and 0s are generated.

5.9. Consider the polynomial $X^4 + X + 1$.
(a) Sketch the shift-register representation.
(b) Is the polynomial reducible?

5.10. Show that the polynomial $X^3 + 1$ is reducible.

5.11. For Problem 5.9:
(a) Find the sequence generated if all the shift registers are initially loaded with 1s.
(b) If the register is initially loaded with (0, 0, 0, 1), show that a shifted version of the sequence is generated.

5.12. Find the length, N, of the sequence generated by $X^{24} + X^7 + X^2 + X + 1$. If the clock rate is 1.0 megabits/s, what is the period of the sequence in minutes?

5.13. For the PN code generator shown, find the sequence generated if the generator starts with a 1 at the output of each stage.

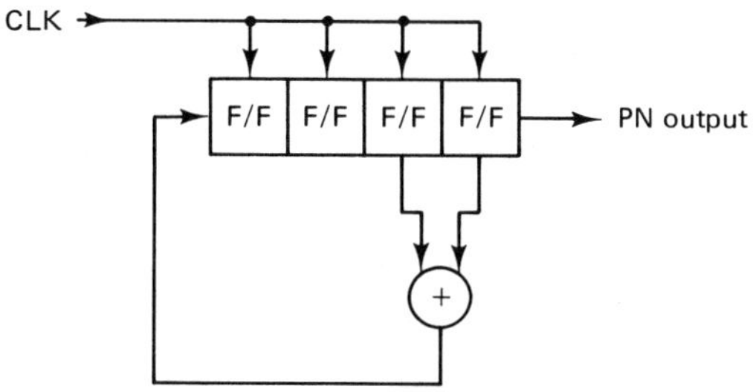

5.14. Given the sequence (1, 1, 1, −1, −1, −1, −1). The sequence is added modulo 2 with itself shifted two places. Compare the result of this operation with the original sequence.

5.15. A linear shift register with 13 stages is available. How many distinct sequences can be generated with this register?

5.16. Given the sequence (+1, −1, −1, +1, −1, +1, +1.) Find the autocorrelation function.

6

PHASE-LOCKED LOOPS

6.0 INTRODUCTION

Phase-locked loops (PLLs) were introduced in Chapter 3 (see Section 3.3.1.2) and simple trigonometry was used to illustrate the basic operation. These devices are of particular importance in the design of digital communication systems and it is important for the communications technologist to have a thorough understanding of these devices. PLL design is fairly straightforward if all factors are considered. The majority of conventional PLLs (Figure 6.1) perform one of the three functions:

- If N = unity, the PLL is a tracking filter.
- If N is greater than unity, the device is a frequency multiplier.
- If N is a selectable value, the PLL is a frequency synthesizer.

The functions just mentioned all find uses in digital systems. Under the tracking category, we can further subdivide its uses as follows:

- Bit synchronizers
- Frequency modulation/demodulation
- Phase modulation/demodulation
- AM modulation/demodulation (including (Double Sideband Suppressed Carrier) DSB-SC and SSB)

In many instances, the advent of integrated circuits has standardized the design for users. PLL principles were discussed in Chapter 3. Therefore, we begin our discussion with an analysis of the characteristics of PLLs.

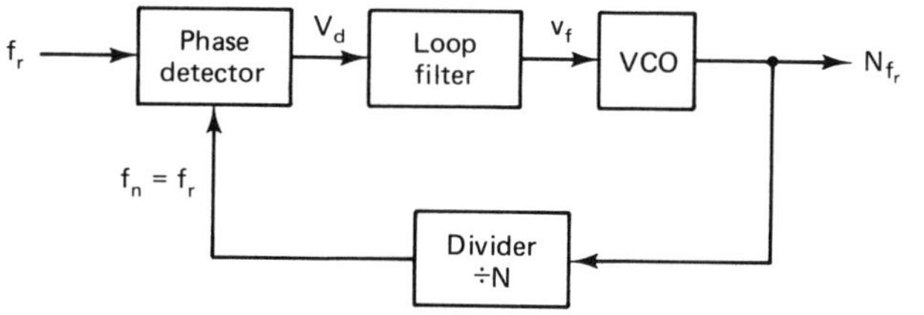

Figure 6.1 Basic PLL.

6.1 PLL CHARACTERISTICS

An elementary loop consisting of a phase detector, a low-pass filter, and a VCO is illustrated in Figure 3.13 and is repeated below as Figure 6.2 on page 136. Note the loop consists of a phase detector, a low-pass (loop) filter, and a VCO. The input signal has phase $\theta_i(t)$ and the phase of the VCO output signal is $\theta_o(t)$. The output of the phase detector is described by equation (3.18) and is given by

$$\begin{aligned} V_d &= K_d \sin(\theta_i - \theta_o) \\ &\approx K_d(\theta_i - \theta_o) \end{aligned} \tag{6.1}$$

where K_d is the gain of the phase detector with dimensions of volts per radian (refer to Figure 6.1). The phase-error voltage, V_d, is filtered by the loop low-pass filter, whose filter transfer function in the frequency domain is given by $F(s)$.

The output of the filter is an error voltage, V_f, which is used to control the

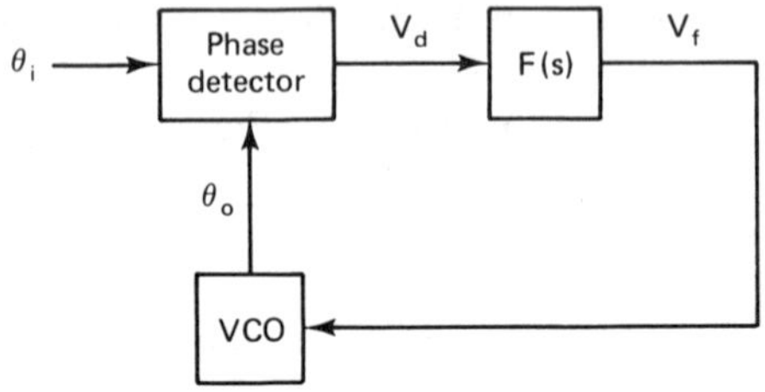

Figure 6.2 PLL block diagram.

frequency of the VCO. The deviation of the VCO from center frequency is given by

$$\Delta\omega = K_o V_f \tag{6.2}$$

where K_o is the VCO gain constant with dimensions of radians per second per volt. Since frequency is the time rate of change of phase, equation (6.2) may be expressed as

$$\frac{d\theta_o}{dt} = K_o V_f \tag{6.3}$$

Equation (6.3) may be transformed to the frequency domain by taking the Laplace transform. That is,

$$S \cdot \theta_o(s) = K_o V_f(s) \tag{6.4}$$

then

$$\theta_o(s) = \frac{K_o V_f(s)}{s} \tag{6.5}$$

Equation (6.5) illustrates that the output phase of the VCO is proportional to the integral of the filtered error voltage, V_f. We can write the transfer functions of the phase detector and loop filter as

$$V_d(s) = K_d[\theta_i(s) - \theta_o(s)] \tag{6.6}$$

and

$$V_f(s) = F(s)V_d(s) \tag{6.7}$$

Next, combining equations (6.5) through (6.7) results in the basic PLL equations:

$$A(s) = \frac{\theta_o(s)}{\theta_i(s)} = \frac{K_o K_d F(s)}{s + K_o K_d F(s)} \tag{6.8}$$

and

$$\frac{\theta_e(s)}{\theta_i(s)} = \frac{\theta_i(s) - \theta_o(s)}{\theta_i(s)} = \frac{s}{s + K_o K_d F(s)} \tag{6.9}$$

where

$K_o K_d F(s)$ = Forward loop gain
K_o = VCO constant (radians per second per volt)
K_d = Phase detector gain constant in volts per radian
s = Laplace operator
$F(s)$ = Transfer function of the loop filter
$A(s)$ = Loop transfer function

6.1.1 First-Order Loop

The order and, therefore, the frequency response of the loop is indicated by the order of equation (6.8). We see from this equation that it is directly dependent upon the filter transfer function, $F(s)$. Consider first a loop where $F(s) = 1$. That is, the low-pass filter is removed and the phase-detector output is connected directly to the VCO. Equation (6.8) reduces to

$$A(s) = \frac{1}{1 + s/K_o K_d} \tag{6.10}$$

This equation describes a first-order loop. Replacing s by $j\omega$ gives the Bode frequency response plotted in Figure 6.3. The natural frequency of the loop is denoted

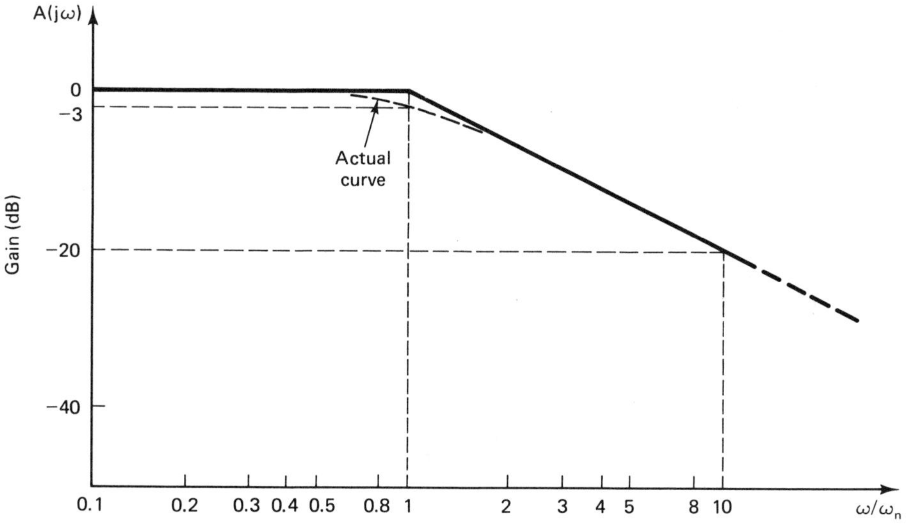

Figure 6.3 First-order loop Bode frequency response.

by ω_n and occurs at the break or corner frequency of the plot. The response of the actual circuit is down -3 dB at this frequency. We define the bandwidth of the loop as that frequency where the normalized loop gain decreases to -3 dB. Note that for a first-order loop, the natural frequency is equal to the bandwidth of the loop. From equation (6.10), the natural frequency of the loop is seen to be equal to the forward loop gain. That is,

$$\omega_n = K_o K_d \tag{6.11}$$

Thus loop bandwidth and loop gain are not independently selectable. The selection of one parameter determines the other.

6.1.2 Second-Order Loops

Fundamental characteristics of a PLL, such as capture range, capture time, loop bandwidth, and transient response, are controlled primarily by the loop filter. Two widely used filters are shown in Example 3.2. These filters are simple and the performance is satisfactory for many purposes. An active filter (see Figure 6.4) provides better tracking performance. The transfer function for this filter is

$$F(j\omega) = \frac{R_2 + 1/j\omega C}{R_1} = \frac{1 + j\omega R_2 C}{j\omega R_1 C} = \frac{1 + j\omega\tau_2}{j\omega\tau_1} \tag{6.12}$$

or

$$F(s) = \frac{1 + \tau_2 s}{\tau_1 s} \tag{6.13}$$

where $\tau_1 = R_1 C$ and $\tau_2 = R_2 C$. Substituting equation (6.13) into equation (6.8)

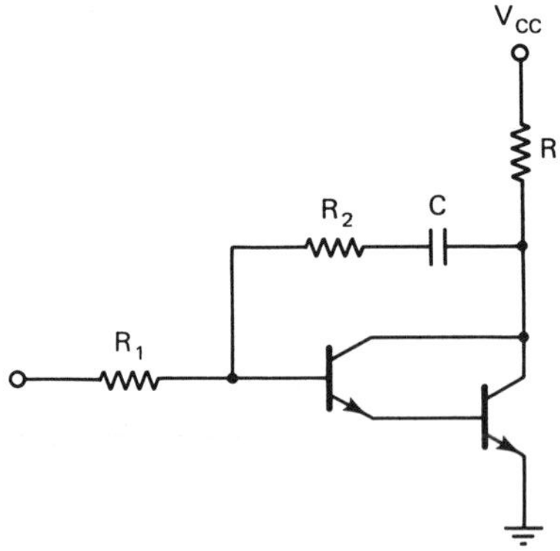

Figure 6.4 Active loop filter.

yields the following expression for the transfer function:

$$A(s) = \frac{K_oK_d(1 + \tau_2 s)/\tau_1 s}{S + K_oK_d(1 + \tau_2 S)/\tau_1 s} = \frac{1 + \tau_2 s}{1 + \tau_2 s + s^2\tau_1/K_oK_d} \tag{6.14}$$

Note that in contrast with equation (6.10), this is a second-order equation. We can extract useful information from this equation by comparing it with the standard form for a second-order system. Thus

$$A(s) = \frac{1 + \tau_2 s}{1 + (2\delta/\omega_n)s + s^2/\omega_n 2} \tag{6.15}$$

In this equation, δ is the damping factor and ω_n remains the natural frequency, as before. If we compare equations (6.14) and (6.15), we see that these quantities can be expressed as

$$\omega_n = \sqrt{\frac{K_oK_d}{\tau_1}} \tag{6.16}$$

and

$$\delta = \sqrt{\frac{K_oK_d}{\tau_1}}\,\frac{\tau_2}{2} \tag{6.17}$$

Note that for a given forward loop gain, K_oK_d, ω_n is independently variable by virtue of the filter time constant, τ_1.

Both ω_n and δ may be chosen from the transient response (Figure 6.5). With these two values defined, the loop filter can be readily designed. Resister values can be calculated using equations (6.16) and (6.17). An example illustrates the procedure.

EXAMPLE 6.1

A PLL is desired that will lock up in less than 2 ms. A peak overshoot of less than 20% is desired. The phase-detector gain, K_d, is 6.2 V/rad and the VCO gain, K_o is 10^7 rad/s/V. Design the loop filter using the configuration of Figure 6.4.

Solution From Figure 6.5, $\delta = 0.8$ will produce less than 20% overshoot. Also, we see that the loop settles to within 5% of normal at $\omega_n t = 4.5$. Thus

$$\omega_n = \frac{4.5}{t} = \frac{4.5}{2 \times 10^{-3}} = 2.25 \times 10^3 \text{ rad/s}$$

Let $C = 1\ \mu$F; then

$$R_1 = \frac{K_oK_d}{\omega_n^2 C} = \frac{(10 \times 10^6)(0.2)}{(2.25 \times 10^3)^2(10^{-6})} = 395\ \text{k}\Omega$$

and

$$R_2 = \frac{2\delta}{\omega_n C} = \frac{(2)(1.8)}{(2.25 \times 10^3)(10^{-6})} = 711\ \Omega$$

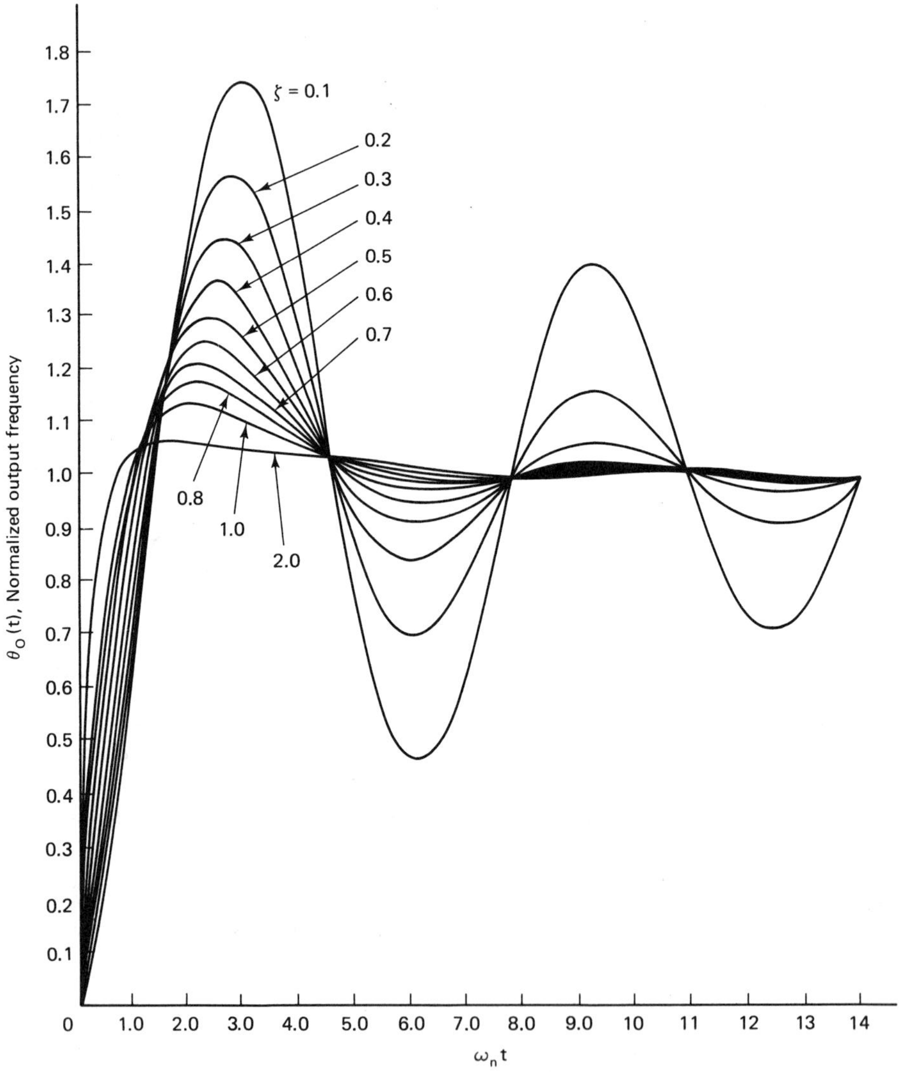

Figure 6.5 Second-order step response.

6.1.3 Frequency Response of a Second-Order Loop

Example 6.1 illustrates how the loop natural frequency, ω_n, and the damping factor, δ, affect the transient response of a PLL. It is of interest to examine the steady-state frequency response in terms of these same parameters. Refer to equation 6.15. The Bode frequency response plot for this equation is shown in Figure 6.6.

We define the bandwidth of the second-order loop as that frequency where the normalized loop gain drops to -3 dB. This same definition was applied to the first-order loop. With some manipulation, the bandwidth can be obtained in the conventional manner from equation (6.15). Because the first-order zero ($S_z = -1/\tau_2$) occurs at a much higher frequency than the second-order pole, we can ignore the zero for purposes of calculating bandwidth. Then, in terms of frequency, we can express equation 6.15 as

$$A(\omega) = \frac{1}{(1 - \omega^2/\omega_n^2) + j2\delta(\omega/\omega_n)} \tag{6.18}$$

The -3-dB frequency implies a voltage gain of 0.707. Thus setting $|A(j\omega)| = 0.707$ we have

$$\frac{1}{\sqrt{\left(1 - \dfrac{\omega^2}{\omega_n^2}\right)^2 + \left(2\delta\dfrac{\omega}{\omega_n}\right)^2}} = 0.707 \tag{6.19}$$

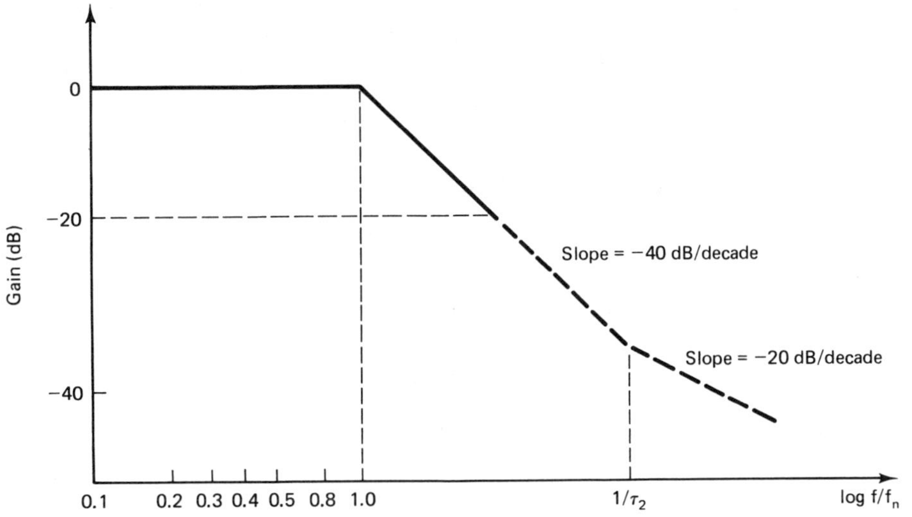

Figure 6.6 Second-order loop Bode frequency response.

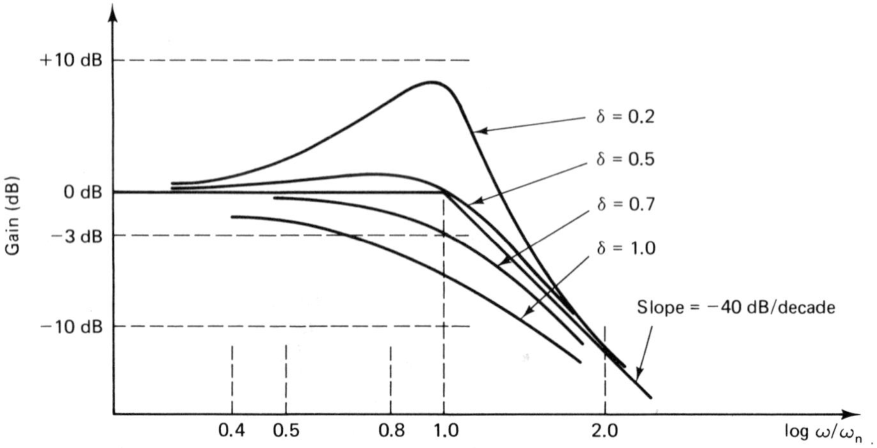

Figure 6.7

Squaring both sides of equation (6.19) and expanding yields

$$\left(\frac{\omega}{\omega_n}\right)^4 + (4\delta^2 - 2)\frac{\omega}{\omega_n} - 1 = 0 \tag{6.20}$$

Solving for ω using the quadratic formula, we have for the −3-dB frequency

$$\omega = \omega_{-3\,\text{dB}} = \omega_n[1 - 2\delta^2 + \sqrt{(2\delta^2 - 1)^2 + 1}]^{1/2} \tag{6.21}$$

Using this equation, for any damping factor δ and natural frequency ω_n, the loop bandwidth, $\omega_{-3\,\text{dB}}$, can be calculated. The actual frequency response as a function of δ and ω_n is shown in Figure 6.7. Note from this figure that for only one damping factor, $\delta = 0.707$, is the bandwidth equal to the natural frequency ω_n. For values of δ greater than 0.707, the bandwidth is less than ω_n. The converse is true for δ less than 0.707. Typical values are tabulated in Table 6.1.

6.1.4 Tracking and Acquisition

Example 6.1 illustrated a direct application of the use of δ and ω_n in designing the loop filter. It is of interest at this point to extend our understanding of the relation of the parameters on tracking and acquisition characteristics of PLLs.

TABLE 6.1

δ	$\dfrac{\omega_{-3\text{dB}}}{\omega_n}$
0.500	1.82
0.707	2.06
1.000	2.48

Consider first the transfer function, or conversion gain of the VCO alone. The units for VCO conversion gain were given as radians per second per volt. The transfer function for a practical VCO is shown in Figure 6.8. This figure illustrates the range of frequencies that the VCO can be tuned over by the control voltage, V_f, from the loop filter. This frequency range, ω_o(min) to ω_o(max) is the *lock range* or *tuning range* of the PLL. After lock has been achieved, the VCO will track the incoming frequency as long as this frequency is within the lock range.

The input may be at any frequency ω_i prior to lock. The multiplying (mixing) action of the phase detector generates sum and difference frequencies ($\omega_i \pm \omega_o$). The sum frequency, $\omega_i + \omega_o$, is removed by the loop low-pass filter. If, now, the difference frequency $\omega_i - \omega_o$ is outside the band edge of the loop filter, it will also be removed. The PLL cannot achieve lock. As the input frequency ω_i approaches the VCO free-running frequency ω_c, the difference frequency $\omega_i - \omega_c$ becomes smaller and enters the passband of the loop filter. The filter applies some of the DC component of the difference frequency to the VCO. This DC drives the VCO frequency, ω_o, toward the input frequency, ω_i. The difference frequency $\omega_i - \omega_o$ decreases and a greater DC component is generated. This positive feedback action continues until the VCO frequency equals the input frequency. The loop is now in lock and will track the input frequency within the VCO lock range. The range of input frequencies ω_i centered about the VCO free running frequency ω_c such that the difference frequency $\omega_i - \omega_o$ allows lock is called the *capture range* of the loop. These characteristics are illustrated in Figure 6.9. Referring to this figure, the lock range is that range of frequencies from ω_1 to ω_4. The capture range is that range of frequencies from ω_2 to ω_3. For the VCO not in lock and $\omega_i < \omega_1$, the VCO output ω_o is equal to its center frequency ω_c. When the input frequency ω_i reaches frequency ω_2, the input control voltage V_f locks the VCO to this frequency. The VCO will now track any frequency in the range ω_1 to ω_4. For any frequency greater than ω_4 or less than ω_1, the PLL will lose lock and the VCO will return to center frequency. A similar procedure occurs if ω_i is greater than ω_4. The capture frequency is ω_3 and the lock range is again

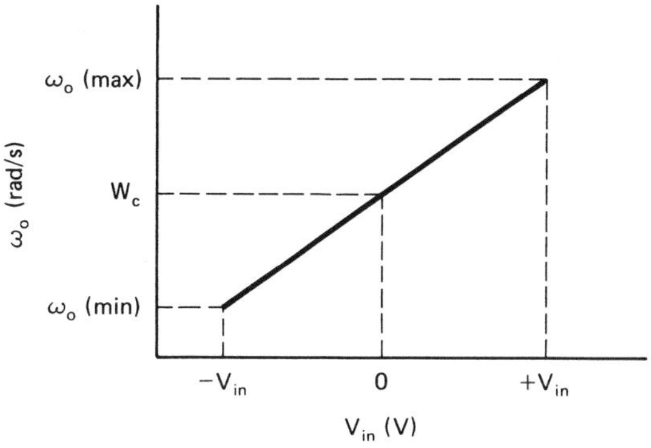

Figure 6.8 VCO with limited frequency range.

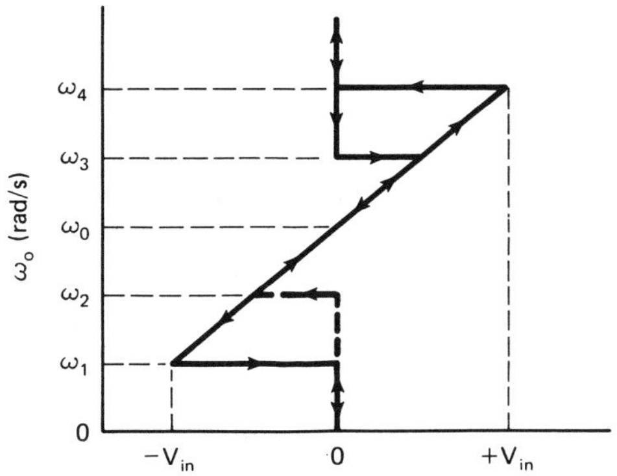

Figure 6.9 VCO capture and lock characteristics.

from ω_1 to ω_4. In order that the loop be capable of tracking over the entire lock range, the loop filter voltage V_f must be of maximum absolute value at the extremes of the range.

The DC or average voltage applied to the VCO control input determines the lock range. This presumes that the circuit capability of the VCO itself is not exceeded. The capture range, however, is a function of the DC component of the difference sideband passed by the loop filter. Therefore, the capture range is a function of the AC frequency response of the PLL. A frequency plot for two PLLs having different AC frequency response characteristics is shown in Figure 6.10. The capture range for each respective loop is equal to $\omega_c \pm \omega_{L1}$, or $\omega_c \pm \omega_{L2}$, where ω_c is the VCO center frequency.

The AC frequency response also affects the PLL output. A quantitative discussion serves to illustrate this. Consider the nature of the VCO output signal. The parameter that varies as a function of the control signal is the VCO frequency, ω_o. The changes that occur are as follows:

- Rate of change of the VCO frequency.
- The range of frequencies over which the output changes for a specified input.

We shall see how the AC transfer function relates to these two characteristics.

Consider a carrier frequency ω_i applied to the PLL input and the loop in lock. If the carrier is now frequency modulated with a signal of frequency, ω_s, the VCO frequency will also vary at the rate, ω_s. For a signal frequency, ω_s, the VCO will vary over a frequency-deviation range that is determined by the amplitude of the input AC signal. The limiting factor in determining Δf_{max} is the lock range of the PLL. For a specified input amplitude of the modulating signal, ω_s, a varying phase difference between the VCO output and ω_i will occur at the ω_s rate. A varying input control voltage to the VCO is generated as a result. At the same time, the average DC value of the VCO input control voltage remains constant. As long as the closed-loop response of the PLL can respond to the ω_s rate, the peak value of the VCO input control voltage will be proportional to the am-

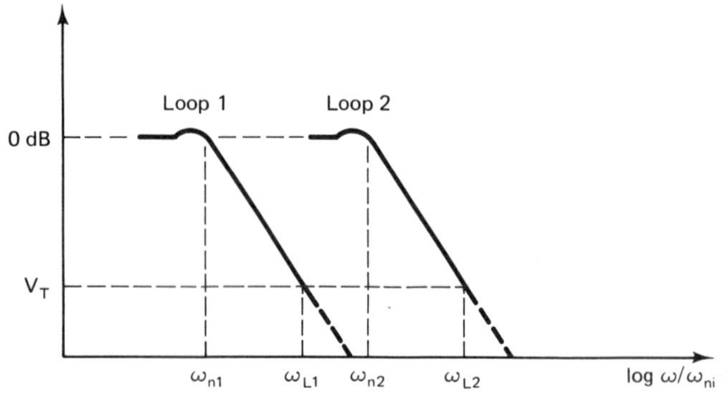

Figure 6.10 VCO relative capture ranges.

plitude of the input signal, ω_s. This will result in a variation of the VCO output frequency that is constant and correlates with the range of input frequencies, ω_s.

Now, as the input signal frequency, ω_s, increases, the PLL becomes less able to track the instantaneous changes in ω_i. Consequently, the PLL is less able to track instantaneous variations in phase difference. The end result is a reduction in the instantaneous variation of the VCO control voltage, and there is less variation in the VCO output frequency. The frequency response is expressed in terms of relative phase angles between the VCO output and the carrier input (see equation (6.8)). Since the relative phase angle determines the VCO control voltage, the PLL AC frequency response can be determined by measuring the rms value of the VCO control voltage as a function of the modulating frequency, ω_s.

At this point we have some feel for the parameters that affect the capture and lock characteristics. Once the loop is in lock, it must track the input signal. In order to illustrate tracking, it is necessary to examine the phase error, θ_e, that results from an input, ω_i. After the loop has locked, there must be a corresponding output cycle for each input cycle. Generally speaking, the phase error should be small. An exception exists where the PLL is used as a discriminator. In this particular situation, the phase error is the output information and close tracking is not desired.

Phase error in the frequency domain is expressed by equation (6.9); that is

$$\theta_e(s) = \frac{1}{s + K_o K_d F(s)} = \frac{1}{\dfrac{1 + K_o K_d F(s)}{s}} \theta_i(S) \qquad (6.22)$$

We can use standard servo theory to rewrite this equation as

$$\theta_e(s) = \frac{1}{1 + G(s)H(s)} \theta_i(s) \qquad (6.23)$$

where $G(s) = 1/s$ and $H(s) = K_o K_d F(s)$. Now $G(s)H(s)$ is the open-loop transfer function, and for any PLL it may be expressed as

$$G(s)H(s) = \frac{K_o K_d F(s)}{s} \qquad (6.24)$$

PLLs are further classified according to type based on the number of poles of the loop transfer function at the origin. Typical loop transfer functions for types 1, 2, and 3 are:

$$\text{Type 1:} \quad G(s)H(s) = \frac{K}{s(s + a)} \qquad (6.25)$$

$$\text{Type 2:} \quad G(s)H(s) = \frac{K}{S^2(s + a)} \qquad (6.26)$$

$$\text{Type 3:} \quad G(s)H(s) = \frac{K(s + a)(s + b)}{s^3} \qquad (6.27)$$

Loop filter configurations for second- and third-order PLLs are illustrated in Figure 6.11. In terms of standard servo theory, the expression $1 + G(s)H(s)$ is referred to as the *characteristic equation*. The order of the loop refers to the highest degree of this polynomial.

Using the preceding information, we can now investigate the tracking characteristics of the different types of PLLs. Steady-state phase errors are those that remain after transients have died away. We can evaluate these errors by use of the final value theorem of Laplace transforms; stated simply, this is

$$\lim_{t \to \infty} y(t) = \lim_{S \to 0} sy(s) \tag{6.28}$$

If we apply this theorem to equation (6.22), the result is

$$\lim_{t \to \infty} \theta_e(t) = \lim_{s \to 0} \frac{s^2 \theta_i(s)}{s + K_o K_d F(s)} \tag{6.29}$$

For a second-order loop, the phase error can be written as

$$\theta_e(s) = \frac{s\theta_i(s)}{s^2 + 2\delta\omega_n + \omega_n^2} \tag{6.30}$$

Let the input frequency change linearily with time at a rate of $\Delta\dot{\omega}$ radians per square second. This could represent either a doppler frequency or sweep-frequency modulation. By the final value theorem, the acceleration error, θ_a, of the

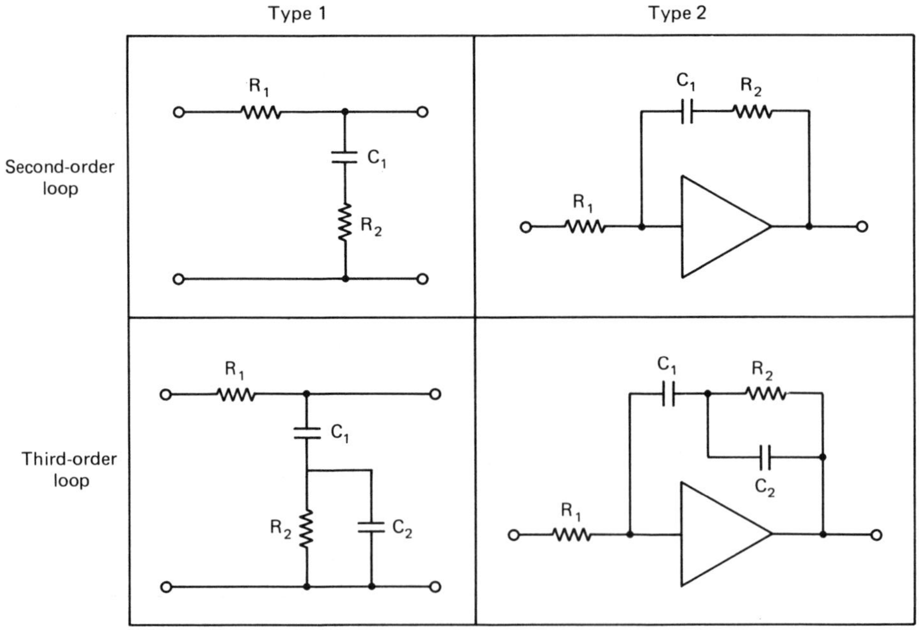

Figure 6.11 Loop filter configurations.

TABLE 6.2 Steady-State Phase Errors for Various System Types

	Type 1	Type 2	Type 3
Step position	Zero	Zero	Zero
Step velocity	Constant	Zero	Zero
Step acceleration	Continually increasing	Constant	Zero

PLL is

$$\theta_a = \lim \theta_e(t) = \lim \frac{\Delta\dot{\omega}}{s^2 + 2\delta\omega_n + \omega_n^2} \qquad (6.31)$$
$$= \frac{\Delta\dot{\omega}}{\omega_n^2} \text{ rad}$$

In words, this says that the second-order PLL will track the linearly changing frequency with a constant phase error θ_a. The acceleration error can be reduced to zero by changing the loop to a third-order loop. The necessary filter configuration is shown in Figure 6.11. Steady-state phase error versus system type is shown in Table 6.2. This table can be used in selecting a PLL type for a specific input signal.

The design of a PLL typically involves selecting the type of loop required, determining the necessary AC bandwidth, and also establishing the required stability. These design constraints are illustrated in Example 6.2.

EXAMPLE 6.2

A frequency synthesizer is desired with the following specifications:

- Lockup time between channels (to 5%), 1.0 ms
- f_{out}, 2 to 3 MHz
- Frequency steps, 100 KHz
- Overshoot, less than 20%

Solution Refer to Figure 6.1. From the specifications, we first obtain the range for N.

$$f_r = f_{in} = 100 \text{ KHz}$$

Then

$$N_{max} = \frac{f_{o(max)}}{f_r} = \frac{3.0 \text{ MHz}}{0.1 \text{ MHz}} = 30$$

$$N_{min} = \frac{f_{o(min)}}{f_r} = \frac{2.0 \text{ MHz}}{0.1 \text{ MHz}} = 20$$

Overshoot is specified to be 20%. The VCO range must accomodate this. The end limits of the VCO are computed as follows:

$$f_o(\text{max}) \geq 3.0 + (0.2)(3.0) = 3.6 \text{ MHz}$$

$$f_o(\text{min}) \leq 2.0 - (0.2)(2.0) = 1.6 \text{ MHz}$$

The range of the VCO is from 1.6 to 3.6 MHz. This corresponds to a tuning range of 3.6/1.6 = 2.25:1.

Next, based on the lockup time and overshoot, we can select the damping factor and the loop natural frequency. From Figure 6.5, a damping factor δ equal to 0.8 will produce a peak overshoot of less than 20%. From this same figure, we see that for $\omega_n t = 4.5$, the response will settle to less than 5%. Thus, the required loop natural frequency, w_n, is

$$\omega_n = \frac{4.5}{t} = \frac{4.5}{10^{-3}} = 4.5 \times 10^3 \text{ rad/s}$$

To complete the design, we must design the loop filter (see Figure 6.4). Design of the filter requires that the gain of the VCO and the phase detector be known. For this design, an MC 4324/4024 VCO and an MC 4344/4044 phase detector are chosen. From specification sheets for these devices the gain constants are:

$$K_d = 11 \times 10^6 \text{ rad/s/V}$$

and

$$K_d = 0.1 \text{ V/rad}$$

With a divider in the loop, the loop transfer function (equation 6.8) becomes

$$A(s) = \frac{K_o K_d K_n(s)}{s + K_o K_d K_n F(s)}$$

The expressions for R_1 and R_2 become (see example 6.1)

$$R_1 = \frac{K_o K_d}{N \omega_n^2 C}$$

$$R_2 = \frac{2\delta}{\omega_n C}$$

Choose $R_1 = 1 \text{ k}\Omega$; then

$$C = \frac{K_d K_o}{N_{\text{max}} \omega_n^2 R_1} = \frac{(0.1)(11 \times 10^6)}{(30)(4.5 \times 10^3)^2 (10^3)} = 1.8 \ \mu\text{F}$$

$$R_2 = \frac{2\delta_{\text{min}}}{\omega_n C} = \frac{2(0.8)}{(4.5 \times 10^3)(2 \times 10^{-6})} = 177.7 \ \Omega$$

We can compute δ_{max} as follows:

$$\delta_{max} = \delta_{min} \sqrt{\frac{N_{max}}{N_{min}}} = 0.8 \sqrt{\frac{30}{20}} = 0.98$$

Refer to Figure 6.5. We see that values of δ up to 0.98 meets the requirements for settling time.

In this chapter, we have considered the essentials of PLL design—a tall order indeed. Chapter 7 examines the problem of bit synchronization in digital communication systems. We will have an opportunity to return again to PLLs and expand our basic knowledge of the design and use of these remarkable circuits.

PROBLEMS

6.1. A PLL uses a frequency divider (see Figure 6.1) such that the VCO output frequency, $f_o = 15f_r$ during a lock condition, that is $f_s = f_r$, $\theta_i - \theta_o \approx 0$. Now let f_r be changed to $f_r + \Delta f$ such that $\theta_i - \theta_o = 0.1$ rad. What is the change in phase angle, θ_o, at the output of the VCO?

6.2. Show that for a loop with a sinusoidal phase detector, the lock (hold-in) range, $\pm\Delta\omega_H$, is given by $\pm\Delta\omega_H = \pm K_v$, where K_v is the forward loop gain.

6.3. In Problems 6.2, the sinusoidal phase detector is replaced with a triangular phase detector (see Figure 7.16). What is the lock range ($\pm\Delta\omega_H$)? What is the lock range if a sawtooth detector is used?

6.4. It is stated that the capture range is equal to $\omega_c \pm \omega_L$. For a second-order loop, it has been shown that the capture range, $\Delta\omega_L$, is given by

$$\Delta\omega_L = \pm \frac{K_v \tau_2}{\tau_1 + \tau_2}$$

where K_v is the forward loop gain. Show that the capture range is given approximately by

$$\Delta\omega_L \approx \pm 2\delta\omega_n$$

6.5. Design a frequency synthesizer to the following specifications:

- $f_o = Nfr$, 1 to 17 MHz
- Frequency steps, 50 kHz
- Overshoot less than 15%
- Lock-up time between channels (to 5%), 2.0 MS

6.6. In Figure 6.1 with $N = 1$, if f_r is a sine wave and f_n is a square wave, find the expression for V_d, the phase detector output for (a) $f_r \neq f_n$ and (b) $f_r = f_n$.

6.7. The PLL is set up with $N = 25$ and $Nf_r = 2.5$ MHz. Assume a locked condition, that is, $\theta_i = \theta_o \approx 0$ and $V_f \approx 0$. A phase shift, $\Delta\theta$, is inserted between the divider and the phase detector using a voltage-controlled phase shifter. Explain in detail the effect on $\theta_i - \theta_n$ and θ_o, where θ_n is at the divider output. What effect does this have on the VCO output frequency, Nf_r?

6.8. Refer to Figure 6.4. The use of this filter results in a second-order system. The filter performs one integration. Where in the loop is the second integration performed? Is the integration with respect to frequency or phase or both? Explain.

6.9. A passive filter that is often used in PLLs is as shown.

(a) Calculate the transfer function of this filter.

(b) Calculate the PLL transfer function using equation (6.8) and show that it can be expressed in a standard form similar to equation (6.15).

(c) Define the quantities ω_n and δ. Under what conditions does the PLL transfer function closely approximate that obtained with the active filter transfer function given by equation (6.13).?

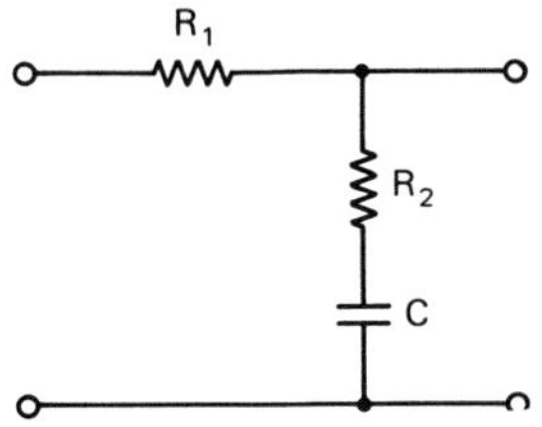

6.10. Verify Table 6.2 for a third-order loop using a type 1 filter.

6.11. A type 2 filter is used in a third-order loop.

(a) Derive the filter transfer function.

(b) Sketch the magnitude of the transfer function $|H(j\omega)|$ and the phase shift versus frequency.

7

DIGITAL SYNCHRONIZATION CIRCUITS

7.0 INTRODUCTION

Synchronization poses a fundamental problem in digital communication systems. In a digital data communication system, there exists a hierarchy of synchronization problems to be considered. For example, if a carrier type system is involved, we are concerned with *carrier synchronization* first. The receiver must generate a reference carrier with a phase that closely matches that of the received signal. This reference signal is used in the data receiver to perform coherent demodulation, thus creating a baseband data signal. The next problem in the hierarchy is that of synchronizing the receiver symbol clock with the baseband data-symbol sequence. This process is called *bit synchronization*.

Beyond the bit synchronizer, other synchronization problems occur. For example, depending on the system type, we may encounter sync problems such as *word*, *frame*, and *packet* synchronization. We commonly associate the first two with PCM systems and the latter with a data communication system wherein packets of data are transmitted. These three synchronization types are distinguished from carrier and bit synchronization schemes by the fact that the problem is usually solved by specially designing the message format. For example, this often involves inserting repetitive bits or words into the data sequence. These words are easily recognizable and can serve as sync words. Often the bit pattern is chosen so that it is unlikely to occur as data.

In systems where the transmitted signal contains an unmodulated sinusoidal component, the standard approach to achieve carrier synchronization is to use a PLL. The PLL is designed with a narrow loop bandwidth such that it will lock onto and track the carrier and not be disturbed by sideband components of the modulated signal. For systems in which the carrier is totally suppressed, it must be recreated from the sideband data. For example, a Costas loop is capable of accomplishing this for PSK systems (see Chapter 3). For modulation formats that exhibit a high bandwidth efficiency—large bits per hertz figure of merit—the accuracy requirements for both carrier and bit synchronization become increasingly severe. In today's digital systems, increasing pressure is being felt to increase bandwidth efficiency. A byproduct, then, is an increased emphasis on synchronization techniques. In the following, we will consider first techniques for carrier synchronization.

7.1 CARRIER SYNCHRONIZATION

In digital communication systems where a modulated carrier is used to transmit the data, a phase-coherent reference carrier is needed to demodulate coherently the carrier signal. In previous chapters we have, to some extent, already considered elements of this problem. We can illustrate the problem by a consideration of the major modulation formats such as FSK, PSK, or DSB-SC. Optimum system performance is obtained with coherent demodulation. The coherent demodulator requires a reference carrier, which is phase coherent with the modulated carrier.

The basic DSBSC system is obtained by multiplying the carrier signal, $\cos \omega_c t$, by the digital data stream, $d(t)$, using a balanced modulator. We can express this operation mathematically as

$$e_{DSBSC} = d(t) \cdot \cos \omega_c t \tag{7.1}$$

The second modulation type is PSK. This is a special case of angle modulation in which the phase is shifted in accordance with the data. That is,

$$\begin{aligned} e_{PSK} &= A \cos[\omega_c t + M_p d(t)] \\ &= A \cos[M_p d(t)] \cos \omega_c t - \\ &\qquad A \sin[M_p d(t)] \sin \omega_c t \end{aligned} \tag{7.2}$$

where M_p is the phase modulation index. PSK is similar to the DSB-SC signal. In fact, if the data is binary, $d(t) = \pm 1$, and $M_p = \pi/2$ rad, then PSK is equivalent to DSB-SC. Thus carrier-synchronization techniques developed for one system would also be applicable to the other system.

7.1.1 Pilot Carrier Synchronization

A residual carrier signal is available for synchronization purposes in a DSB-AM system. Isolation of the unmodulated carrier from the composite spectrum implies the use of narrow-band filtering. This is necessary to separate the carrier from the data sidebands. Premodulation filtering of the data makes this an easier task by minimizing sidebands near the carrier. An alternative choice is to decrease the modulation index. Both of these techniques degrade data detection performance; consequently neither is a desirable solution.

The actual technique commonly used to enable carrier synchronization is the addition of a pilot carrier to DSB-SC modulation. The pilot carrier can be located in a region where the data sidebands are small. This technique also allows the carrier level to be adjusted independently of the data modulation and in some cases allows one pilot carrier to be used with a set of several DSB-SC or PSK data signals at different carrier frequencies. In a power-limited communication system, the carrier power is a major part of the overall transmitter power. For a given total RF output power, this reduces the overall system performance but makes the solution of carrier synchronization a trivial problem. A carrier filter consisting of a band-pass filter is commonly employed. For a typical band-pass filter, the rise time may be approximated as:

$$t_r \approx \frac{0.7}{B} \tag{7.3}$$

where B is the -3dB bandwidth. The sync acquisition time in carrier cycles is then

$$N = t_r \cdot f_c = \frac{0.7}{(B/f_c)} \text{ Hz} \tag{7.4}$$

where f_c is the carrier frequency. For coherent demodulation, a perfectly stable carrier is desired. For example, any phase jitter is passed on to the recovered baseband data as noise. The recovered carrier signal at the output of the filter may be expressed as

$$y(t) = s(t) + n(t) \tag{7.5}$$

where $s(t)$ is the signal and $n(t)$ is noise. The addition of noise will change the zero crossing of the sinusoidal carrier, thus introducing jitter.

We can estimate the timing jitter at the zero crossings of the recovered carrier by considering the rate of change of voltage at the crossing and the rms value of noise voltage. Let the carrier be represented as

$$s(t) = \sqrt{2}A \cos \omega_c t \tag{7.6}$$

where A is the rms value. Then the rate of change with respect to time is

$$\frac{ds}{dt} = \sqrt{2}A\omega_c \sin \omega_c t \tag{7.7}$$

At the zero crossing $\sin \omega_c t = 1$ and the maximum rate of change is

$$\left.\frac{ds}{dt}\right|_{\max} = \sqrt{2}A\omega_c \tag{7.8}$$

If the rms noise voltage at the output of the carrier filter is σ_n, the rms value of the timing jitter is approximately:

$$\Delta t_{\text{rms}} \approx \frac{\sigma_n}{(\Delta s/\Delta t)} = \frac{\sigma_n}{\sqrt{2}A\omega_c} \tag{7.9}$$

The jitter normalized to the carrier period is given by

$$\frac{\Delta t_{\text{rms}}}{T} = \frac{1}{(2\sqrt{2}\pi A/\sigma_n)} = \frac{1}{(2\sqrt{2}\pi \text{ SNR})} \tag{7.10}$$

where $T = 1/f_c$ and $\text{SNR} = A/\sigma_n$. Note that the timing jitter is proportional to the pilot carrier SNR measured at the output of the carrier recovery filter. Equation 7.10 is plotted in Figure 7.1 on page 156.

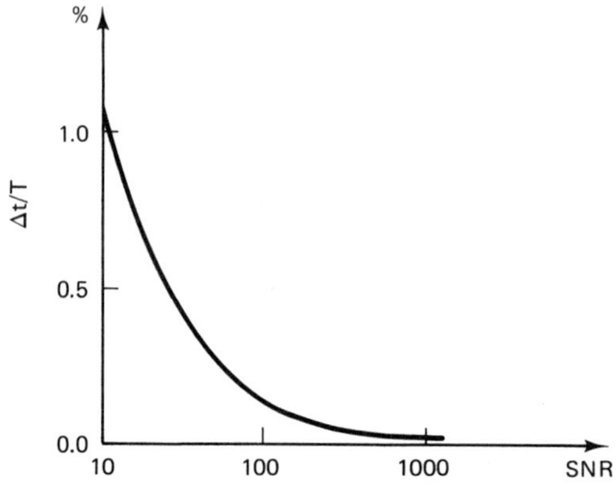

Figure 7.1 Normalized carrier phase jitter versus SNR.

7.1.2 Carrier Extraction in Suppressed Carrier Systems

Both DSB-SC and PSK modulation result in a suppressed carrier system (see equations (7.1) and (7.2)). If devoting power to pilot carriers is not desired, the carrier must be reconstructed from the available received spectrum. When the modulating data, $d(t)$, is random, the received signal spectrum does not contain any discrete spectral components. One approach to deriving a carrier synchronization signal is to pass the signal through a nonlinear element to generate a carrier component. Variations of this technique were introduced in Chapter 3.

One technique for extracting a carrier is simply to square the signal. For example, using equation (7.1), we have

$$\begin{aligned} r(t) &= d^2 \cdot \cos^2 \omega_c t \\ &= d^2[1 + \cos 2\omega_c t] \end{aligned} \tag{7.11}$$

Although $d(t)$ is random, the quantity $d^2(t)$ fluctuates around a constant mean value, which leads to a discrete spectral value of $2\omega_c$. A narrow-band-pass filter or PLL can be used to provide a stable reference carrier (see Figure 7.2). This technique results in a 180° phase ambiguity problem (see Chapter 8).

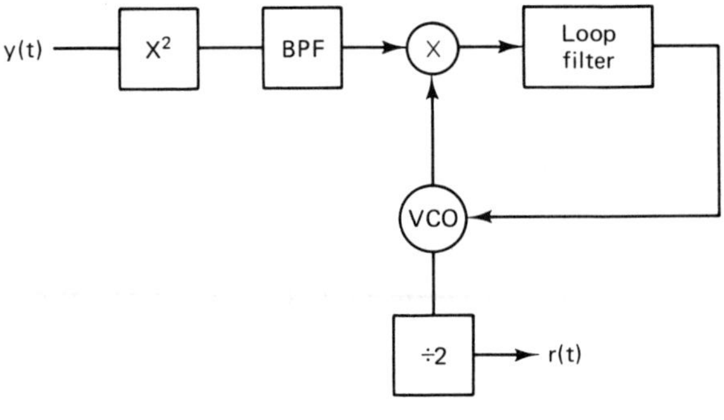

Figure 7.2 Squaring loop carrier synchronization circuit.

A second approach to carrier reconstruction for PSK and DSB-SC is use of a Costas loop. This technique is discussed in Section 3.3.3.1. This particular technique multiplies the received signal by in-phase and quadrature reference carriers. The multiplier outputs are low-pass filtered to remove twice frequency components. The filtered signals are again multiplied and the product low-pass filtered to produce an error signal for the loop VCO. A mathematical analysis is given in Chapter 3.

A third type of carrier synchronizer configuration is known as a decision-directed loop (Figure 7.3). Operation of the loop is easily explained if we assume it is already in lock. That is, the VCO is already operating at the correct frequency with a small phase angle, θ. Refer to the figure. The output of the inphase VCO reference may be expressed as:

$$r_1 = \cos(\omega_c t + \theta) \tag{7.12}$$

For small θ, $r_1 \approx \cos \omega_c t$. The quadrature reference is described by

$$r_2 = \sin(\omega_c t + \theta) \tag{7.13}$$

The inphase reference r_1 coherently demodulates the carrier and the matched filter provides an optimum estimate of the data d. That is, the output of the multiplier is

$$\begin{aligned} Z(t) &= y(t) \cdot r_1(t) \\ &= d(t) \cos \omega_c t \cdot \cos \omega_c t \\ &= d(t)\,[1 + \cos 2\omega_c t] \end{aligned} \tag{7.14}$$

The twice-frequency term is removed by the matched filter. The filter output is sampled and a comparator completes the estimate of the data, $\hat{d}$. This estimate then remodulates the carrier to produce r_3. That is,

$$r_3 = \hat{d}(t) \cdot \sin \omega_c t \tag{7.15}$$

The data estimate $\hat{d}$ is delayed by 1 bit period from the input data by the integrating

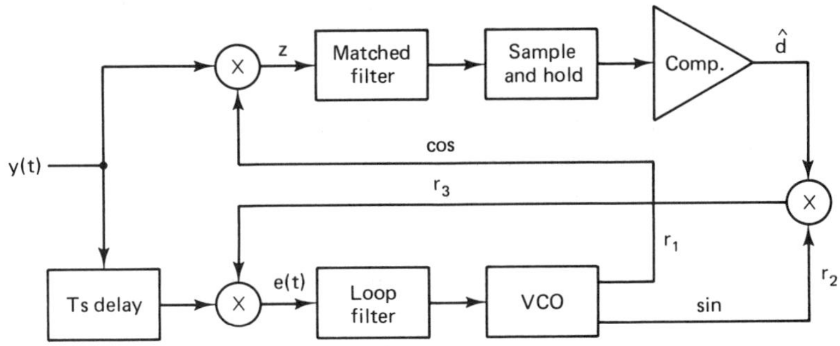

Figure 7.3 Decision-directed loop.

action of the matched filter. A compensating delay of T seconds is placed in the quadrature path. The output, $e(t)$, to the loop filter is

$$\begin{aligned} e(t) &= r_3 \cdot d \cdot \cos \omega_c t \\ &= d \cdot \hat{d} \sin \omega_c t \cdot \cos \omega_c t \cdot \cos \theta \\ &\quad + d \cdot \hat{d} \cos^2 \omega_c t \cdot \sin \theta \end{aligned} \tag{7.16}$$

The loop filter eliminates twice-frequency terms and averages the product of the data and its estimate ($d \cdot \hat{d}$). This provides a proportional phase-error term, $f(t)$, to the VCO.

$$\begin{aligned} f(t) &= K_1 \sin 2\,\theta \\ &\approx K_2 \theta, \qquad \text{for small } \theta \end{aligned} \tag{7.17}$$

This phase error term drives the loop to lock.

All three types of carrier synchronizers can be shown to be equivalent to PLLs. This suggests a design approach based on such parameters as the acquisition performance of the equivalent PLL. Noise performance of the three loops is considerably more complex. Considering that the synchronizer is designed to track a modulated carrier, it is reasonable to consider as a lower bound the noise performance of a PLL tracking an unmodulated carrier. It is shown in the literature that the noise performance of the squaring and Costas loop are identical. The performance of the decision-directed loop depends upon the implementation. This loop can approach that of the squaring and Costas loops. The theoretical tracking jitter for carrier synchronization loops in terms of loop SNR is illustrated in Figure 7.4. A design approach for a Costas loop is discussed in the next section.

7.1.2.1 Costas loop design. Consider the design of a Costas loop for a binary PSK signal. The loop is illustrated in Figure 3.22 and is repeated here for reference (Figure 7.5, see page 160). The design approach is to first find the equivalent PLL. We begin with equation (3.25). The input to the loop filter is

$$e_f = K_m \frac{A^2}{4} \sin[2(\theta_i - \theta_o)] \tag{7.18}$$

where

A = rms value of the PSK signal
K_m = multiplier gain
θ_i = input signal phase
θ_o = VCO output signal phase

We can express the output of the loop filter, $f(t)$, as

$$f(t) = H(\rho)\left[K_m \frac{A^2}{4} \sin 2(\theta_i - \theta_o) \right] \tag{7.19}$$

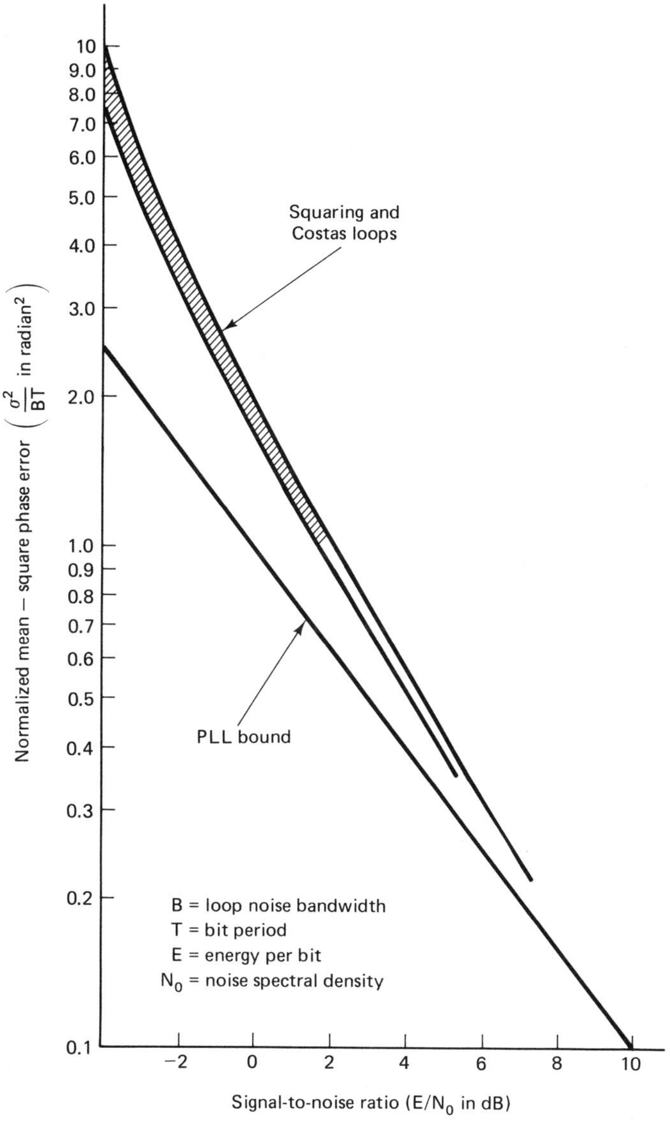

Figure 7.4 When normalized mean-squares phase error is plotted as a function of SNR, it can be seen that the squaring and Costas loops possess identical phase-error performance of carrier-tracking loops.

where $H(\rho)$ is an operator that describes the effect of the loop filter. The phase of the VCO is the integral of frequency, which is, in turn, proportional to the input voltage $f(t)$. Thus

$$\theta_o = K_o f(t)dt$$

where K_o is the VCO sensitivity in radians per second per volt. The rate of change

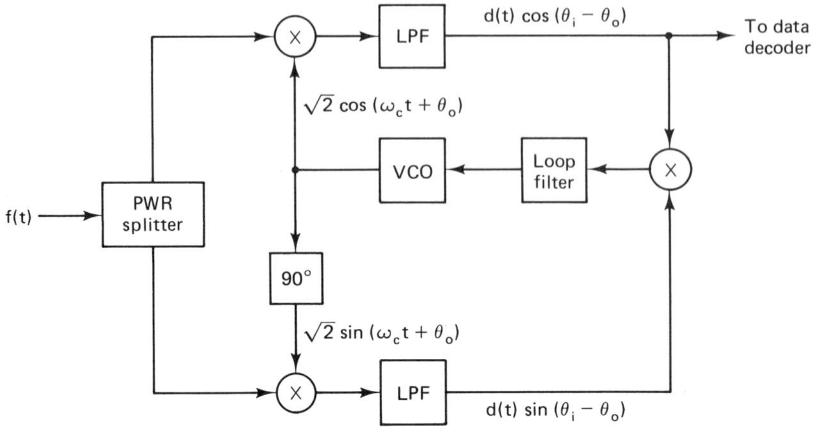

Figure 7.5 Costas carrier recovery loop.

of phase is given by

$$\begin{aligned}\frac{d\theta_o}{dt} &= K_o f(t) \\ &= K_o H(\rho)\left[K_m \frac{A^2}{4} \sin 2(\theta_i - \theta_o)\right]\end{aligned} \tag{7.20}$$

At this point, rather than trying to solve this equation, we can develop a block diagram equivalent to the equation (see Figure 7.6). For small phase error, $\sin 2(\theta_i - \theta_o) \approx 2(\theta_i - \theta_o)$. Thus the equivalent gain of the phase detector, K_d, is

$$K_d = \frac{A^2 K_m}{4} \tag{7.21}$$

If we assume a loop filter with transfer function, $H(s) = (1 + \tau_2 s)/\tau_1 s$, the loop

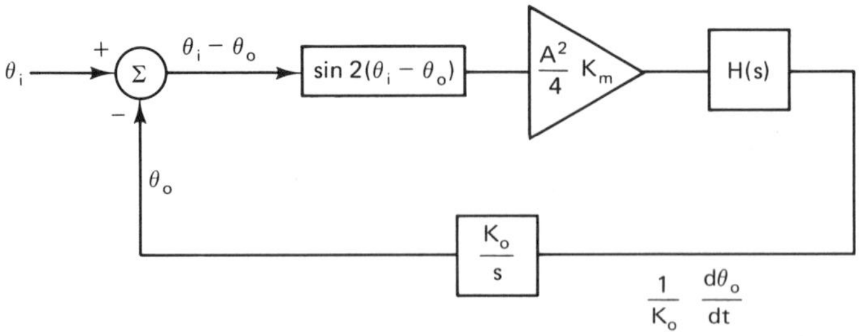

Figure 7.6 Equivalent PLL block diagram for a Costas loop.

transfer function is (see equation (6.15))

$$A(s) = \frac{1 + \tau_2 s}{1 + (2\delta/\omega_n)\delta + s^2/\omega_{n^2}} \tag{7.22}$$

The loop natural frequency, ω_n, and damping factor, δ, are

$$\omega_n = \sqrt{\frac{K_o K_d}{\tau_1}} \tag{7.23}$$

and

$$\delta = \sqrt{\frac{K_o K_d}{\tau_1}} \frac{\tau_2}{2} = \sqrt{\frac{K_o K_d}{4\tau_1}} \tag{7.24}$$

Note from equation (7.21) that the loop parameters ω_n and δ are a function of the input signal level, A, by virtue of the equivalent comparator gain.

As illustrated by the design examples in Chapter 6, selection of the loop parameters, ω_n and δ, starts the design process. These, in turn, are derived from specifications of the design. The damping factor is normally selected in the range 0.5 to 1.0, whereas the loop width, ω_n, is selected based on the sync-acquisition and jitter requirements. At any rate, ω_n must be less than the data rate in order that data frequency components not perturb the loop error signal (be passed by the loop filter). An example illustrates the design process.

EXAMPLE 7.1

Design a Costas loop to recover the carrier from an 1800-Hz PSK signal, which is modulated by 1200-bits/s binary data. Carrier synchronization is to be accomplished in 10 bit periods. The input signal level is 100 mV(rms).

Solution From Figure 6.5, the loop settles to within 5% for $\omega_n t = 0.5$. Ten bit periods corresponds to 15 carrier cycles. Thus

$$t = (15)\left(\frac{1}{1800}\right) = 8.33 \text{ ms}$$

Then

$$\omega_n = \frac{4.5}{t} = \frac{4.5}{8.33 \times 15^3}$$

$$= 540 \text{ rad}$$

or

$$f_n = 85.94 \approx 86 \text{ Hz}$$

For design purposes, choose a VCO with a sensitivity of 1000 Hz/V. Thus

$$K_o = 6280 \text{ rad/s-V}$$

Then

$$K_d = \frac{A^2 K_m}{4} = (0.1)^2 \frac{K_m}{4} = 0.0025 K_m$$

The multiplier gain, K_m, and the loop filter remain as design variables. From equation 7.24

$$\delta = \omega_n \frac{\tau_2}{2}$$

or

$$\tau_2 = \frac{2\delta}{\omega_n}$$

From Figure 6.5, a value of $\delta = 0.8$ allows the loop to settle within 5% for $\omega_n t = 4.5$. Then

$$R_2 = \frac{2(0.8)}{540} = 0.002963 \text{ s}$$

Choose $C = 1\ \mu\text{F}$, then

$$R_2 = \frac{0.002963}{10^{-6}} = 2963\ \Omega$$

Using equation (7.23), we have

$$\tau_1 = \frac{K_o K_d}{\omega_{n^2}}$$

and

$$K_d = 0.0025 K_m$$

A design trade-off exists between τ_1 and K_m. Chose $K_m = 50$; then

$$\tau_1 = \frac{(6280)\ (0.125)}{540^2} = 0.002692 \text{ s}$$

and

$$R_1 = \frac{0.002692}{540^2} = 2692$$

A circuit implementation using MC 1496s as the multipliers is shown in Figure 7.7. The loop multiplier gain consists of the actual multiplier gain along

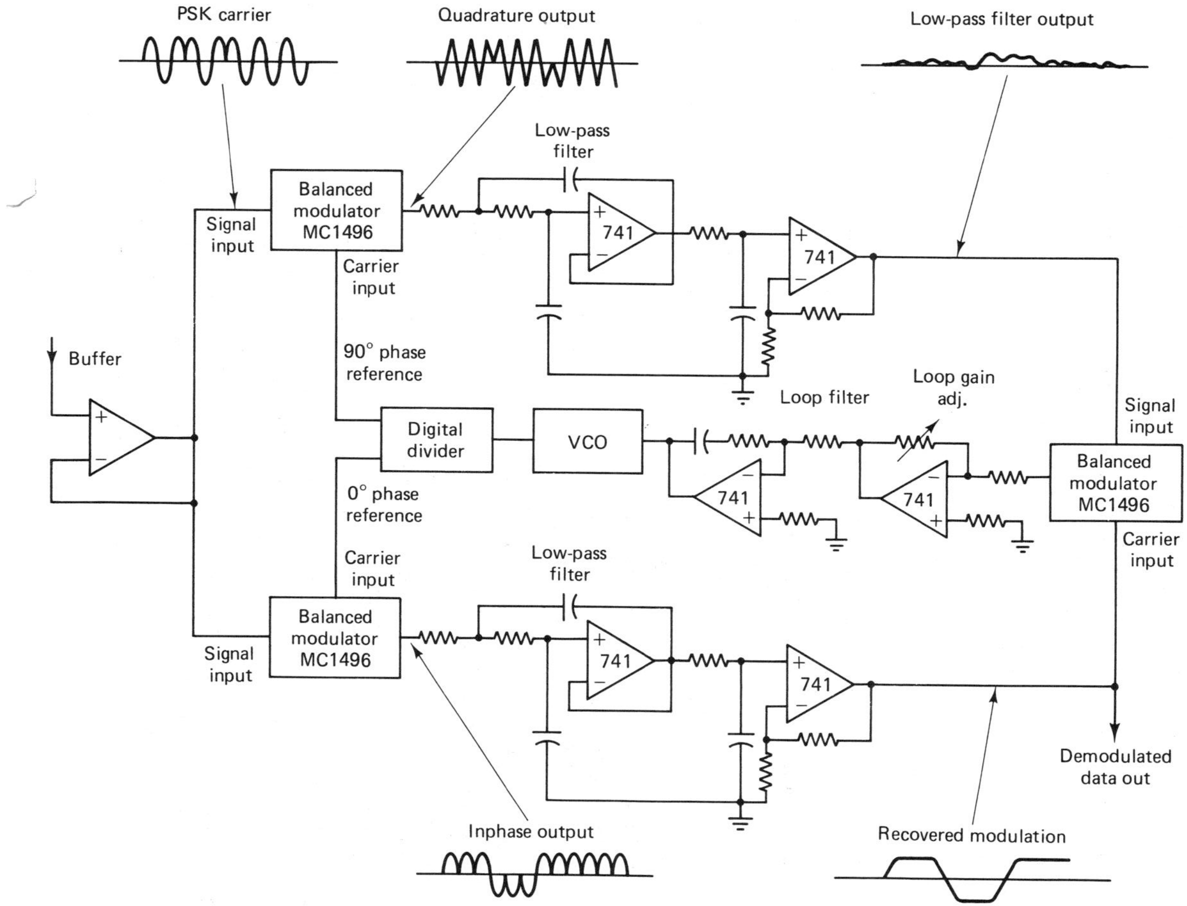

Figure 7.7 Block diagram of the complete loop shows typical waveforms when the loop is near lock.

with the gain of a cascaded amplifier. The low-pass filters in the arms following the two product multipliers should be wide enough to pass the data but reject second harmonic components. In order to preserve the time-domain characteristics of the data, linear-phase (Bessel) filters are used. For design purposes, a cutoff frequency equal to the bit rate is sufficient. Third-order filters will then provide 20-dB attenuation of the second harmonic terms. The output of the inphase channel is fed to bit synchronization and detection circuits to form the complete PSK demodulator. These techniques are discussed in the next section.

7.2 BIT SYNCHRONIZATION

Digital codes and signaling schemes are inextricably linked to bit synchronizers. Bit-error performance as well as the ability to extract timing depends on the choice of code. Some common PCM codes are illustrated in Figure 3.2. For example, if we chose to transmit data as NRZ PCM, a long string of continuous 1s or 0s will create timing problems because of a lack of data transitions. On the other hand, a biphase level code always contains a transition at midbit time, thus exhibiting a self-clocking property. A clock can be extracted from this type of signal using a thresholding or logical operation.

The reconstruction of a serial data stream that is corrupted by noise and delayed by an unknown time begins with the extraction of a timing signal, which can be used to reconstruct the clock. For example, zero crossovers of a received signal contain timing information. Detecting a zero crossing appears at first to be a simple enough problem. Difficulties do occur, though, if the data is noisy. Noise displaces the zero crossings and, consequently, timing jitter is introduced. Problems such as this lead to design trade-offs. The primary choice is usually between acceptable jitter performance versus time of acquisition.

The regeneration of a clock for NRZ PCM was introduced in Chapter 1 for asynchronous transmission (see Figures 1.5 and 1.6). As noted, the free-running multivibrator is tuned to oscillate at a frequency slightly lower than the expected received symbol rate. Now, if the free-running frequency differs from the symbol rate by Δf hertz, any phase error between the true symbol rate and the local clock (multivibrator) will increase with the time between transitions as

$$\Delta\theta = 2\pi(\Delta f)(T) \text{ rad} \tag{7.25}$$

where T is the free-running time. We can express the time, T, as

$$T = \frac{N}{f_B} \tag{7.26}$$

where N is the number of bit periods and f_B is the symbol rate. If the timing errors must be limited to ϕ radians, the maximum allowable difference, Δf, may be expressed as

$$\Delta f \leq \frac{\Delta\theta}{2\pi T} = \frac{\phi}{2\pi T} \text{ Hz} \tag{7.27}$$

Combining equations (7.26) and (7.27) we have

$$\frac{\Delta f}{f_B} \leq \frac{\phi}{2\pi N} \tag{7.28}$$

This expression allows a determination of the required local stability of the clock in an asynchronous receiver. For convenience it is plotted in Figure 7.8.

EXAMPLE 7.2

Data is transmitted using the format of Figure 1.5. If the maximum timing error is limited to 22.5°, What is the maximum deviation of the local clock from the input symbol rate?

Solution With this format, 8 bit periods can occur without a data transition. From Figure 7.8, the local clock must be within $\pm 0.78\%$ of the input symbol rate. Thus

$$\Delta f = (0.0078) f_B \text{ Hz}$$

The solution we have just examined provides an inexpensive solution to the problem of generating a local clock. It suffers, however, from the problem of

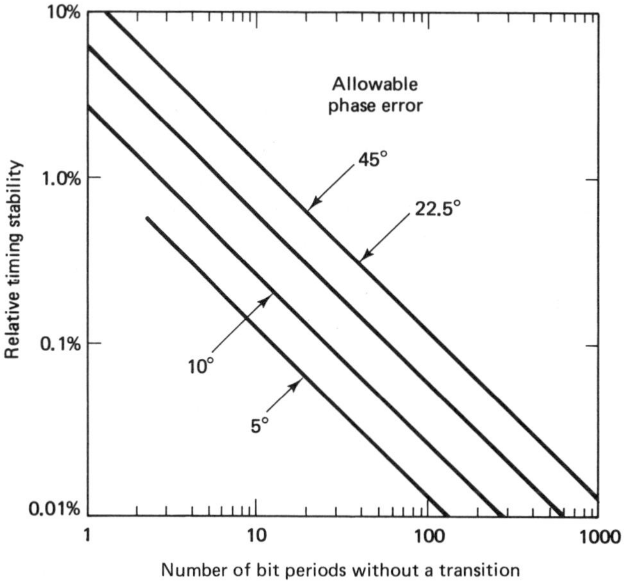

Figure 7.8 To find the local stability in an asynchronous receiver, select the appropriate curve for the maximum allowable phase error. Next, determine a point on this curve from the number of bits that can occur without data transition. The vertical axis determines maximum allowable deviation of the local clock.

frequency drift and external component selection. For this particular data format it is well to consider existing commercial solutions. The universal asynchronous receiver/transmitter (UART) is designed for just this problem. An investigation of a UART receiver shows that a receiver-register clock signal must be supplied to the UART. The clock must be at a frequency equal to 16 times the received bit-rate frequency, f_B. Start-bit-detection logic initiates receiver-register clocking at approximately mid-bit time. The inherent timing resolution of the UART is $f_B/16$. At this point we may ask ourselves how accumulated phase error (again using a 22.5° limit) can remain less than the inherent timing resolution for the maximum 11-bit message. An example illustrates.

EXAMPLE 7.3

Data is transmitted and received using UARTs. The format of Figure 1.5 is used with an 11-bit message. What is the maximum deviation of the receiver clock from the input symbol rate?

Solution The inherent resolution of the receiver clock is:

$$\text{Res} = \frac{360}{16} = 22.5^\circ$$

From Figure 7.8, the local clock must be within $\pm 0.58\%$ of the bit period.

Systems of this type rely on a single bit to initiate synchronization. This introduces a problem if the data is noisy. The probability of either correctly or incorrectly initiating the bit rate clock exists. Overcoming this limitation requires more time dedicated to synchronization. Consequently, in synchronous systems tens or hundreds of bits may be devoted to initial bit synchronization prior to the beginning of a continuous data bit stream. The design of synchronization circuits for systems of this type is more closely related to synchronous system design than to that of the asynchronous systems we have been discussing. Greater flexibility in design may be achieved in the design of synchronization circuits for asynchronous systems with long message lengths or with continuous synchronous systems. In this situtation, care is required in evaluating design trade-offs.

7.2.1 Noise Induced Design Trade-Offs

Circuitry for extracting a clock for a synchronous system using a polar RZ, self-clocking code was shown in Figure 1.7. In the absence of noise, the reconstructed clock from this circuit may be used directly. A more sophisticated design is indicated if the signal is noisy. The effect that additive noise has on timing jitter can be assessed by examining a noisy signal near the limiter decision threshold. By additive noise, we mean a noise voltage that adds linearly with the signal voltage. Refer to Figure 7.9. If the limiter in the circuit of Figure 1.7 makes an

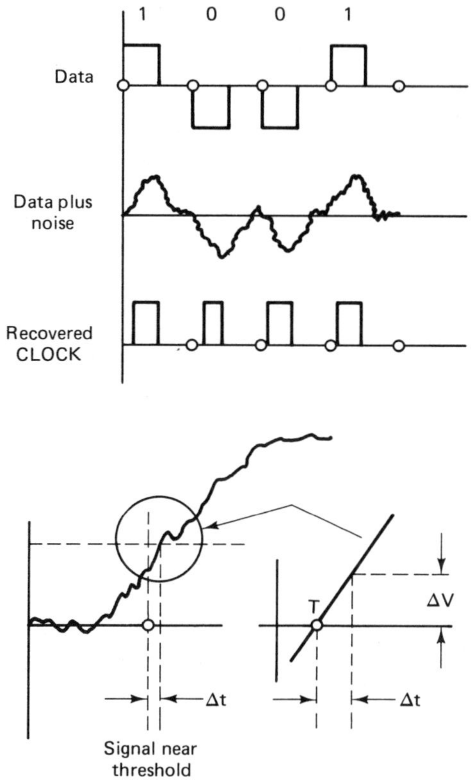

Figure 7.9 A received signal, corrupted with noise, produces a reconstructed clock with noise-induced jitter.

ideal decision at the threshold, the reconstructed signal will contain jitter due to the time displacement of the signal by the noise voltage. If the signal has a finite linear rate of change at the threshold level, a noise voltage of amplitude $-\Delta V$, occurring at time $(T + \Delta t)$, will displace the intended zero crossing of the signal at Δt seconds.

In terms of statistics, noise voltage—when plotted—is a probability density function, with a given rms value. Therefore, the displacement of the zero crossings from the true value, is also a probability density function. The rms value, $\sigma_{\Delta t}$, may be expressed as

$$\sigma_{\Delta t} = \frac{\sigma_{\Delta V}}{(dv/dt)} \text{ s} \tag{7.29}$$

where dv/dt is the time rate of change of the signal at the threshold, T, in volts per seconds and $\sigma_{\Delta V}$ is the rms value of noise voltage in volts. We can approximate the rate of change of the signal as

$$\frac{dv}{dt} \approx \frac{V_{PP}}{t_r} \tag{7.30}$$

where V_{pp} is the peak-to-peak signal voltage and t_r is the rise time. Using this expression, equation (7.29) may be written as

$$\sigma_{\Delta t} \approx \frac{t_r}{(V_{pp}/\sigma_{\Delta V})} = \frac{t_r}{\text{SNR}_{\text{in}}} \tag{7.31}$$

where SNR_{in} is the input SNR. Thus the timing jitter is inversely proportional to this ratio. Note the similarity of this expression to equation (7.9) for carrier phase jitter. This approximation is useful only for small noise levels. Multiple zero crossings can occur if the noise level becomes a significant fraction of the signal level. In this case, the timing signal becomes useless. This situation is analogous to the threshold effect in FM receivers. The effect is a strongly nonlinear function of the input SNR. Obviously, from equation (7.31), timing jitter improves if we can effect an improvement in the SNR of the demodulated but as yet not decoded data signal. Typically this involves baseband and/or predetection filtering. The implications of this approach are examined next.

7.2.2 Jitter-Reduction Techniques

An intuitive technique for reducing jitter is to rectify the signal and then pass it through a band-pass filter tuned to a frequency equal to the symbol rate (see Figure 7.10). The output of the band-pass filter is a sine wave of frequency $(1/T)$ whose rms value may be expressed as

$$A_{\text{rms}} = \left(\frac{2}{T\sqrt{2}\pi}\right) \mid H(f_o) \mid V \tag{7.32}$$

where $H(f_o)$ is the gain of the band-pass filter at the fundamental frequency $f_o = 1/T$, and V is the peak-to-peak signal level at the output of the rectifier.

In order to see how this circuit reduces timing jitter, it is necessary to analyze its performance in the presence of noise. We can express the output of the rectifier, $y(t)$, as

$$y(t) = [s(t) + n(t)]^2 = s^2 + 2sn + n^2 \tag{7.33}$$

where $s(t)$ is the time-varying input signal and $n(t)$ is the noise. Note from this expression that at the input to the bandpass filter we have the desired signal (s^2) and two undesired terms—both functions of the noise. If the noise is small relative

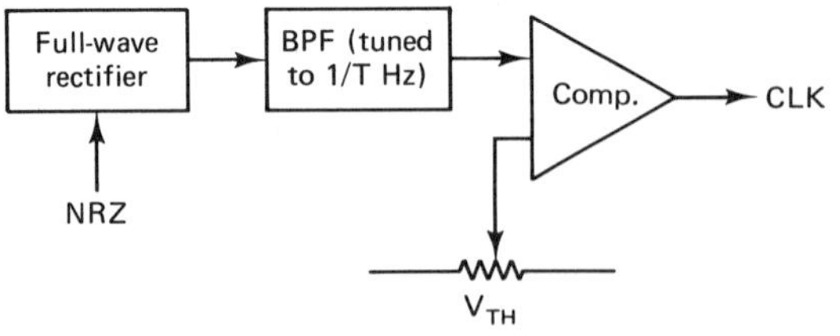

Figure 7.10 Reducing timing jitter may be accomplished by passing the rectified signal through a band-pass filter tuned to the frequency of the symbol rate. Remember, any improvement in jitter performance results in a longer acquisition time.

to the signal, we can approximate equation (7.33) as

$$y(t)\ s^2 + 2sn \tag{7.34}$$

Jitter performance is dominated by the product of signal and noise, *sn*.

In order to quantify the effect of signal times noise on jitter it is necessary to determine values for each at the output of the band-pass filter. More precisely, we are interested in the SNR at the filter output. We can describe noise in terms of its power spectrum. The input noise is *white*; consequently it is characterized by a flat *power spectral density* of N_o watts per hertz. Statistically, it has a Gaussian probability density. The noise spectrum is illustrated in Figure 7.11. Consider a bandwidth, Δf, of 1 Hz in the vicinity of the bit rate. We can represent this elemental cycle of noise as a sine wave with random phase and power of N_o watts. Specifically, for frequency f_k, the noise term N_k is

$$N_k = N_o \sin(\omega_k t + \theta) \tag{7.35}$$

The rms value of noise is $\sqrt{N_o}$; thus, the peak value is $\sqrt{2}\sqrt{N_o}$. Equation (7.35) may be expressed as

$$N_k = \sqrt{2}\sqrt{N_o} \sin(\omega_k t + \theta) \tag{7.36}$$

The product of signal times noise produces terms containing the sum and difference frequencies between the signal and noise frequencies. The band-pass filter is tuned to the fundamental frequency, $1/T$. Thus only terms whose frequencies fall in the vicinity of $1/T$ appear at the filter output. The signal may be represented as

$$s = \sqrt{2}\sqrt{P_s} \sin \omega_o t \tag{7.37}$$

Thus $s \cdot n$ is given by

$$\begin{aligned} s \cdot n &= (\sqrt{2}\sqrt{P_s} \sin \omega_o t)\ (\sqrt{2}\sqrt{P_s} \sin(\omega_k t + \theta) \\ &= 2\sqrt{N_o}\sqrt{P_s}\ [\sin(\omega_o - \omega_k)t - \theta] + \sin(\omega_o + \omega_k)t + \theta] \end{aligned}$$

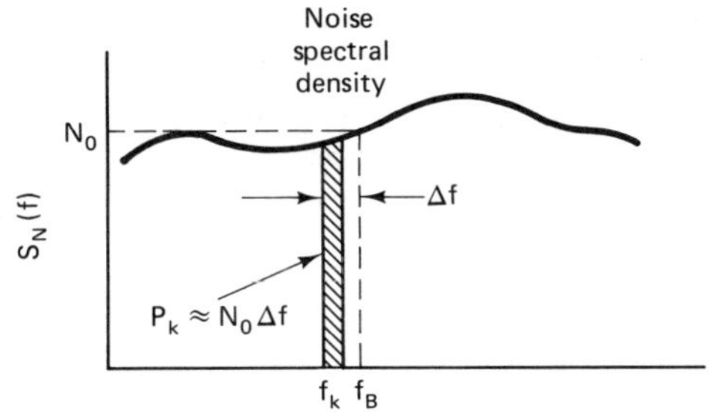

Figure 7.11 Noise in elemental bandwidth Δf, at f_k, can be approximated by a sine wave of random power equal to the noise power in the bandwidth.

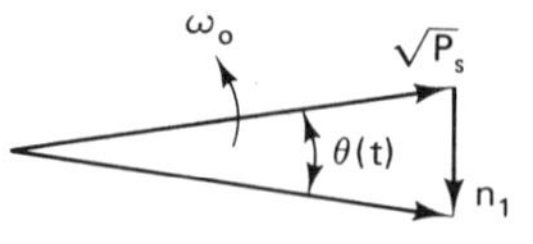

Figure 7.12 Phasor diagram of a small interfering signal.

The value of this product at the filter output is

$$\begin{aligned}(s\cdot n)_o &= \sqrt{2}\,\sqrt{P_s}\int N_o\,|\,H(j\omega)\,|^2\,df \\ &= P_s N_o B_N\end{aligned} \tag{7.38}$$

where $H(j\omega)$ is the filter transfer function and B_N is the equivalent noise bandwidth (ENBW) of the band-pass filter. When white noise is passed through a narrow-band filter, the filter output can be represented by

$$n(t) = n_1(t)\cos\omega_o t + n_2(t)\sin\omega_o t \tag{7.39}$$

The resultant output noise signal is a sinusoidal voltage with time-varying amplitude $R(t) = \sqrt{n_1^2(t) + n_2^2(t)}$ and phase $\theta_n(t)$. Comparing this to the signal (equation (7.37)), we see that the noise has a componentinphase with the signal and another component, $n_1(t)\cos\omega_o t$, in phase quadrature with the signal. Referring to Figure 7.10, we see that the limiter eliminates the amplitude variations and only the phase variations, $\theta_n(t)$, remain. This phase variation represents the timing jitter. If the noise term is small relative to the signal, we can represent the result as in Figure 7.12. The phase perturbation is given by

$$\begin{aligned}\theta_n(t) &= \tan^{-1}\frac{n_1}{\sqrt{P_s}} \\ &\approx \frac{n_1}{\sqrt{P_s}} = \frac{1}{\text{SNR}}\end{aligned} \tag{7.40}$$

If we compare this result to those arrived at earlier, we note the not surprising fact that timing jitter is inversely proportional to the SNR. Consequently, we can reduce the timing jitter by reducing the synchronization circuit bandwidth, that is, by reducing the noise power into the comparator.

The price paid for improved jitter performance is an increase in acquisition time. If we suddenly apply a signal to the input of the circuit shown in Figure 7.10, the output of the band-pass filter, in response to a step input, exhibits a rise time that is inversely proportional to the filter bandwidth. If we assume that the synchronizer requires a full rise time to achieve synchronization, the sync-acquisition time, T_{ACQ}, is approximately

$$T_{\text{ACQ}} \approx \frac{1}{B_n}\ \text{s} \tag{7.41}$$

where B_n is the filter bandwidth. In terms of bit periods, the acquisition time is

$$N_{\text{ac}} \approx \frac{f_B}{B_n} \tag{7.42}$$

This corresponds to equation (7.4) for carrier synchronization acquisition. The bandwidth, B_n, is inversely proportional to the Q of the filter; thus the ratio f_b/B_n is proportional to the filter Q. For a bandpass filter, $Q = f_o/B_n$. Increasing Q decreases timing jitter by increasing the SNR. A corresponding increase in sync-acquisition time occurs. This represents a fundamental design trade-off. In the next section we extend these results to codes that are not self-clocking. These codes may be considered to represent random data sequences. Considerable attention must be paid to sync acquisition, especially where missing data transitions are involved.

7.3 BIT SYNCHRONIZATION FOR RANDOM BASEBAND SIGNALS

In the previous section we considered techniques for synchronization of self-clocking codes. We wish now to extend these ideas to cover the entire range of random baseband signals. A logical first step in designing a bit synchronizer for a particular random data stream is to examine the power spectrum of the signal as a check to see if there are any periodic frequencies which are related to the bit rate. Power spectrums for some popular codes are shown in Figure 3.3. We illustrate three of these in Figure 7.13, wherein a random data pattern is assumed. A signal is called *periodic* if the designer can replace the time variable t with t

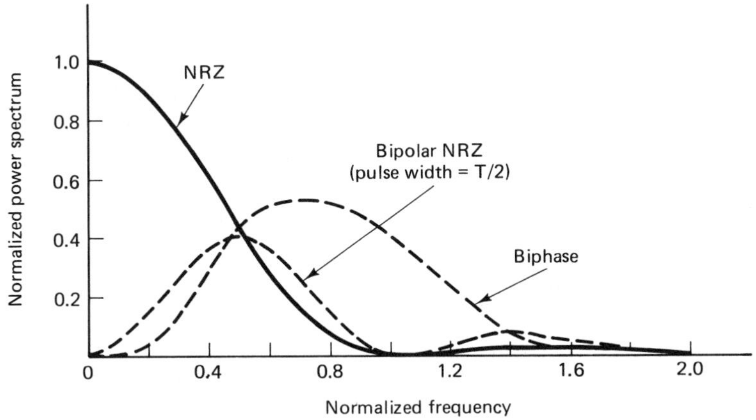

Figure 7.13 In designing a synchronizer, first examine the power spectrum of the signal to discover any periodic components that might be related to the symbol rate. Unfortunately, the three commonly used codes shown here assume a random data pattern that prevents the simple filtering of signals to extract a timing signal.

+ T and obtain the same result, T being the repetition period. Referring to Figure 7.13, a continuous spectra is indicated, and this implies a lack of periodicity. Thus there are no deterministic frequency components in the spectrum. In other words, a signal does not exist that can be filtered to extract a timing signal. Also, since the data is assumed to be random, any operation that processes the signal linearly will not produce any periodic components. The solution is to process the signal nonlinearly to produce new frequencies.

In general, a nonlinear operation can be described by a power series. We are familiar with this, for example, in representing the forward-biased characteristics of a diode by a power series. Since the constant and linear terms in a power series do not represent new frequencies, this leads us immediately to the squared term. A squaring circuit is suggested. The output of a square-law device (full-wave rectifier) yields a periodic component whose frequency either equals the symbol rate or is an integral multiple of it. We can use this component for creating a bit-synchronization signal. There are exceptions, though, where the result is not usable. For example, if NRZ PCM is squared where a positive pulse is a 1 and negative pulse is a 0, the result is a constant dc level. The squared signal is periodic but does not have any transitions that can be used to produce a timing signal.

A second scheme for extracting a timing signal is to design a transition detector that will produce an output of the same polarity for every data transaction. We can accomplish this by first limiting the input signal and then using the positive- and negative-going edges of the limited signal to trigger separate monostable multivibrators (see Figure 7.14(a)). Referring to this circuit, we see that both one-shot outputs are ORed together to produce an output pulse train with positive pulses occurring at each data transition. When a single transition occurs at each symbol period, the output is periodic with period equal to the symbol rate. If there are no data transitions, the output is zero. The device, then, is equivalent to a gated periodic signal.

Other techniques also exist for detecting data transitions. For example, we can differentiate the input signal with an RC network and then rectify the differentiated pulses. Another technique is simply to delay the input signal by a time Δt and then multiply the signal and the delayed version (see Figure 7.14(c)). Still another technique is illustrated in Figure 7.14(d). The signal is delayed using a clocked F/F. The exclusive-or circuit then performs the equivalent of multiplication. A little thought will convince us that all these circuits are equivalent to a gated periodic signal. The transition density of the input signal determines the gating action.

The nonlinear action of the circuits generates either a periodic or gated periodic signal, which is derived from the input random signal. We can use this signal to extract the required timing clock. In the following we consider two techniques for accomplishing this. The first method utilizes a narrow-band, band-pass filter. The second technique relies on a PLL. Both techniques are, at this point, familiar to us.

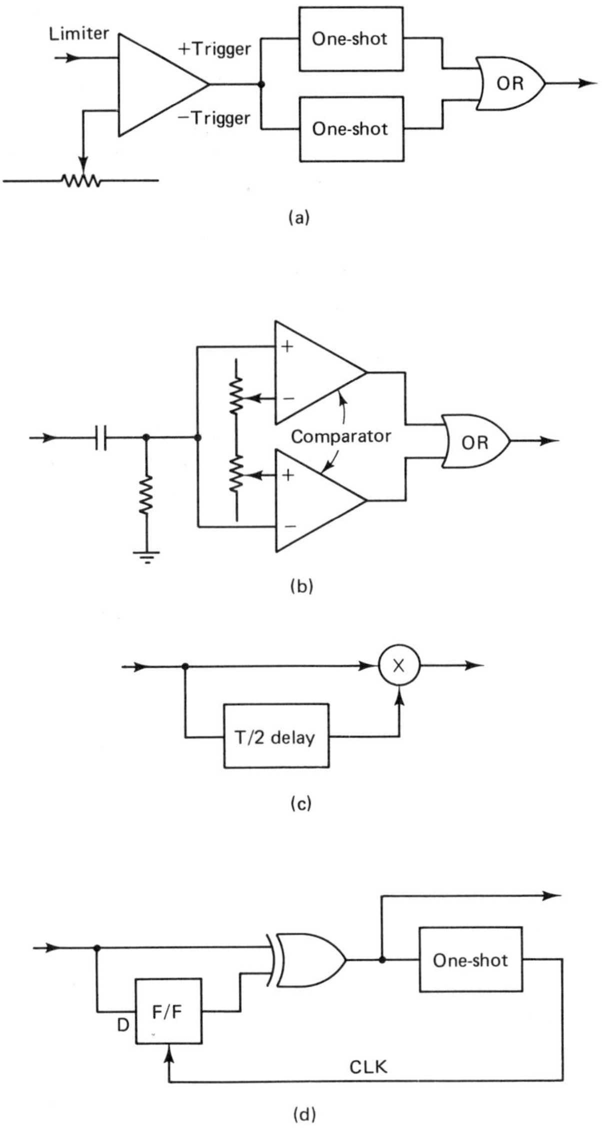

Figure 7.14 A transition detector produces an output of the same polarity for every data transition. Various techniques (a) limit input and use edges to trigger the first monostable (while negative-going edges trigger the second) and (b) differentiate the input and then rectify the pulses or delay the input by ΔT and then multiply this delayed input by the original signal. This last process can be done by (c) analog or (d) hybrid analog-digital circuits.

7.3.1 Band-Pass Filter Bit-Sync Derivation

Use of a band-pass filter for carrier synchronization was considered in Section 7.1.1. It is also useful as a follow-on device for the circuits illustrated in Figure 7.14. The filter output in this case becomes the bit-synchronization signal. When a band-pass filter with center frequency equal to the symbol rate is used to filter a periodic signal, the zero crossings of the signal will cause the filter to ring in conjunction with the input signal. The synchronization signal must lie within the narrow passband of the filter. If the synchronization signal does not exhibit a transition at each symbol period, the filter will continue to ring (i.e., coast) through these periods. The actual length feasible is dependent upon filter Q (bandwidth). The narrower the bandwidth, the longer the filter will ring. The trade-off here is the required synchronization time—the filter time constant increases. On the other hand, a decrease in phase jitter is achieved (see Section 7.2.2). For very wide bandwidths, the filter will tend to oscillate at the input rate. However, the oscillations will decay rapidly as soon as the input is removed. It is important that the input frequency and the center frequency of the filter closely coincide. For different frequencies, the output signal exhibits a fixed time delay due to phase shift characteristics of the filter. Thus, variations in the time delay will occur as a function of the relative frequency difference.

The speed with which we can acquire synchronization with a band-pass filter depends markedly upon bandwidth. Filter rise time is given by equation (7.3). Acquisition time, then, can be expressed in numbers of bit periods, N, as

$$N = Kt_r(\text{BR}) \text{ bits} \tag{7.43}$$

where BR is the bit rate, t_r is filter rise time in seconds, and K is a constant greater than or equal to 1.0. The constant K depends on the filter type. Typically, K varies between 1.0 and 2.0.

EXAMPLE 7.4

How many bit periods are required for acquisition if the filter bandwidth is 10% of the bit rate? Assume $K = 1.5$.

Solution

$$N = Kt_r(\text{BR}) = K\left(\frac{0.7}{B}\right)\text{BR}$$

$$= (0.7)(K)\left(\frac{\text{BR}}{B}\right) = (0.7)\,(1.5)\left(\frac{1}{0.1}\right)$$

$$= 10.5 \text{ bits}$$

For this particular filter 10.5 bits are required to achieve synchronization. A block diagram of the complete bit synchronizer is shown in Figure 7.15.

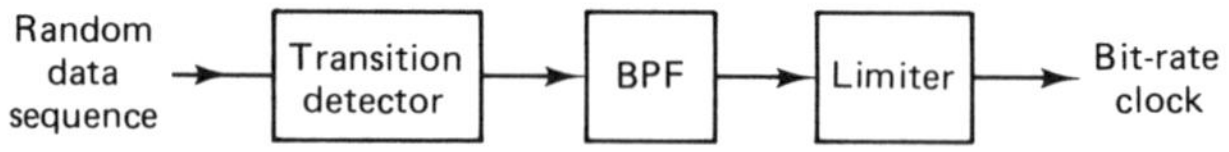

Figure 7.15 Bit synchronizer block diagram.

7.3.2 PLL Bit-Sync Derivation

An alternative to the band-pass filter in the bit synchronizer is use of a PLL. The PLL also receives its input from the transition detector. It may be shown from the transfer function of a PLL (equation (6.8)), that, like the narrow-band filter, the PLL also exhibits a very narrow bandwidth. We have also seen previously (Chapter 6), that a PLL will track input frequency variations. Thus the PLL will provide a continuous clock, afford a large measure of noise rejection because of the narrow bandwidth, and will not get out of tune, as a band-pass filter may.

The characteristics of PLLs were discussed in Chapter 6 and the phase detector output is given by equation (6.1). An interesting feature of phase detector characteristics that affects the operation of a PLL lies in the periodicity of the phase-detector output function (see Figure 7.16). The phase error at the detector output repeats at integral multiples of 2π rads of phase. If for some reason a disturbance (noise voltage) enters the PLL and causes the phase error to exceed $\pm\pi/2$ rad, the PLL will skip one or more cycles. This phenomenon ultimately determines synchronization threshold in a bit synchronizer that uses a PLL.

In bit synchronizers that use PLLs, we can define acquisition time as the approximate time it takes the PLL to achieve frequency lock plus phase lock. The time required for frequency lock is the time it takes for the PLL to reduce the initial frequency offset to zero. After frequency lock is obtained, a finite period of time is required to reduce phase error to an acceptable value. A phase error of less than $\pm 10°$ is considered to represent an acceptable phase error.

An analysis of frequency lock time is extremely difficult. Considerable analysis has been devoted to this problem, and the time for frequency lock is given approximately by

$$T_{FL} \approx \frac{4\Delta f^2}{B_L^3} \tag{7.44}$$

where Δf is the difference between the incoming frequency and the VCO frequency. The quantity B_L is the equivalent square noise loop bandwidth of the

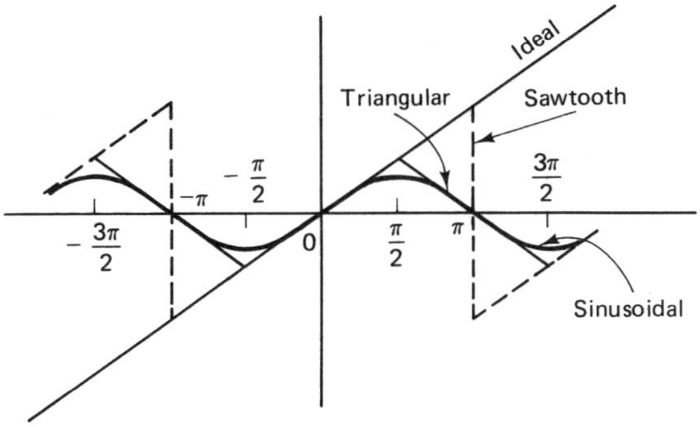

Figure 7.16 Typical phase detector characteristics show periodicity with detector-output error identical for errors differing by integral multiples of 2π rad. When a disturbance enters the PLL and causes phase error to exceed $\pm\pi$ rad, the PLL will skip one or more cycles. For a triangular phase comparator, the PLL skips cycles if the phase comparator exceeds $\pm\pi/2$ rad.

PLL. The *loop noise bandwidth* may be calculated by

$$B_L = \int_0^\infty | A(j\omega) |^2 \, df \text{ Hz} \tag{7.45}$$

where $A(j\omega)$ is the loop transfer function (see equation 6.8).

The time required to achieve phase lock is derived from the response of the PLL to an input step phase change. The mean time to phase lock normalized to loop noise bandwidth as a function of the damping factor for a sawtooth phase comparator is shown in Figure 7.17. The mean phase-lock time, however, is not a strong function of the damping factor. In fact, it changes only about 10 percent for a damping factor between 0.6 to 1.0. For a second-order loop, time to phase lock is approximately

$$T_{PL} \approx \frac{1.3}{B_L} \tag{7.46}$$

The total sync-acquisition time is then approximately the sum of frequency lock time plus phase-lock time, or

$$T_{\text{ACQ}} \approx T_{FL} + T_{PL} \approx \frac{4\Delta f^2}{B_L^3} + \frac{1.3}{B_L} \tag{7.47}$$

Expressed in terms of number of symbols or bit periods, equation (7.47) becomes

$$N \approx 4\left[\frac{(\Delta f/B_L)^2}{(B_L/\text{BR})} + \frac{1.3}{(B_L/\text{BR})}\right] \tag{7.48}$$

Equation (7.48) is plotted in Figure 7.18. Note that in the absence of frequency offset ($\Delta f = 0$), the acquisition time is dependent primarily on the time to phase lock and thus on PLL bandwidth. The mean time to phase lock is rarely greater than $1.5/B_L$ and practically never larger than $2.0/B_L$. Equation (7.47) assumes a large SNR. That is, the loop noise is less than the equivalent $\pm 10°$ phase error.

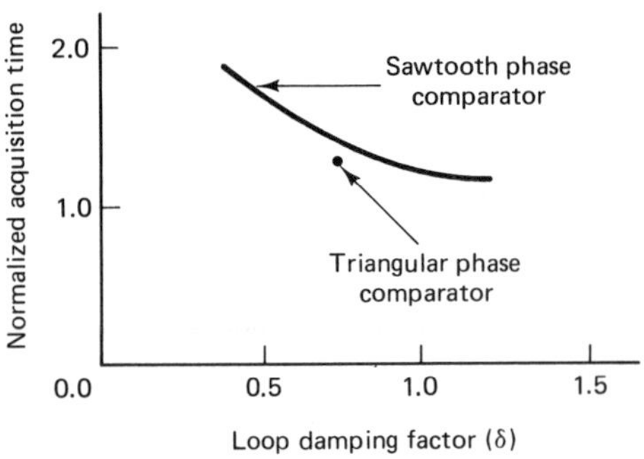

Figure 7.17 Normalized acquisition time (mean time to phase lock normalized to loop-noise bandwidth) is shown as a function of damping factor.

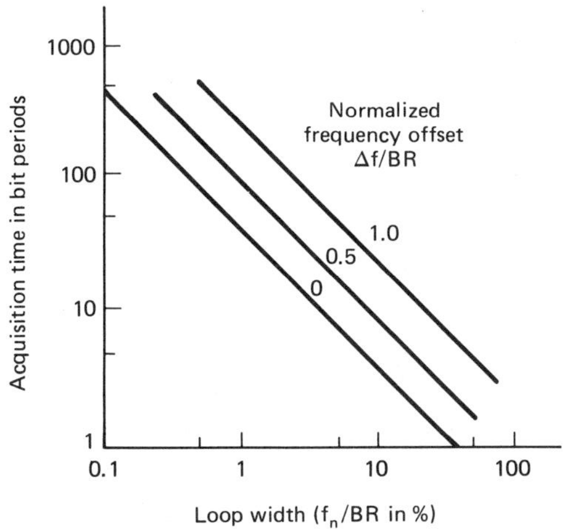

Figure 7.18 Acquisition time versus PLL loop bandwidth is normalized to bit periods for convenience of the designer. The bit period is the basic time quantity.

7.3.2.1 Input noise and synchronization. A block diagram of the bit synchronizer is shown in Figure 7.19. The squaring circuit generates the frequency component that the PLL tracks. The squaring circuit generates three terms (see equation (7.33)), and for a large SNR this reduces to equation (7.34). The input noise exhibits itself as the product of additive noise and a deterministic signal. For a large SNR, we would expect a relatively small phase error.

For a low SNR, the noise squared term cannot be neglected. Input noise to the loop results in phase jitter at the output. Referring to Figure 7.16, if the phase error exceeds $\pm\pi/2$, a loss of lock results in a loss of bit synchronization. A variety of definitions exists for sync threshold. The definition that we employ here is that SNR at which the probability of clock slippage occurs is equal to 10^{-6}. Since the probability of bit slippage depends on the probability of losing lock, we can approach the problem in terms of the probability of phase error exceeding a given threshold.

The *loop noise bandwidth* of a PLL is given by equation (7.45). For a second-order loop, the one-sided noise bandwidth is

$$B_L = \frac{\omega_n}{2}\left(\delta + \frac{1}{4\delta}\right) \text{ Hz} \tag{7.49}$$

The output phase jitter (fluctuation of VCO phase) is the input phase, θ_i, modified by the closed-loop transfer function of the PLL, $A(j\omega)$. Thus, the mean square output phase jitter is

$$\overline{\theta_{no}^2} = \int_0^\infty \phi \,|\, A(j\omega)\,|^2\, df = \phi \int_0^\infty |\, A(j\omega)\,|^2\, df = \phi B_L \tag{7.50}$$

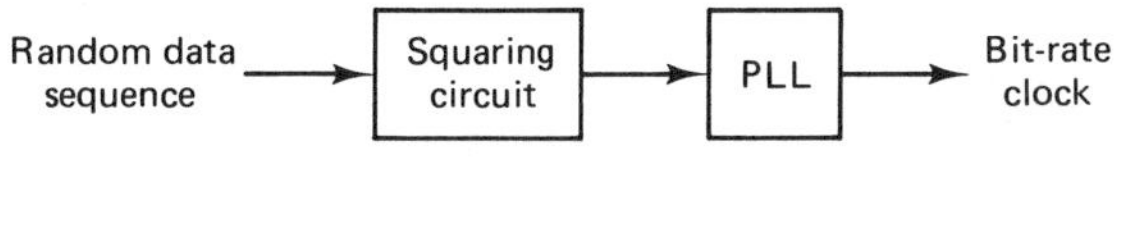

Figure 7.19 Bit synchronizer incorporating a PLL.

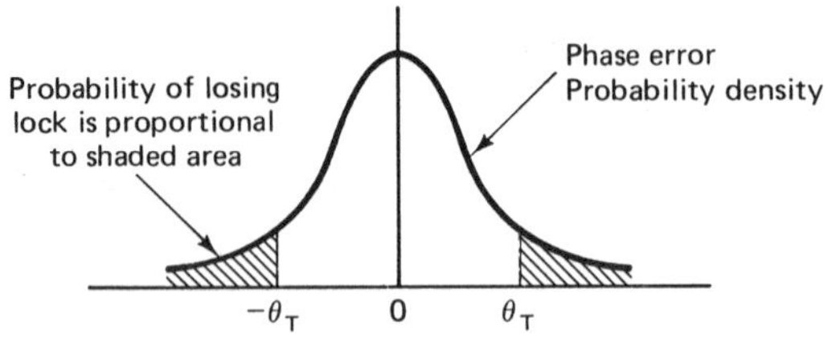

Figure 7.20 Probability of phase error exceeding the threshold, θ_T.

where ϕ is the noise spectral density in radians squared per hertz. It can be shown that the output phase, θ_o, and the loop SNR are related by

$$\overline{\theta_{no}^2} = \frac{1}{2(\text{SNR})_L} \tag{7.51}$$

where $(\text{SNR})_L$ is the loop SNR.

The probability that phase error exceeds a given threshold is illustrated in Figure 7.20. The probability of losing lock is the probability that the magnitude of the phase error exceeds the threshold, θ_T. Thus

$$\text{Prob}\ (\mid \theta \geq \theta_T \mid) \approx \text{erfc}(\theta_T/\sqrt{\text{SNR}_L/2}) \tag{7.52}$$

where erfc is the complementary error function. A theoretical analysis based on the Fokker-Plank equations results in expressions for the probability of bit slippage that behave asymptotically the same as equation (7.52). In practice, the highly nonlinear nature of the bit-slippage process is such that the approximation described by equation (7.52) based on Gaussian (white) noise is a usable approximation to actual performance.

The phenomenon of cycle slipping by the PLL exhibits a characteristic threshold effect. The probability of a cycle slip is infinitesimal for a loop SNR_L greater than a particular level. Below this SNR_L, the loop quickly loses lock. The threshold, for a loop with a triangular phase comparator, is theoretically on the order of 8 to 12 dB. The synchronization threshold is defined as that SNR_L where the loop starts slipping cycles. In practice this is an optimistic range, and a more realistic range is an SNR_L of 15 to 20 dB. A plot of typical sync threshold versus loop bandwith is shown in Figure 7.21.

Input timing jitter will also influence performance of the bit synchronizer. If the input signal contains frequency jitter, the loop phase error is (see equation (6.9)):

$$\theta_e(s) = \frac{\theta_i(s)}{s + KF(s)} \tag{7.53}$$

For a second-order PLL, the relative response of the loop to input jitter is

$$\frac{\theta_e(s)}{\theta_i(s)} = \frac{s}{s^2 + 2\delta\omega_n + \omega_{n^2}} \tag{7.54}$$

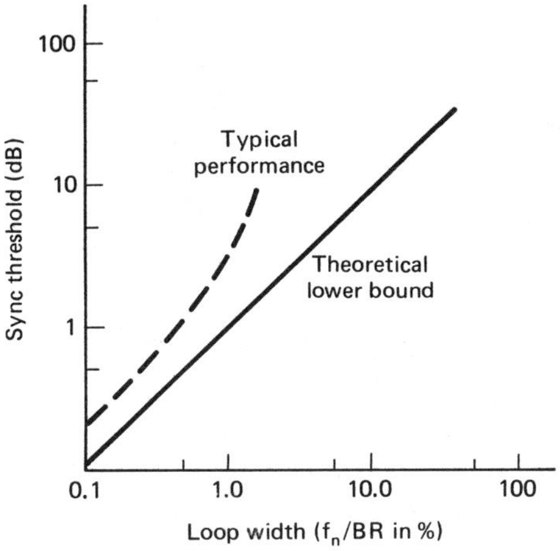

Figure 7.21 Typical synchronization threshold, as a function of loop width, has a theoretical lower bound (100 percent transition density) that is rarely achieved in practice.

This response is equivalent to a band-pass filter with center frequency equal to the loop natural frequency. Therefore, the loop phase error will be maximum for jitter frequencies near the loop natural frequency. Expressed mathematically, the maximum phase error, in radians, for a jitter frequency f_n with peak amplitude Δf hertz is:

$$\theta_e = \frac{\Delta f}{2} \delta f_n \text{ rad} \tag{7.55}$$

Maximum jitter amplitude (expressed as a percent of bit rate) is shown in Figure 7.22 as a function of normalized loop bandwidth. In this figure, the maximum allowable phase error is assumed to be 1 rad.

For a large SNR, the number of consecutive transitions that can be missing depends upon frequency error in the loop at the last data transition. In the absence

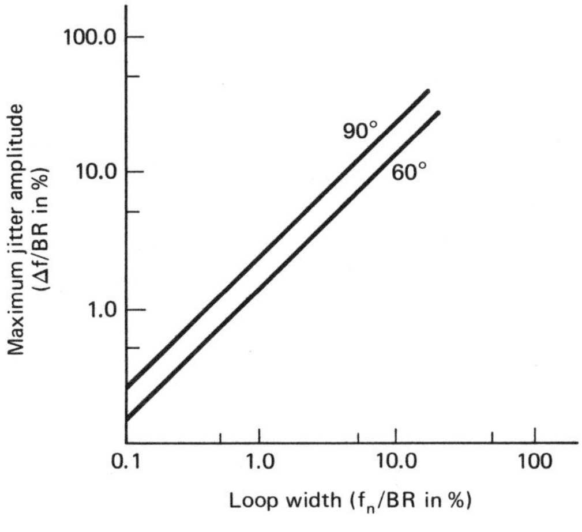

Figure 7.22 Maximum jitter amplitude is a function of loop width, with maximum phase error the running parameter. In this graph we assume jitter frequency equal to the loop's natural frequency.

of data transitions, any loop frequency error will integrate to drive the loop out of lock. For a low-input SNR, extraneous zero crossings induced by noise increase loop error deterioration. At each data transition, the phase detector injects a signal plus noise into the loop. If there is no data transition, only noise circulates in the loop. Decreasing data transition density directly decreases the energy component of the signal. Consequently, synchronizer threshold performance deteriorates at less than 3 dB for each halving of transition density. An application of the theory discussed is given in Example 7.5.

EXAMPLE 7.5

Design a bit synchronizer that will be capable of extracting a timing clock from a noisy NRZ PCM signal. The synchronizer must acquire synchronization within 100 bit periods. After lock is achieved, it must maintain lock down to an SNR of 0 dB. The nominal bit rate is 10 kilobits/s ±10% with an instantaneous jitter amplitude of ±0.10% of the bit rate. A multiple-phase output clock is desired.

Solution

(a) *System parameters:* The first step in the design process is to establish system parameters for the synchronizer. In Figure 7.18, the loop width to ensure acquisition in 100 bit periods or less with no frequency offset must be greater than 0.4%. Since the nominal bit rate may be offset up to ±10% (statement of problem), the acquisition loop width should be made wider to accomodate this offset. As a design margin, choose a loop width of 1% for acquisition.

Next, refer to Figure 7.21. From this figure, note that a loop width of less than 0.4% is required to maintain synchronization at 0 dB SNR with 100% transition density data. If the transition density is less than 100%, the loop width must be made less than 0.4%. In fact, reliable synchronization requires a loop width on the order of 0.1%.

The foregoing indicates that a switched synchronizer design is in order. During acquisition, the synchronizer is switched to the 1% position. After lock has been achieved, the loop is narrowed to 0.1%. In the tracking mode this corresponds to a loop natural frequency, f_n, of 0.1% times 10 KHz = 10 Hz. From Figure 7.22, the maximum jitter amplitude at the loop natural frequency (10 Hz) is approximately ±0.22%. For jitter frequencies less than or greater than f_n, the loop can tolerate much greater jitter amplitudes. Thus the tracking loop will be able to track short-term data jitter.

(b) *Synchronizer circuit design:* Refer to Figure 7.19. The first requirement is a nonlinear element to extract a timing component from the NRZ signal. A data transition detector of the type shown in Figure 7.14(d) is suitable. A PLL with an exclusive-or phase

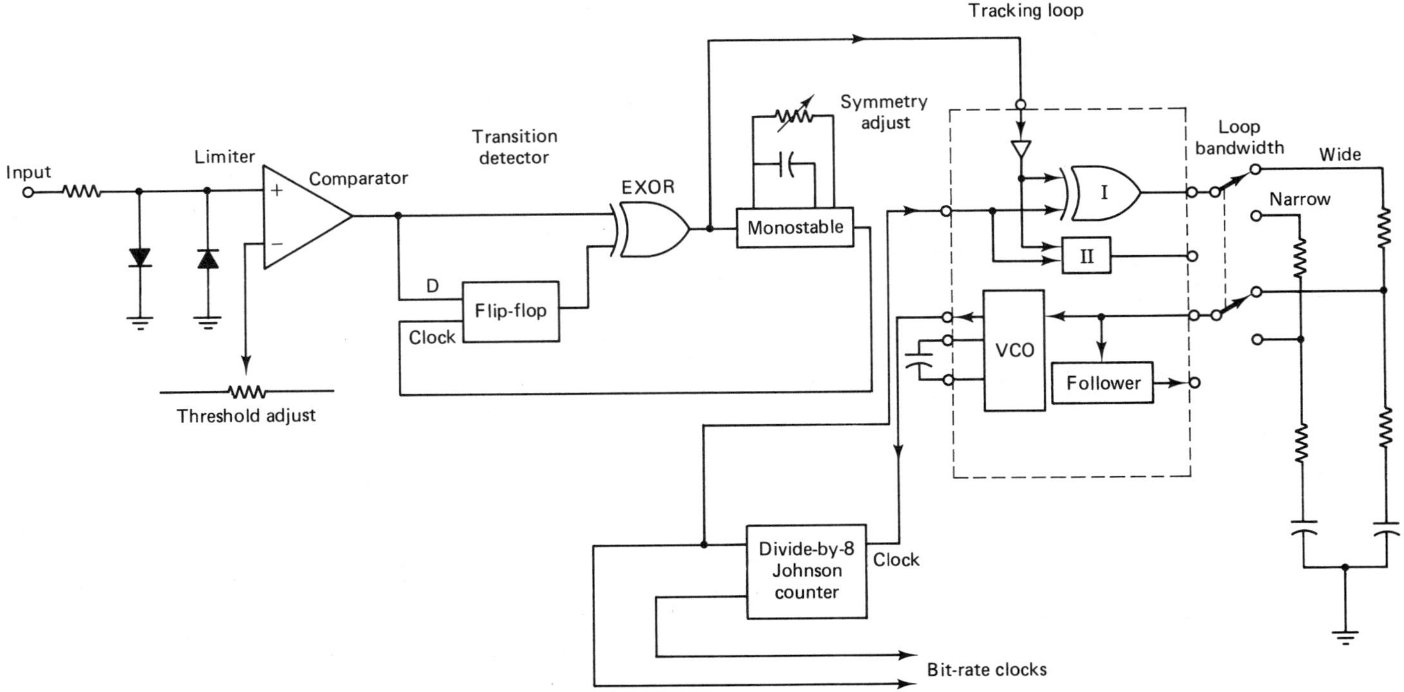

Figure E.5 Actual circuit implementation for a NRZ bit synchronizer employs a manual switch for changing loop bandwidths. An electronic switch could just as easily detect when the loop is in lock and switch bandwidths automatically.

comparator can be used to provide the basic timing loop (see the accompanying figure). The loop VCO operates at an integral multiple of the bit rate in order to provide clock outputs of different phases with respect to the input data. In addition to a clock in phase with input data transitions, a bit synchronizer must provide a clock with a leading edge occurring at midbit time. The Johnson counter provides a convenient method for obtaining a number of different clock phases. The monostable multivibrator in the transition detector is adjusted for one-half a bit period (50 Ns). This output clocks the F/F at midbit period and produces a pulse train at the output of the exclusive-or with a fundamental frequency equal to the bit rate. Waveform symmetry is adjusted by adjusting the duration of the one shot.

The output of the exclusive-or is also fed to one input of the phase comparator. This type of phase comparator provides a triangular phase-error transfer function (see Figure 7.16). The switchable loop filter provides the two loop bandwidths. Although this is shown as a manual switch, a lock-detect circuit can be used to make this function automatic (see Example 3.3). The tuning capacitor is selected to set the VCO frequency to eight times the bit rate (80 kHz). A symmetrical reference clock from the Johnson counter is available for application to the phase comparator and a midbit time clock is available directly from the counter.

This example illustrates fundamental techniques involved in design of a bit-rate synchronizer. Additional factors must be considered in practice. For example, unless the filter capacitors are precharged, switching transients may cause the loop to lose lock. In addition, the designer must consider such factors as drift, stability, reliability, and environmental requirements.

At this point, we have discussed design techniques for the RF clock and the associated bit synchronizer. The next requirement in a digital receiving system is a regulated frame clock. A frame clock is necessary to divide the bit stream into data frames, which then permit further decoding. In the next section, we discuss techniques for accomplishing this function.

7.4 FRAME-SYNCHRONIZATION TECHNIQUES

The problem of frame synchronization is best introduced with an example. The following extreme example illustrates the variety of problems, requirements, and parameters that may arise in a frame sync problem. Consider a digital system that transmits a sequence of frames 2 bits in length (see Table 7.1). Note that the only

TABLE 7.1

1	00
2	01
3	10

frame not acceptable is the binary tuple 11. However, a 11 sequence can occur as an overlap of frames 2 and 3. That is,

$$\underbrace{0\ \overbrace{1}\ }_{2}\ \underbrace{\overbrace{1}\ 0}_{3}$$

If there are no errors in the received data, a simple frame-sync algorithm for this system is: Frame sync occurs between adjacent 1s in the bit stream and in synchronism with every second bit clock before and after. Note that a necessary condition is that the bit synchronizer is already synchronized.

In the previous section, we considered the performance of bit synchronizers in the presence of noise. If noise is present, it can manifest itself in the following ways:

- *Substitution error:* The bit synchronizer mistakes a 0 for a 1, or vice versa.
- *Deletion error:* The bit clock skips a beat and a data bit is lost.
- *Insertion error:* The bit clock has an extra pulse and creates an extra bit.

Each of these error types can play havoc with the frame-sync process. The effect is illustrated in Figure 7.23. Transmission errors (1) and (2) cause loss of lock or

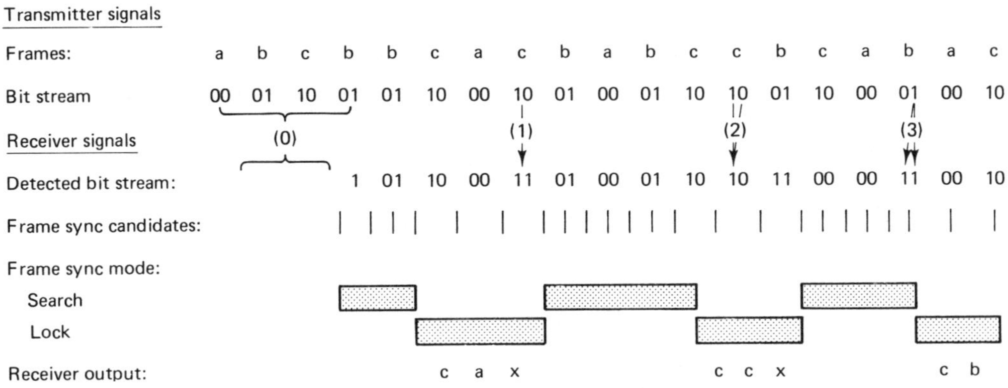

Figure 7.23 Frame sync example for four anomalies. (0) Bits lost during bit sync search. (1) Substitution error. (2) Deletion error. (3) Insertion error. (Note that X denotes an unacceptable frame.)

loss of frame sync. This puts the clock control back in the search mode. The third transmission error type results in an erroneous frame sync. In this example channel errors are difficult to combat or compensate for.

As noted in the previous example, if frame sync is lost, the synchronization circuitry enters the search mode. After the synchronization pattern is detected, a verification strategy is initiated. This may be as simple as requiring that a number of successive frame syncs be detected in the expected time slot. Many specific frame-sync strategies have been proposed. A frame for the Bell System T1 digital carrier communication system is shown in Figure 7.24. Each T1 frame is composed of 8-bit samples from each of 24 channels plus 1 sync bit, for a total of 193 bits per frame. The sync bit is alternately 0 and 1 in successive frames.

For voice with a bandwidth of 4 kHz, this means 8000 frames per second in order that the sampling theorem be adhered to. Thus, the BR is

$$\begin{aligned} \text{BR} &= (193 \text{ bits/frame})\ (8000 \text{ frames/s}) \\ &= 1{,}544{,}000 \text{ bits/s} \end{aligned}$$

In determining a frame-sync pattern for the 193rd bit, it was found that a 0 1 0 1 0 1 . . . bit pattern never occurs in digitized voice. A sequence such as this would imply a 4-kHz component in the voice signal. The low-pass filters through which the voice signal passes prior to being digitized would filter it anyway. In the framing sequence, the receiving equipment examines bits separated by 193 bits. When a 1 0 1 0 1 0 . . . pattern is found, this is assumed to be the framing pulses. If all 192 positions are examined, the time to frame is approximately 50 ms. When a loss of synchronization occurs, this event is detected by the framing circuitry in 0.4 to 0.6 ms. In the worst case, time to reframe is 50 ms.

In this chapter, we have considered synchronization problems associated with a digital system. Obviously, if several systems must communicate, then network synchronization must be ensured. If time delays exist in the communication paths, this imposes significant conditions on the synchronizing devices, particularly if the time delays are variable. The advent of time-division multiple access

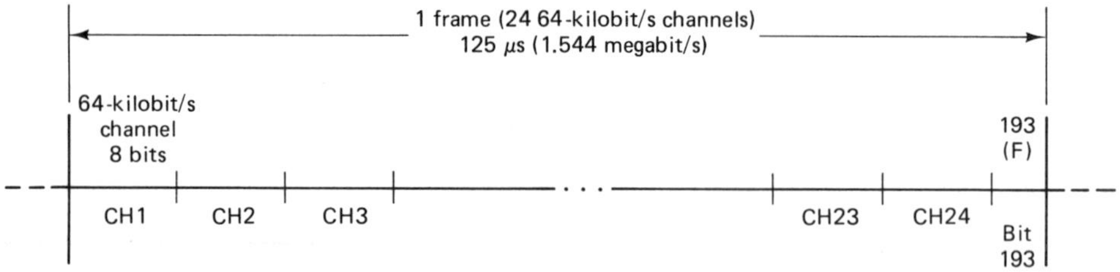

Figure 7.24 Bell system T1 PCM frame.

(TDMA) satellite-switched systems forces all Earth stations to synchronize to the sync window in the satellite. In Chapter 11 we again return to the problem of synchronization.

PROBLEMS

7.1. A PCM system transmits RZ pulses over a channel with Nyquist bandwidth. It is designed to operate at a BER of 10^{-6} when no timing errors are present. For a 10% error in bit timing, how much of an increase in SNR is required to maintain the same error probability? With perfect timing, what percent jitter is present in the reconstructed data? (*Hint:* See Figure 1.4.)

7.2. An equivalent baseband phase-tracking loop, such as shown in Figure 7.6, has the following parameters.

$$K_m = K_o = 2$$

$$F(\omega) = \begin{cases} 200 & |\omega| \leq 20{,}000\pi \\ 0 & |\omega| > 20{,}000\pi \end{cases}$$

If the modulated signal is

$$s(t) = A\cos[\omega_o t + \theta(t)]$$

and

$$\theta(t) = \begin{cases} 100\pi t + 20\pi & t \geqslant 0 \\ 0 & t < 0 \end{cases}$$

what value of A is required to keep the final steady-state phase error, $\theta_e(\infty)$, less than $\pi/6$?

7.3. For the PLL of Example 6.1, find the output phase jitter if the input phase noise has a uniform spectral density of 10^{-8} rad^2/Hz.

7.4. A crystal-controlled oscillator often serves as the reference clock in electronic equipment. The maximum allowable drift rate is specified in terms of parts per day, where the term *parts* refers to the fractional frequency error of the oscillator. For example, a 1-GHz oscillator with an error of 1 Hz has a fractional error of $1/10^9 = 10^{-9}$ parts, or hertz per hertz. Thus parts per day means hertz error per hertz per day. If the total allowable clock error excursion is $\pm 500\ \mu$s in 30 da, find the maximum allowable drift rate for a reference oscillator in parts per day. Assume the crystal frequency is 10^{12} Hz.

7.5. Find the clock error in microseconds, 30 da after calibration, if the reference oscillator has an error of 10^{-11} parts. Assume that the drift is zero and the crystal frequency is 10^{+10} Hz. (*Hint:* See Problem 7.4.)

7.6. A clock is driven by an oscillator with frequency 10^{10} Hz whose drift is zero. The oscillator has a frequency offset of $+5$ Hz. We can write this as a frequency error of $+5$ parts in 10^{+10}.

(a) If the initial clock error is 0, what is the error after 24 h?

(b) If in addition to the frequency offset stated, the oscillator has an aging rate of -10 parts in 10^{+10} per day, what is the clock error after 24 h? Assume the initial clock error is 0 at the time of clock setting.

7.7. Explain how an eye diagram can be used to measure data transition jitter.

7.8. Clock recovery for a bit sync is shown.

(a) For the signal shown, draw the signal at point X.

(b) With reference to the spectrum of the signal at point X, what are the bandwidth requirements of the PLL?

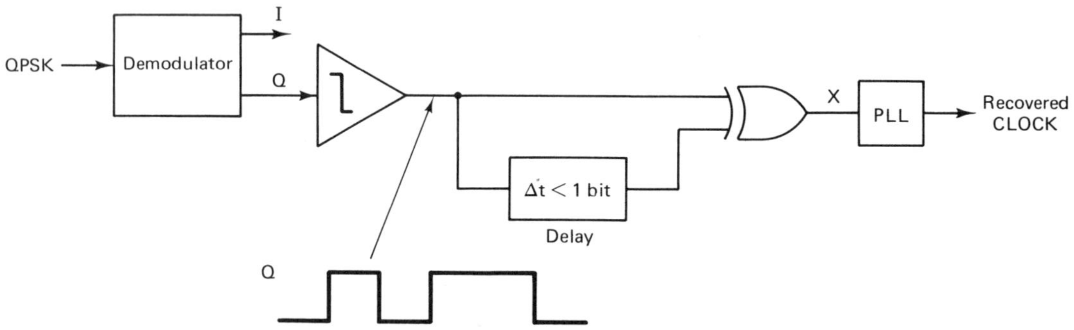

7.9. **(a)** Prove equation (7.9).

(b) If the signal is a 10-kHz sinusoid with peak value 2 V, what is the rms timing jitter if the SNR is 15 dB?

7.10. In Figure 7.2, assume that $y(t)$ is a BPSK signal.

(a) Use mathematics to show how the squaring loop recovers the carrier reference, $r(t)$.

(b) Demonstrate that the squaring loop does or does not have a 180° phase ambiguity.

7.11. Refer to Figure 7.10. Assume a noise-free Manchester-encoded PCM signal to present at the input. Draw waveforms at various points in the bit synchronizer for the input signal shown.

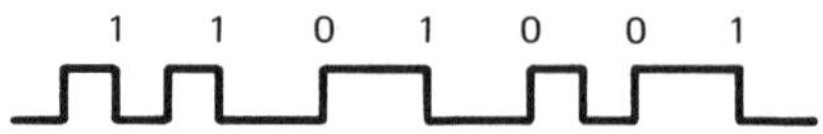

7.12. Would you expect the bit synchronizer of Figure 7.10 to work better for a Manchester-coded input or a bipolar input (See problem 7.11)?

7.13. Use mathematics to show that the extended version of the Costas loop shown may be used for QPSK carrier recovery. Assume a locked condition.

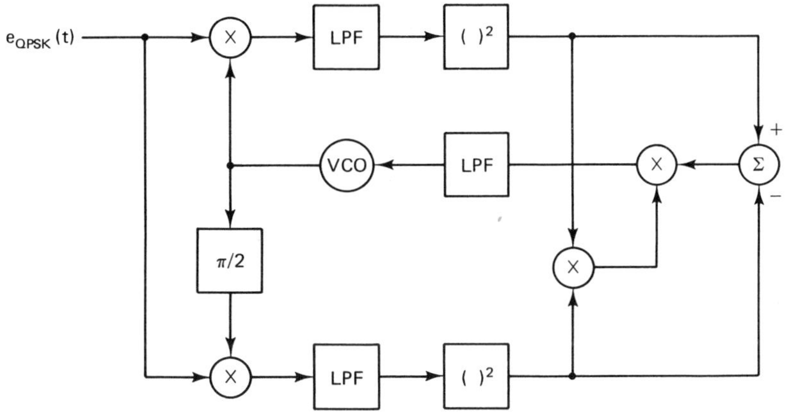

7.14. Show that the decision-directed loop is equivalent to a PLL.

7.15. Refer to Figure 7.2. Prove whether square-law recovery will or will not work for QPSK.

8

DIGITAL RADIO SYSTEMS

8.0 INTRODUCTION

Digital radio differs from analog radio not in the radio transmission technique but rather in the scheme of modulation used. In Chapter 3 we considered digital modulation or signaling techniques. We saw, specifically, that these techniques take advantage of digital phase modulation and a combination of amplitude and phase-shift keying. The primary components of a digital communication system are shown in Figure 8.1. All the digital modulation schemes discussed previously can be described in terms of radio-frequency carriers (sinusoids), which are modulated by low-frequency (baseband modulation) signals that convey the information. In designing a system we are free to filter, weight, or otherwise shape these baseband signals prior to modulating the carrier in order to achieve the most desirable results. One objective of such signal processing could be bandwidth control of the transmitted signal. At the receiver the baseband information is recovered by the process of detection or demodulation. Coherent detection requires a sinusoidal reference signal, which is phase locked in both frequency and phase to the received carrier. Techniques for accomplishing this were discussed in Chapters 6 and 7. Noncoherent detection schemes do not require a phase reference. Following the detector a baseband decision device (bit synchronizer) is required (see Chapter 7) to decode the baseband signal into a stream of data bits.

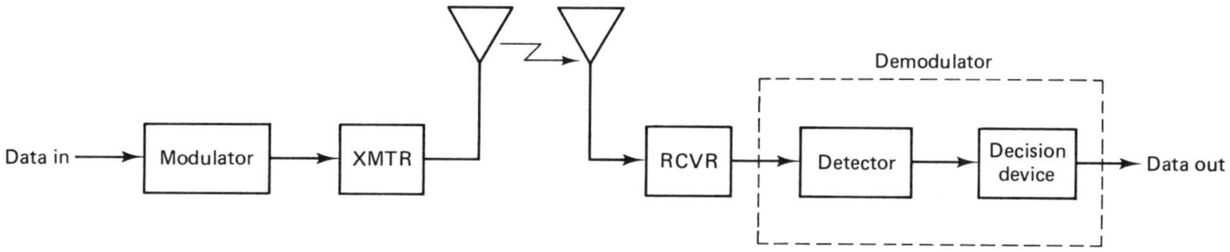

Figure 8.1 Digital radio system.

It is apparent to us at this point that many factors are involved in the initial design phase of a digital radio system. Referring to Figure 8.1, it is clear that, neglecting the data itself, one of the first choices to be made is the modulation technique to be used. We need to be able to, for example, compare the many modulation techniques available to us. Following this is the specification of the transmitter. An unlimited bandwidth is never available to us; therefore, we are driven to the design of a bandlimited digital radio system. With these choices made, design specifications for the receiver follow. We begin our discussion, then, with a comparison of digital radio modulation techniques.

8.1 COMPARISON OF DIGITAL RADIO MODULATION TECHNIQUES

The simplest digital AM modulation technique is double-sideband (DSB) AM modulated by a binary signal. Mathematically we can represent this operation as

$$e_{DSB}(t) = \frac{A}{2}[1 + m(t)] \cos \omega_c t \tag{8.1}$$

where ω_c is the carrier angular frequency (radians per second) and $m(t)$ is the modulating signal. Consider the case where $m(t)$ is an NRZ binary data waveform $[m(t) = \pm 1]$. For 100 percent modulation this results in an on-off keying (OOK) modulation. An OOK waveform can be demodulated either coherently or incoherently. The increase in performance is so slight, however, that coherent detection is not employed for this type of modulation when it is used for digital radio channels.

It is recognized that the carrier itself carries no information. Consequently, transmission efficiency is improved by resorting to the use of DSB-SC modulation. The general form is

$$e_{DSB\text{-}SC}(t) = Am(t) \cos \omega_c t \tag{8.2}$$

For the case where $m(t)$ equals 0 or 1, we again have OOK modulation. When $m(t) = \pm 1$, this results in binary PSK (see Section 8.1.2). For applications where spectral efficiency is important, we can make use of a single sideband. The SSB signal is usually generated by use of a band-pass filter to suppress the upper or lower sideband. If a band-pass filter is used with smooth roll-off rather than a sharp cutoff at or near the carrier frequency, this procedure results in a vestigial sideband (VSB) signal, wherein one sideband is completely present but the other sideband is only partially present.

Another AM alternative is QAM. QAM is produced by summing two DSB-SC signals where the premodulation carrier phases are 90° apart. Mathematically this operation is

$$e_{QAM}(t) = A[M_I(t) \cos \omega_c t + M_Q(t) \sin \omega_c t] \tag{8.3}$$

where $M_I(t)$ and $M_Q(t)$ are either independent or interdependent binary data signals. For example, the incoming data stream can be divided into two separate data streams, each at one-half bit rate. QAM exhibits the advantage of being as efficient in required power and bandwidth as SSB but without the stringent filtering requirements. In addition, if $M_I(t)$ and $M_Q(t)$ are three-level duobinary signals (+1, −1, or 0) coded in a manner to minimize ISI caused by filtering; the result is quadrature partial response (QPR) modulation (see Section 3.3.6.2). Both QAM and QPR require coherent detection. Any errors in phase tracking cause interference between the I and Q channels, thus degrading performance. If $M_I(t)$ and $M_Q(t)$ have values of ± 1, QAM is identical to QPSK (see Section 8.1.2).

The two techniques will differ, however, if $M_I(t)$ and $M_Q(t)$ are not rectangular pulses.

8.1.1 Frequency-Modulation Techniques

In terms of binary signaling, the simplest FM technique is FSK. We make use of two frequencies separated by Δf hertz, where Δf is twice the carrier deviation. This shift, Δf, is commonly called the *shift interval* and is small relative to the carrier f_c. It is common practice in FSK schemes to specify the frequency spacing in terms of the modulation index, m_f. We define the modulation index as

$$m_f = \frac{\Delta f T}{2} \tag{8.4}$$

where T is the symbol duration. It is equal to the inverse of the data rate, $1/f_b$, for binary schemes. In common with a number of other modulation schemes, FSK can be demodulated either coherently or noncoherently. For example, noncoherent demodulation can be effected by two passband filters followed by envelope detectors and a decision device. Using this approach, the filter spacing must be at least $\Delta f = 2(m_f/T)$, where $m_f > 1$. This requirement will prevent significant overlap of the filter passbands. An alternative approach that will limit the constraint on the modulation index is to use a discriminator to convert the frequency variations to amplitude variations and then employ an AM envelope detector. Coherent demodulation is readily accomplished using a PLL.

There is at present considerable interest in modified versions of FSK. These schemes are based on preventing abrupt changes in phase at the bit transitions. The idea then is to obtain continuous-phase FSK (CPFSK). This technique results in rapid spectral roll-off with improved transmission efficiency. A second modified technique is MSK, sometimes called fast-frequency-shift keying. MSK is a special case of CPFSK where the modulation index m_f is 0.5 and coherent detection is used. The performance for this particular technique is identical to coherent PSK plus retaining the superior spectral properties of CPFSK.

8.1.2 Phase-Modulation Techniques

Digital phase-modulation (PM) schemes, for the most part, require coherent detection. Basically, there are three versions of BPSK. Of these, the most straightforward approach is BPSK. Using this technique the carrier phase is shifted either 0° or 180°. A precise phase reference is needed for demodulation. Some phase reference–extraction techniques exhibit 180° phase ambiguities. An example of this is a Costas loop. To combat this ambiguity, a modified form of PSK called *differentially encoded PSK* (DE-PSK) is often used (see Section 3.3.4). The performance of DE-PSK is slightly inferior to that of PSK. This is intuitively reasonable. For example, with DE-PSK, no carrier phase transition may correspond to a space and a $\pm 180°$ transition may correspond to a mark. Thus, a bit-decision error on the current bit will induce another error on the subsequent bit.

DPSK is the third version of binary PSK. The information for DPSK is also differentially encoded. With DPSK, no effort is made to extract a coherent phase reference. Using this technique, the signal from the previous bit interval becomes the phase reference for the current bit interval. The performance of DPSK is less than that of DE-PSK because the phase reference is not smoothed over many bit intervals, relying instead on a single bit interval. QPSK schemes were also discussed in Chapter 3. Coherent QPSK involves encoding 2 bits at a time into one of four carrier phases spaced 90° apart. Just as in the binary case, the data can be differentially encoded and differentially detected with an accompanying loss in performance. We denote this scheme as DQ-PSK. In recent years a modified version of QPSK called SQPSK has come into use. This scheme offers advantages over conventional QPSK with regard to spectral efficiency, synchronization and sideband efficiency (see Chapter 3).

The requirements for high-speed data transmission have led to an increasing need for bandwidth conservation. This has resulted in the use of hybrid AM/PM techniques called amplitude and phase-shift keying (APK). The phase-amplitude space for 16-ary APK is shown in Figure 8.2. APK is a form of QAM wherein $M_I(t)$ and $M_Q(t)$ in equation (8.3) are multilevel (rather than binary) baseband signals.

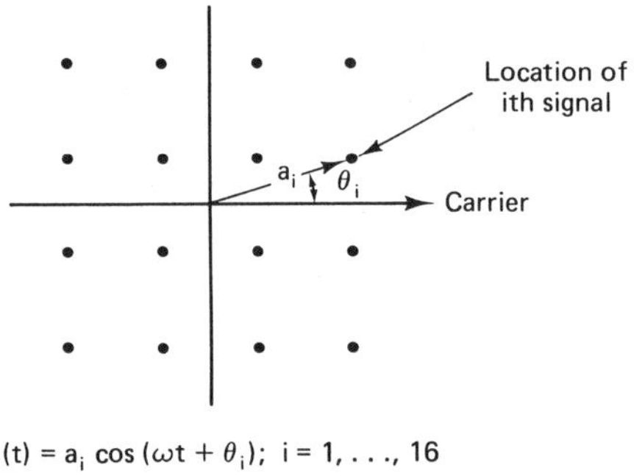

Figure 8.2 16-ary APK phase-amplitude space.

8.1.3 Comparison of Modulation Techniques

We see from the preceding discussion and also from Chapter 3, the variety of modulation and signaling schemes available for consideration in an initial design. An intelligent choice requires that we compare techniques under conditions characteristic of digital radio channels. A performance measure that is often used is the baseband E_B/N_0 ratio (the ratio of average signal energy per bit to noise power per hertz, as measured at the receiver input) required to achieve a particular BER. For the following discussion, we assume a BER of 10^{-4}. For most general-purpose digital radio applications, this error is sufficient. The ideal performance of a number of modulation schemes is shown in Table 8.1.

TABLE 8.1 IDEAL PERFORMANCE OF REPRESENTATIVE MODULATION SCHEMES

Type	Modulation scheme	E_B/N_0 (dB)*
AM	OOK—coherent detection	11.4
	OOK—envelope detection	11.9
	QAM	8.4
	QPR	10.7
FM	FSK—noncoherent detection ($d = 1$)	12.5
	CP-FSK—coherent detection ($d = 0.7$)	7.4**
	CP-FSK—noncoherent detection ($d = 0.7$)	9.2**
	MSK ($d = 0.5$)	8.4
	MSK—differential encoding ($d = 0.5$)	9.4
PM	BPSK—coherent detection	8.4
	DE-BPSK	8.9
	DPSK	9.3
	QPSK	8.4
	DQPSK	10.7
	OK-QPSK	8.4
	8-ary PSK—coherent detection	11.8
	16-ary PSK—coherent detection	16.2
AM/PM	16-ary APK	12.4

* For BER of 10^{-4}. d = FM modulation index
** Assumes 3-bit observation interval.

This table compares performance in the presence of AWGN. Referring to the table, note that CP-FSK with coherent detection over a 3-bit observation interval outperforms BPSK. The identical ideal performance is shown for QAM, MSK, and QPSK. This results from the underlying similarities of these schemes. In fact, offset-keyed QAM, MSK, and OK-QPSK differ only in the weighting functions applied to the I and Q channels.

A comparison of modulation schemes can be made based on the spectral characteristics of the modulated signal. Bandwidth efficiency was discussed in Section 3.7. Of particular interest is the extent to which a signal will interfere with signals in adjacent channels. One way to measure this quality is the attenuation of the power spectrum a specified distance from the center frequency. For example, at an arbitrary distance $8/T$ hertz from the center frequency ($T = 1/f_b$), AM modulation schemes are attenuated approximately 25 dB. For PM modulation, the side lobes are down by 33 dB while with continuous phase FM the side lobes are down by approximately 60 dB. Generally speaking, for frequencies with large displacement from the center frequency [large values of $(f - f_c)T$], the spectrum for AM and PM falls off as f^{-2}, whereas that for CP-FM falls off as f^{-4}.

At this point we could infer a significant advantage for FM. A note of caution, however, is in order. First, the figures given for PM assume that abrupt phase transitions occur. Improved spectral characteristics can be achieved by making the phase transitions occur more smoothly. Modifications can be made to AM schemes as well. For example, continuous-phase AM has a spectrum that falls off as f^{-4}. As an additional spectrum conservation measure, side lobes can be reduced by suitable postmodulation filtering. The penalty paid is a decrease in performance. We see, then, that a choice of modulation technique based on spectral efficiency needs to be made on the basis of a rather detailed trade-off study based on both cost and the E_b/N_0 performance criteria.

In Chapter 3 we determined the transmission bandwidth required to transmit data at a specified information rate. We characterized the speed of a modulation technique as being equal to R/W, where R is the data rate and W is the IF bandwidth. Obviously in choosing a modulation scheme, this becomes an important figure of merit. The speeds for each technique together with the E_b/N_0 required for a 10^{-4} BER are listed in Table 8.2. The figures shown in this table are indicative of the results that may be expected. It should be noted, however, that these results do not represent the optimum combination of speed and efficiency that can be obtained.

8.1.3.1 Interference effects. The effects of cochannel and adjacent-channel interference were mentioned in the previous section. We noted, for ex-

TABLE 8.2 RELATIVE SIGNALING SPEEDS OF REPRESENTATIVE MODULATION SCHEMES

Type	Modulation scheme	Speed (bits/s/Hz)	E_B/N_0 (dB)*
AM	OOK—coherent detection	0.8	12.5
	OOK—envelope detection		
	QAM	1.7	9.5
	QPR	2.25	11.7
FM	FSK—noncoherent detection ($d = 1$)	0.8	11.8**
	CP-FSK—coherent detection ($d = 0.7$)		
	FP-FSK—noncoherent detection ($d = 0.7$)	1.0	10.7
	MSK ($d = 0.5$)	1.9	9.4
	MSK—differential encoding ($d = 0.5$)	1.9	10.4
PM	BPSK—coherent detection	0.8	9.4
	DE-BPSK	0.8	9.9
	DPSK	0.8	10.6
	QPSK	1.9	9.9
	DQPSK	1.8	11.8
	OK-QPSK		
	8-ary PSK—coherent detection	2.6	12.8
	16-ary PSK—coherent detection	2.9	17.2
AM/PM	16-ary APK	3.1	13.4

* For BER of 10^{-4}. d = FM modulation index
** Discriminator detection.

TABLE 8.3 PERFORMANCE OF REPRESENTATIVE MODULATION SCHEMES IN THE PRESENCE OF CW INTERFERENCE

Type	Modulation scheme	E_B/N_0 required $S/I = 10$ dB	$S/I = 15$ dB
AM	OOK—coherent detection		
	OOK—envelope detection	~20	14.5
	QAM		
	QPR		
FM	FSK—noncoherent detection ($d = 1$)	14.7	13.3
	CP-FSK—coherent detection ($d = 0.7$)		
	CP-FSK—noncoherent detection ($d = 0.7$)		
	MSK ($d = 0.5$)		
	MSK—differential encoding ($d = 0.5$)		
PM	BPSK—coherent detection	10.5	9.2
	DE-BPSK	11.0	9.7
	QPSK	12.0	10.3
	DQPSK	>20	14.0
	OK-QPSK		
	8-ary PSK—coherent detection	~20	15.8
	16-ary PSK—coherent detection		24
AM/PM	16-ary APK		

* For BER of 10^{-4}. d = FM modulation index

ample, that MSK and CP-FSK schemes exhibit a large advantage in rate of signal strength roll-offs (f^{-4} versus f^{-2}) over AM and PM, if postmodulation filtering is not employed. Given that adjacent-channel interference does occur, we must consider the amount of performance degradation caused by a specified level of interference. One method of measuring this effect is to apply an in-band CW interfering signal 10 or 15 dB down in power from the desired signal and then determine the E_b/N_0 required for a BER of 10^{-4}. This setup can be considered as a model for interfering side lobes from adjacent channels or for the main lobe from a cochannel interferer. Table 8.3 represents the performance for different modulation schemes. Referring to this table, noncoherent FSK and BPSK exhibits the least degradation from ideal performance. The most degradation from ideal is exhibited by 8-ary and 16-ary schemes (see Table 8.1). In compiling this table, data was not available for CP-FSK schemes or for OOK-QPSK.

8.1.3.2 Effects of fading. Signal fading is a problem often encountered over digital radio links. For example, multipath fading is caused when the digital receiver receives identical transmitted signals that have traveled different effective paths (see Figure 8.3). We can classify the interfering waves according to two types. The first type results from the effects of atmospheric propagation, whereas the second results from reflections caused by large surfaces such as the ground or bodies of water. For both types of interference, a frequency-selective fading

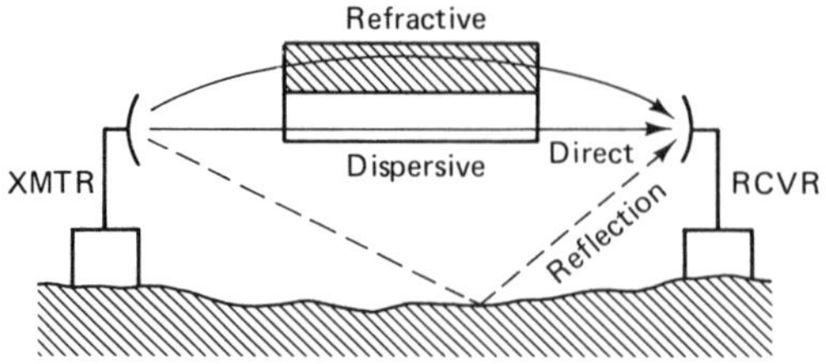

Figure 8.3 Multipath fading is caused when a receiver picks up identical transmitted signals that have traveled paths of differing effective lengths.

phenomenon results. This happens because of phase differences in the received signals at common frequency bands. Common inphase frequency components increase the signal amplitude, whereas out-of-phase components decrease the signal. This creates peaks and notches in the amplitude versus frequency response, as illustrated in Figure 8.4. A well-designed digital radio will include a fading compensation circuit or equalizer to offset these losses. The design of an adaptive transversal equalizer is discussed in Section 8.7.1.

If the fading is caused by two resolvable multipath components, the results tabulated in Table 8.3 can be used. The CW interference signal represents the signal from the interference path. On the other hand, if fading is caused by a large number of equal-amplitude components, a more complex model (Rayleigh fading) is appropriate. Performance results for a Rayleigh fading channel are shown in Table 8.4. The results in this table are tabulated for a BER of 10^{-2}. Comparing these results to those in Table 8.3, we see that the effects of a Rayleigh fading channel are rather severe. Error-control coding can be used to alleviate this. Comparing Tables 8.1 and 8.4, we see, however, that the relative performance of the different modulation schemes does not differ markedly. The results in Table 8.4 assume that for the AM and AM/PM schemes, fading is slow enough that the decision threshold can be continuously adjusted to an optimum value.

The results just discussed can be used as a starting point in solving systems engineering problems. In many cases most of the alternatives can be eliminated, based on economic and physical restraints. An example of the use of the tables is illustrated in Example 8.1.

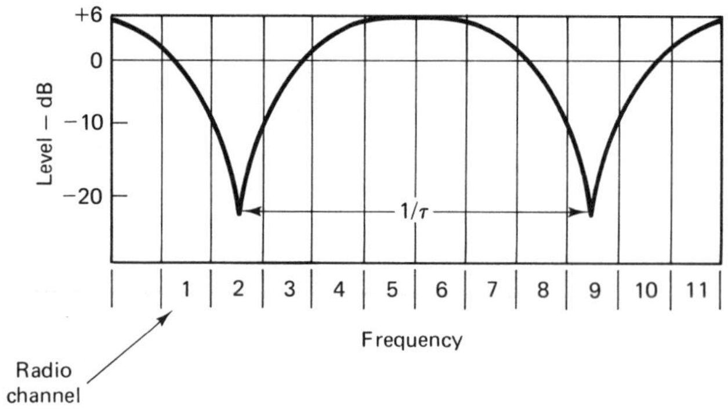

Figure 8.4 When multipath signals combine, instances of large negative phase coincidence create "notches" in the received amplitude-versus-frequency response.

TABLE 8.4 PERFORMANCE OF REPRESENTATIVE MODULATION SCHEMES ON A RAYLEIGH FADING CHANNEL

Type	Modulation scheme	Average E_B/N_0 (dB)*
AM	OOK—coherent detection	17**
	OOK—envelope detection	19**
	QAM	14
	QPR	
FM	FSK—noncoherent detection (d = 1)	20
	CP-FSK—coherent detection (d = 0.7)	13†
	CP-FSK—noncoherent detection (d = 0.7)	18
	MSK (d = 0.5)	14
	MSK—differential encoding (d = 0.5)	17
PM	BPSK—coherent detection	14
	DE-BPSK	17
	DPSK	17
	QPSK	13.5
	DQPSK	20
	OK-QPSK	13.5
	8-ary—coherent detection	16.5
	16-ary PSK—coherent detection	21
AM/PM	16-ary APK	18

* For BER of 10^{-2}. d = FM modulation index
** Assumes optimum variable threshold.
† Assumes three bit observation interval.

EXAMPLE 8.1

We wish to choose the best modulation scheme for transmitting 90-megabit/s data over a 40-MHz radio channel.

Solution The signaling speed, R/W = (90 megabits/s)/(40 MHz) = 2.25 bits/s/Hz. Refer to Table 8.2. Of the modulation schemes listed, QPR is the most efficient. It requires the lowest E_b/N_0 ratio and is the least complex. Eight-ary, 16-ary PSK, and 16-ary APK can also achieve the signaling speed but are more complex. Based on economic factors, then QPR is the choice.

8.2 BANDLIMITED DIGITAL RADIO SYSTEM DESIGN

The theoretical works of Nyquist, Lender, Kretzmer, and others forms the basis for the design of bandlimited digital radio systems. Nyquist published analytical work in 1928, which provided the theoretical foundation for transmission of telegraph signals over a bandlimited cable with zero ISI. In 1963, Lender published the analytical basis for duobinary transmission, and the concept was generalized by Kretzmer in 1966. This early work was directed toward improving data transmission over existing analog facilities. In 1974 the FCC established out-of-band emission regulations for multichannel digital radio, thus providing the framework wherein major development efforts could proceed.

Typically, in a multichannel radio system, the limiting factor on error performance is adjacent-channel interference. The ratio S/I of inband signal power to adjacent-channel interference depends solely on the transmitted power spectrum and the selectivity of the receiver filter. The design objective, then, is to optimize the composite filter function between transmitter and receiver to enhance detection at the receiver. Filters designed to shape a given modulation form into the desired bandlimited signaling pulse do not, in general, correspond to classical Butterworth or Chebyshev designs. The desired shaping filters must meet performance objectives in both the frequency and time domains. Nonlinear power amplification is often used. The point within the system where the filter is introduced depends upon whether the amplifier is linear or nonlinear. If nonlinear, the pulse-shaping filter must follow the amplifier.

8.2.1 Design Concepts

A block diagram of a digital radio system was shown in Figure 8.1. A more detailed diagram is illustrated in Figure 8.5. Refering to this figure, the transmitter functions are incoming data precoding, modulation, pulse shaping, and power amplification. The precoder encodes the incoming data into a modulator control signal and the output of the modulator is carrier pulses, which are shaped by the pulse-shaping network. The power amplifier, discussed previously, is not shown since its location—before or after the pulse shaping network—depends on whether or not it is linear. Particular circuit design fundamentals are involved. For example, if the amplifier is nonlinear, the modulator output is constant amplitude and is

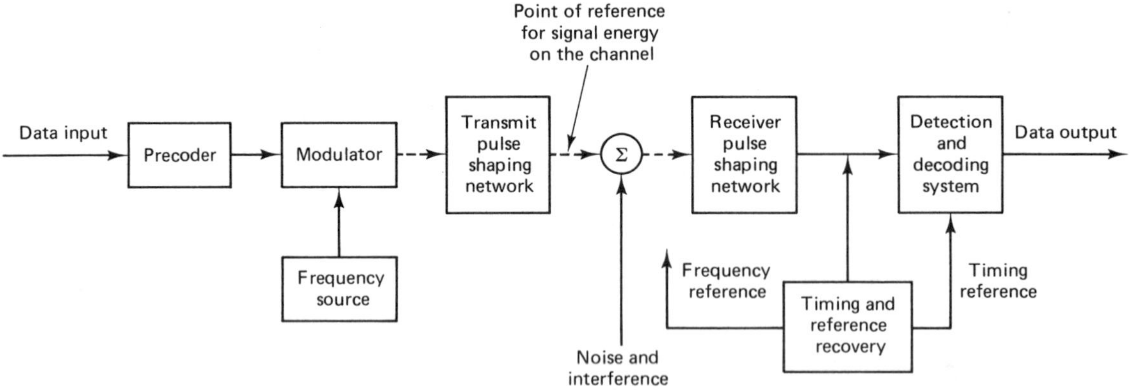

Figure 8.5 Block diagram of digital radio system.

fed directly into the amplifier. The amplifier drives the pulse-shaping network with a signal whose bandwidth is greater than the bandwidth of the network. This results in reflected power, which must not reach the modulator or else distortion of the modulator output occurs. Isolation can be achieved by using a circulator with a resistive load to dissipate the reflected power.

The transmitter pulse-shaping network is designed to suppress signal energy outside the assigned channel bandwidth. The pulse-shaping network in the receiver is designed to shape further the transmitted pulse, reject out-of-band noise, and suppress adjacent-channel interference. In order to recover the data from the pulse shaping circuitry, precise frequency and timing signals and bit synchronizers are required. This topic is discussed in detail in Chapter 7.

Digitally modulated signals such as PSK, MSK, QAM, and those produced by combined amplitude and phase modulation can be expressed in terms of DSB-SC quadrature notation (see equation (8.3)). We generalize this concept as

$$S(t) = I(t) \cos \omega_c t + Q(t) \sin \omega_c t \tag{8.5}$$

where $I(t)$ and $Q(t)$ are the binary or multilevel inphase and quadrature phase modulator pulse sequences, respectively and ω_c is the carrier frequency. Note that $I(t)$ and $Q(t)$ are pulse sequences whose amplitudes are data dependent. We can express these quantities as

$$\begin{aligned} I(t) &= \sum_n a_n P(t - nT) \\ Q(t) &= \sum_n b_n q(t - nT) \end{aligned} \tag{8.6}$$

In these expressions, T is the symbol interval on the channel and a_n and b_n are amplitude constants that depend on the data values over the time interval $nT < (n + 1)T$. Also, $P(t)$ and $q(t)$ are modulator pulse functions. We can more easily assign physical meaning to equations (8.5) and (8.6) if we consider a digital communication system with more detail specified (see Figure 8.6).

From the figure we see that the input data stream, which consists of full-length rectangular pulses, is converted by a serial-to-parallel converter into two full-length rectangular pulse streams labeled $I(t)$ and $Q(t)$ and $a_n = b_n = \pm A/2$. The baud rate for each is one-half the incoming bit rate. For this particular scheme, $Q(t)$ is delayed by one bit duration and is thus offset from the $I(t)$ string. In a practical design, pulse shaping can be effected by baseband filtering both $I(t)$ and $Q(t)$ and also the modulator output signal $S(t)$. The receiver samples the received signal at the center of each pulse and decides if it is positive or negative based on a decision threshold. Errors are made due to the addition of white Gaussian noise or other types of interference to the received signal.

A key design feature is determination of the composite characteristics of the transmitter and receiver filters, which results in a signal stream at the decision threshold in the receiver that is free of ISI at the sampling instants. With this constraint on the composite filter design, the transmitter filter can be specified

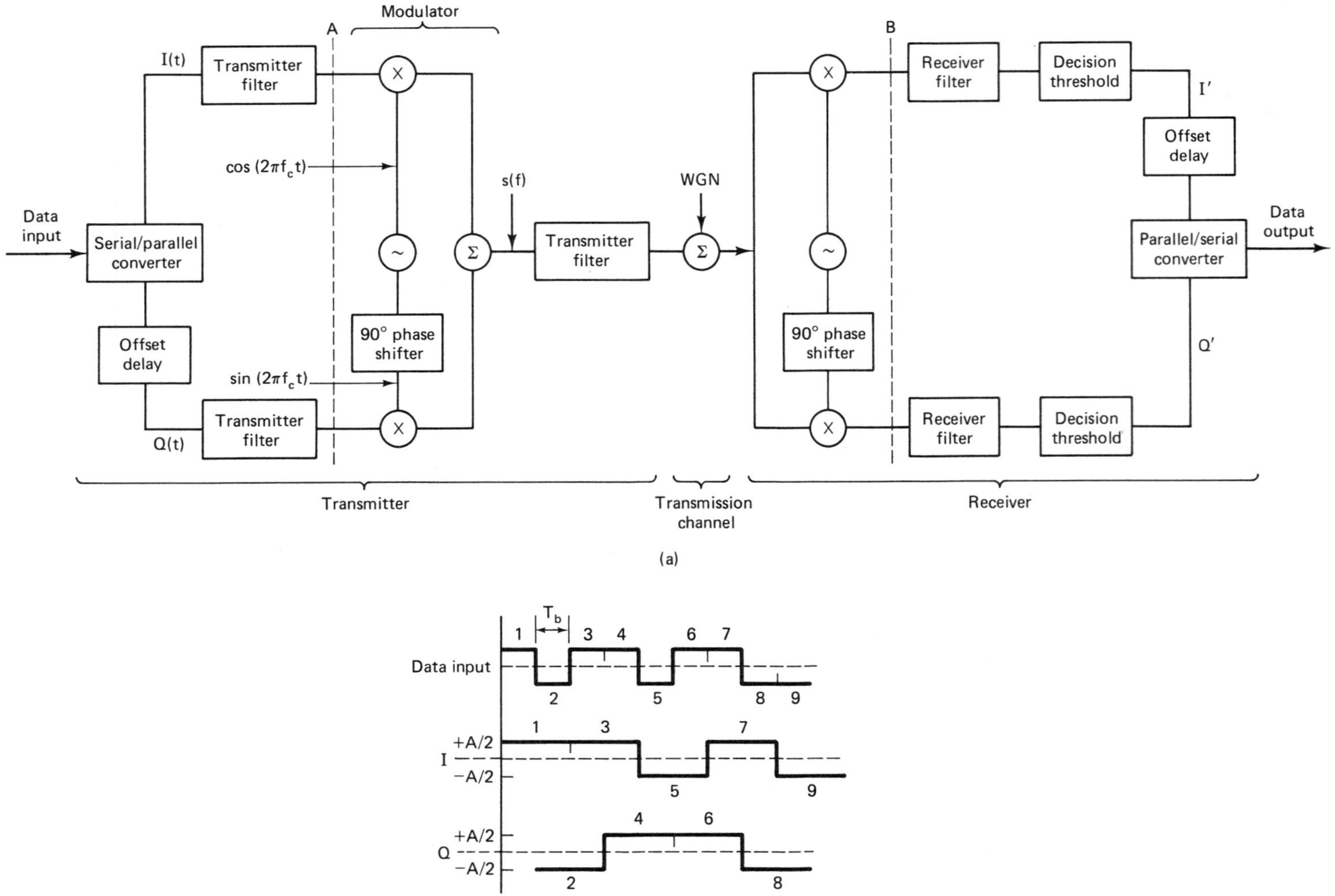

Note: Data input bit number n is converted to I/Q bit number n

(b)

Figure 8.6 (a) Block diagram of QPSK digital communication system. (b) QPSK data streams.

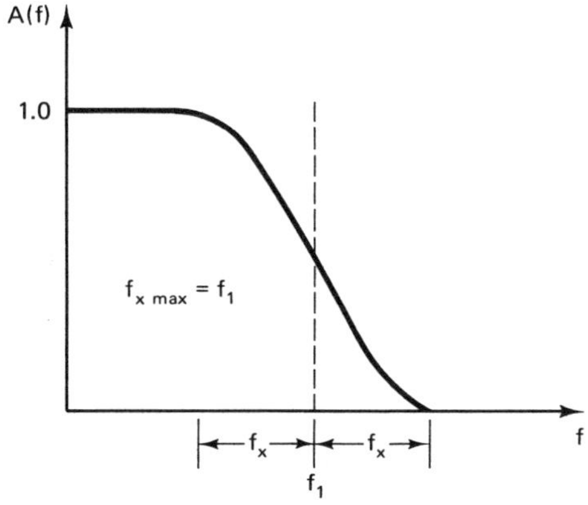

Figure 8.7 Raised cosine filter frequency response.
(© 1979, IEEE)

in order to limit the transmitted signal to the available transmission channel. The receiver filter can then be specified to limit adjacent-channel interference and to optimize the P_e versus the receiver input SNR. From Nyquist's results, we know that a data rate of $2f_1$ can be transmitted through a filter of bandwidth f_1 hertz and recovered without interference between the peaks of the received pulses. This filter must have the property of linear phase and the amplitude characteristics must be either ideal low pass or gradual symmetrical roll-off (see Figure 1.2).

A special case of a gradual roll-off filter is the raised cosine filter. A Bode plot is shown in Figure 8.7. Mathematically, we can characterize the frequency response of the filter, $A(f)$, as

$$A(f) = \begin{cases} 1, & 0 < f < f_1 - f_x \\ \dfrac{1}{2}\left[1 - \sin\dfrac{\pi}{2\alpha}\left(\dfrac{f}{f_1} - 1\right)\right], & f_1 - f_x < f < f_1 + f_x \\ 0, & f > f_1 + f_x \end{cases} \tag{8.7}$$

where

$$\alpha = \frac{f_x}{f_1} = \text{roll-off factor}$$

Here α is the ratio of the excess bandwidth of the composite channel to the Nyquist bandwidth, $1/2T$. The phase shift of the filter is given by

$$\phi(f) = Kf \tag{8.8}$$

where K is a constant. A voltage impulse exhibits a flat infinite bandwidth spectrum and constant phase. Thus the impulse response of a filter is the amplitude and phase response of the filter. This output in the time domain, when the filtering is raised cosine, is a pulse whose shape depends on the roll-off factor α (see Figure 8.8). We previously encountered this pulse shape in Figure 1.2(c). The pulse shape is such that impulses can be transmitted at a rate equal to $2f_1$ without interference

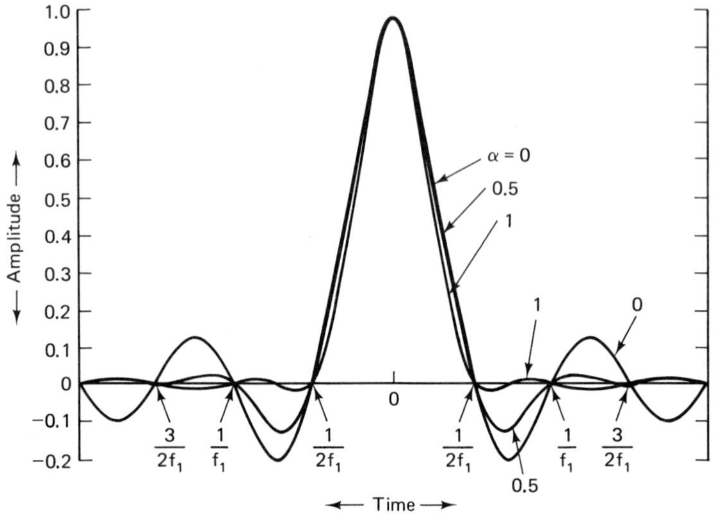

Figure 8.8 Impulse response obtained when filtering is raised cosine for various roll-offs.

between the peaks of the received signals. Sampling time is, of course, the center of the pulse.

An impulse is not necessary to create the responses shown in Figures 8.7 and 8.8. Any pulse that has a spectrum whose amplitude exceeds zero between dc and frequency $f_1 + f_x$ and whose phase remains constant over the same frequency interval can be filtered to yield those same responses. Thus, this pulse can also be transmitted at rate $2f_1$, without ISI. For example, the full-length

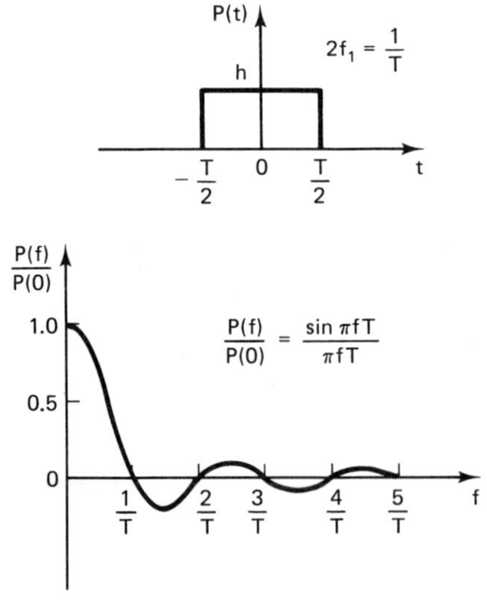

Figure 8.9 Spectral amplitude function of rectangular full-length pulse.

rectangular pulses shown in Figure 8.6 for the offset QPSK system fulfill these requirements.

Refer to Figure 8.6(a). The composite filter transfer function, $F(f)$, of the combined transmitter and receiver filtering is given by

$$F(f) = T_x(f)\cdot R_x(f) \tag{8.9}$$

where $T_x(f)$ and $R_x(f)$ are the transfer functions of the transmitter and receiver filters, respectively. In order that the impulse response at the decision device in the receiver be the same as that in Figure 8.8, the frequency response of the filter $A(f)$ must be modified by the spectral amplitude function of $p(t)$ or $q(t)$. Normally, $p(t) = q(t)$. For the offset system, $p(t) = q(t - T/2)$. The spectral amplitude of a rectangular full-length pulse is shown in Figure 8.9. The frequency response of $A(f)$ must be modified by the factor

$$\frac{P(o)}{P(f)} = \frac{\pi f T}{\sin \pi f T} \tag{8.10}$$

where

$$T = \frac{1}{2f_1}$$

The composite filter transfer function, F(f), becomes

$$F(f) = \begin{cases} \dfrac{\pi f/2f_1}{\sin\left(\dfrac{\pi f}{2f_1}\right)}, & 0 < f < f_1 + f_x \\[2ex] \dfrac{1}{2}\left[1 - \sin\dfrac{\pi}{2\alpha}\left(\dfrac{f}{f_1} - 1\right)\right]\dfrac{\pi f T}{\sin \pi f T}, & f_1 - f_x < f < f_1 + f_x \\[2ex] 0, & f > f + f_x \end{cases} \tag{8.11}$$

The phase function is as before. That is,

$$\phi(f) = Kf, \qquad 0 < f < f_1 + f_x \tag{8.12}$$

Filter values for various values of α are shown in Figure 8.10. Technically speaking, for $\alpha = 0$, the filter is no longer raised cosine but is "brick wall" instead.

Now that we know what the composite filter is, what then are the component filter characteristics? For the specific case of full cosine roll-off ($\alpha = 1$) and white noise at the receiver input, it may be shown that the receiver filter transfer function, $R_x(f)$, necessary to yield optimum P_e versus input SNR for an offset QPSK system, is given by

$$R_x(f) = \cos\left(\frac{\pi f}{4f_1}\right), \qquad f < 2f_1 \tag{8.13}$$

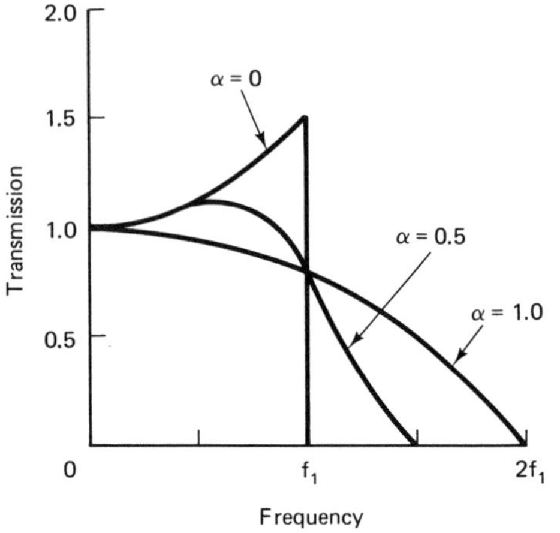

Figure 8.10 Filtering necessary to rectangular pulse train to create raised cosine pulse at filter output.

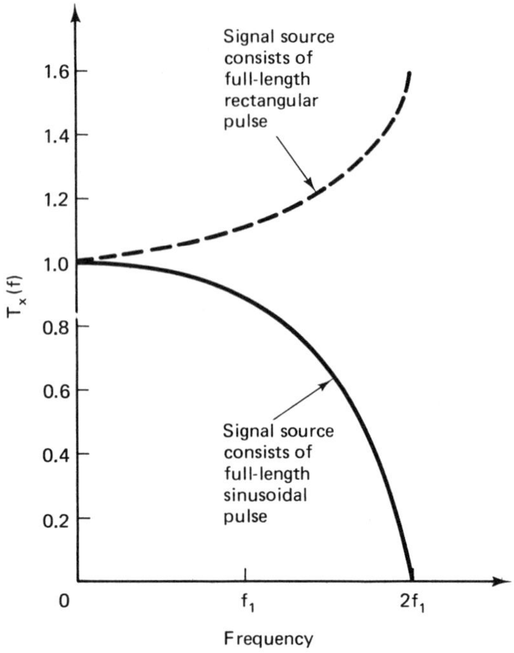

Figure 8.11 Characteristic of transmitter filter.

The system P_e is given by

$$P_e = \frac{1}{2} \operatorname{erfc} \sqrt{\frac{S}{N_b}} \tag{8.14}$$

where S is the average signal power at the filter input and N_b is the noise in the bit-rate bandwidth, f_b. The transmitter function to complete the full raised cosine rolloff spectrum at the receiver decision threshold input is given by

$$T_x(f) = \frac{\pi f/4f_1}{\sin(\pi f/4f_1)} \tag{8.15}$$

For an MSK system, the transmitter filter may be expressed as

$$T_x(f) = \frac{\cos(\pi f/4f_1)}{\cos(\pi f/2f_1)} \left[1 - \left(\frac{f}{f_1}\right)^2\right] \tag{8.16}$$

The corresponding receiver filter for MSK is the same as for offset QPSK (equation (8.13)). Transfer functions of the transmitter filters are shown in Figure 8.11. Note that the transmitter filter for MSK is a gradual roll-off filter and is thus easier to realize in a design than the peaked response of the offset QPSK filter. A block diagram for an MSK digital communication system is shown in Figure 8.12.

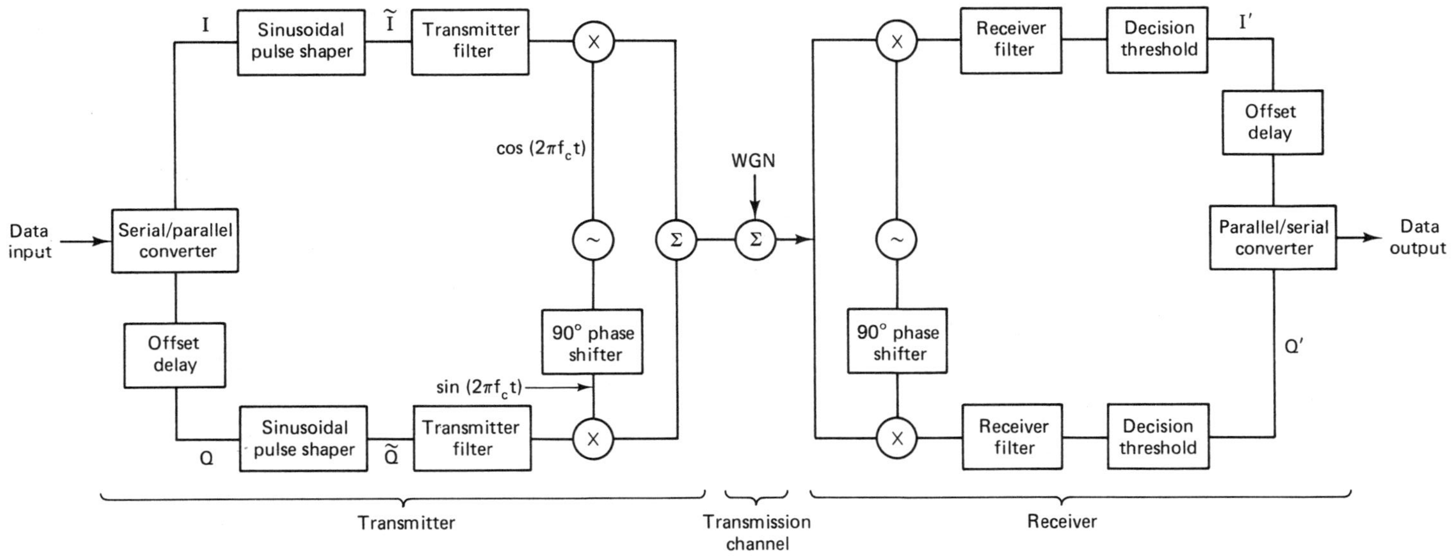

(a)

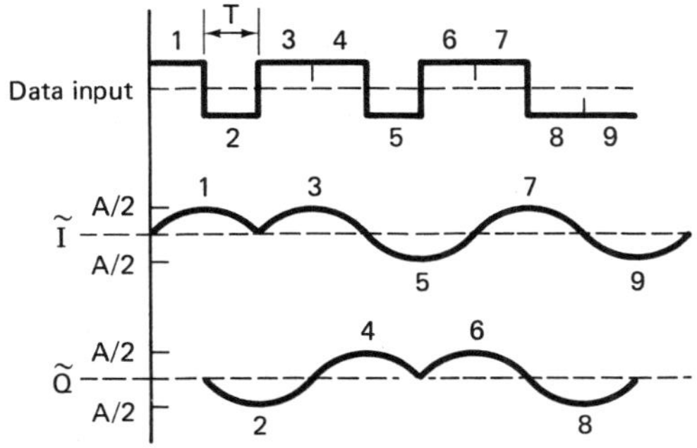

Note: Data input bit number n is converted to I/Q bit number n

(b)

Figure 8.12 Block diagram for an MSK digital communication system.
(© 1979, IEEE)

The data rate for zero symbol interference is $2f_1$ bits per second. Refer to Figure 8.6. Multiplying the I and Q data stream with the carrier results in a double-sided RF spectrum that is twice the baseband spectrum. The Q data stream is inphase quadrature with the I data stream and its RF spectrum is superimposed on that of the I stream. An RF filter of $2f_1$ bandwidth will then pass the composite RF spectrum. A spectral efficiency of 2 bits/s/Hz is achieved.

8.3 SYSTEM REQUIREMENTS

The design of a digital radio communication system begins with two major design issues: (1) selection of modulation methods to obtain the required bandwidth efficiencies and (2) choice of baseband or RF filtering to meet the transmitter emission limitations. The design begins with a consideration of system design specifications. Some of the more important system parameters are listed in Table 8.5. If more than one level of performance is specified, different selectable types of modulation may be necessary.

Suppose, for example, that a system is to be designed with the requirement that for $E_b/N_0 \leq 23$ dB, BER $= 5 \times 10^{-9}$. A bandwidth efficiency of 1.9 bits/s/Hz is desired. The first step is to assemble a list of candidate modulation techniques as in Table 8.6. The profusion of modulation techniques may make this

TABLE 8.5 IMPORTANT SYSTEM PARAMETERS

- Radio frequency band
- Overall bit rate
- Modulation format
- Transmitter power (at the antenna port)
- Noise figure
- Receiver threshold (at BER $= 10^{-y}$)
- System gain (at BER $= 10^{-y}$)

TABLE 8.6 MODULATION CANDIDATES FOR 1.9 bits/s/Hz BANDWIDTH EFFICIENCY

Modulation	Number of signal states	Excess bandwidth, %	E_b/N_0 at 5×10^{-9}, dB*
4-PSK (QPSK)	4	10	12.5
8-PSK	8	60	16.1
QPR	3 × 3	10	14.6
7-PSK/OOK	1 × 7	60	14.3
QAM	4 × 4	100	16.5
16-PSK	16	100	20.6

* Assumes differential encoding and threefold expansion due to scrambling.

the most difficult part of the task. In addition, finding data that has been evaluated for the same filter bandwidths is a problem. In terms of SNR, QPSK is an attractive candidate. It is not, however, attractive at a 10 percent excess bandwidth, since the filter would be difficult to synthesize. Modulations employing 4 bits/symbol, such as 16-phase PSK or QAM, can easily meet the bandwidth requirements. The drawback is modem complexity. This leaves 8-phase PSK, QPR, and 7-phase PSK with an additional off state. Both 8ϕ-PSK and 7ϕ-PSK/OOK provide 1.9 bits/s/Hz using a raised cosine filter with $\alpha = 0.6$. This leaves theoretical link margins of 6.9 dB and 8.7 dB, respectively. Both PSK modulations can be implemented with baseband or RF filtering to obtain the desired raised-cosine response.

Nine-state (3×3) QPR modulation is the quadrative addition of two 3-level partial-response, or duobinary, signals. This technique offers a low bandwidth because of the spectral shaping that results from its adjacent bit correlation (see Section 3.3.6.2). A theoretical margin of 8.5 dB is achieved. Compared to 8ϕ-PSK, QPR has a 1.5-dB link margin and the modem design is simpler. The 10 percent excess bandwidth is not an attractive option. Either choice is a viable design choice.

8.3.1 Waveshape Filtering Method

Filtering in the digital communication system must satisfy two criteria:

- It must yield the desired interference control.
- The output spectrum must meet the FCC emission mask requirement.

The task of the designer is to decide whether the filtering is to be performed in the receiver, at baseband or RF in the transmitter, or a combination of both. We have previously discussed how the filtering should be apportioned between the transmitter and receiver for offset QPSK and MSK digital systems (see equations (8.11) through (8.16)). The ideal filter apportionment is shown in Figure 8.13. Regardless of how the filtering ends up being apportioned, it must meet the −80-dB out-of-band levels of the FCC emission mask. The FCC transmission mask for a 6-GHz digital radio system is shown in Figure 8.14. Two points are critical:

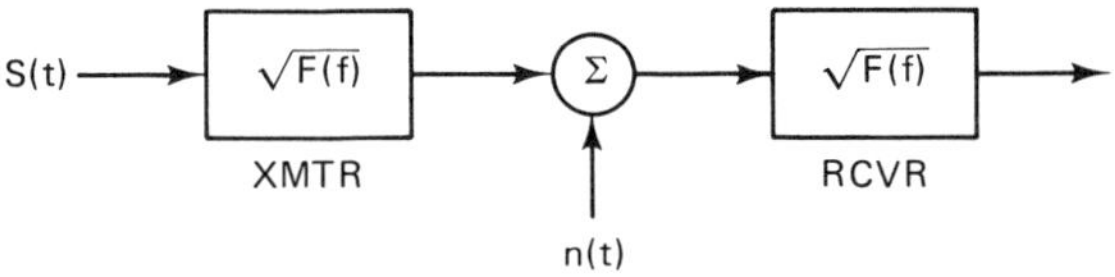

Figure 8.13 Ideal waveshape filter apportionment.

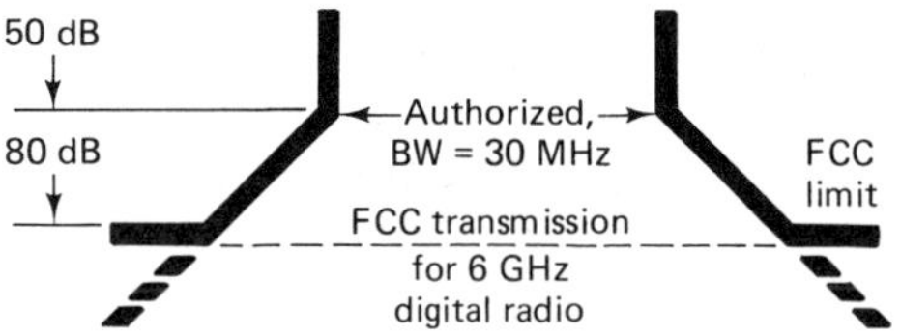

Figure 8.14 Graphical representation of FCC-transmitted spectrum specifications.

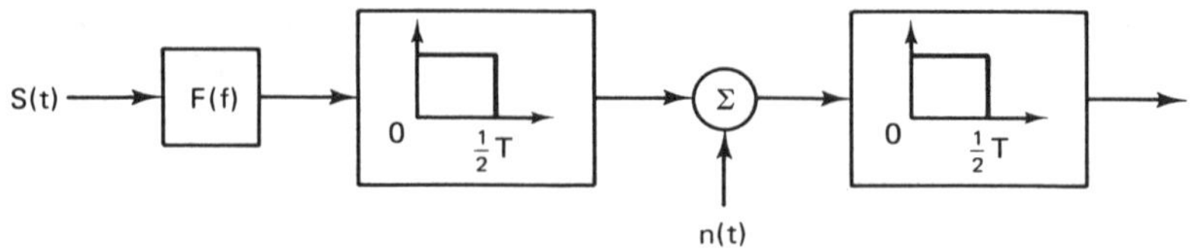

Figure 8.15 QPR waveshape filter apportionment.

(1) The spectrum width at the 50-dB down (from the mean transmitted power) points should not exceed the authorized bandwidth and (2) the output level should be at least 80 dB down outside the frequency range of the FCC mask.

For QPR modulation, it is convenient to place the total pulse-shaping filter in the transmitter. Accompanying this filter is a brick wall response in both the transmitter and receiver (see Figure 8.15). The latter aids in meeting FCC emission levels, that is, the −50-dB step in the FCC mask (Figure 8.14). The penalty paid in SNR performance is a receiver noise bandwidth increase of $\pi/4$ for an SNR of 1.05 dB. This same approach for QPSK with $\alpha = 1$ results in an SNR penalty of 3 dB. This increase in SNR is too large and does not represent a viable design approach.

8.4 DIGITAL RADIO SPECTRUM MEASUREMENTS

A digital modulation system is a complex combination of AM, FM, or PM pulse combinations. This results in a signal with a rather wideband spectrum, which must be shaped as discussed previously. It must also be carefully adjusted and monitored in order to meet the interference and fidelity requirements as specified by the FCC. The primary instrument used for this is the spectrum analyzer. FCC regulation ∮21.106 specifies that for digital modulation below 15 GHz in any 4-KHz band, the center frequency of which is removed from the assigned frequency by more than 50 percent up to and including 250 percent of the authorized bandwidth, the following equation applies:

$$A = 35 + 0.8\,(P - 50) + 10\log_{10} B \tag{8.17}$$

where

A = the decibels of attenuation below the mean output power level of the transmitter
B = authorized bandwidth in megahertz
P = percentage removal from the carrier.

Refer to Figure 8.14. The attenuation must be in no event less than 50 dB. Attenuation greater than 80 dB is not required. A different equation applies for transmisssion above 15 GHz. The FCC specifies authorized bandwidths of 30

MHz at 6 GHz, 40 MHz at 11 GHz, and so on. As we noted previously, two points in the spectrum need to be adjusted and/or verified:

- Spectrum width 50 dB down from the mean transmitted power should not exceed the authorized bandwidth.
- The output level should be at least 80 dB down outside the frequencies in the FCC mask.

EXAMPLE 8.2

For a 6-GHz digital communication system, calculate the required attenuation at a frequency of 5.98 GHz.

Solution

$$P = \tfrac{20}{30} \times 100 = 66.6$$

$$A = 35 + 0.8\,(66.6 - 50) + 10 \log_{10} 30 = 63.05 \text{ dB}$$

In addition to performing spectrum occupancy test, a spectrum analyzer can also be used to adjust for maximum peak output power, spectrum shape and symmetry, comparison of pre- and postoutput filter performance, spurious emissions from the carrier, interference pickup, and so on. When measuring the spectrum to FCC specifications, the occupied bandwidth is measured at the 50-dB points relative to the *mean output-power level*, as illustrated in Figure 8.16. If the power outside the output-filter bandwidth is ignored, then the mean output power is the same as the unmodulated level. The relative level between signal display peak and the mean power level is given by

$$\text{Relative level (dBs)} = 10 \log \frac{B_n}{f_s} \tag{8.18}$$

where B_n is the noise bandwidth (4 KHz) over which power is being measured and f_s is the signaling rate.

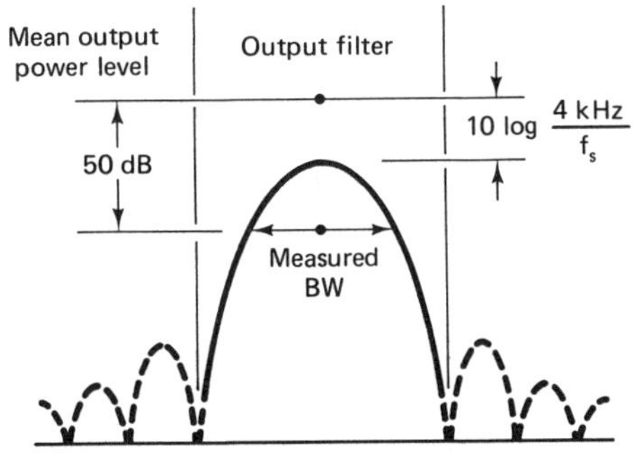

Figure 8.16 Spectrum output level relative to mean output power.

EXAMPLE 8.3

Compute the relative power level if the signaling rate is 30.086 MHZ.

Solution

$$\text{Relative level} = 10 \log[(4 \times 10^3)/(30.086 \times 10^6)] = -38.76 \text{ dB}$$

For this example, the main lobe is 38.76 dB down from the mean output. The output level in a 4-KHz noise bandwidth must be at least 80 dB down from the mean output level outside 250 percent of the specified bandwidth offset. Without an output filter, the side lobes are only about 11 dB below the main lobe. Thus the function of the output filter is to reduce the side lobes to the necessary −80-dB level. Thus in Example 8.3, the side lobes are approximately 49 dB below the unmodulated peak value of the signal. The output filter must attenuate the side lobes an additional 31 dB.

A relatively painless way of checking all the parameters of the FCC specification is to use a transparent CRT overlay (mask) that graphs the requirements of equation (8.17). The mask is similar to Figure 8.14. If the signal (spectrum) falls within the mask, it is within specification.

8.5 RECEIVING SYSTEM DESIGN PARAMETERS

The design of a digital communication system involves trade-offs between the receiver noise figure, receiver sensitivity, third-order intercept level, and dynamic range. Receiver sensitivity is limited directly by noise; therefore, it is of interest first to take a brief look at the physics of noise. Thermal noise is due to the random movement of particles in the effective impedance at the input to the receiver at an equivalent temperature that depends upon the ohmic losses in the antenna and its ambient temperature and upon the noise received from external sources by the antenna. There is no way to avoid the first contribution other than by cooling the entire antenna system down to absolute zero. The noise will be present even if a shielded 50- or 75-Ω resistor representing the antenna impedance is plugged into the receiver input. In this case the equivalent temperature will be equal to the ambient temperature of the resistor. The thermal noise voltage generated by a resistance is given by

$$V = \sqrt{4kTBR} \tag{8.19}$$

where

V = RMS noise voltage
k = Boltzman's constant (1.38×10^{-23})
B = receiver noise bandwidth in hertz
T = temperature in degrees kelvin
R = resistance/impedance in ohms

For example, for a 50-Ω impedance, ambient temperature equal 300°K, and a bandwidth of 3 KHz, the RMS noise voltage is

$$V = \sqrt{(4)(1.38 \times 10^{-23})(300)(3000)(50)}$$

$$= 0.05\ \mu\text{V}$$

System sensitivity is defined as the signal strength that a properly modulated signal must have to provide the desired output in a receiver system. Specifically, it may be expressed as

$$\text{Sensitivity} = kTB + F_s + \text{C/N} + 10 \log_{10} \text{BW} \tag{8.20}$$

where

Sensitivity = system sensitivity (dBm)
kTB = thermal noise level (dBm)
F_s = system noise in decibels
$\frac{C}{N}$ = carrier-to-noise decibel ratio required to produce the desired system output
BW = bandwidth in hertz

The kTB term represents the rms power level of thermal noise present in the input to a receiver system. For a 1-Hz bandwidth, 290 K, 50-Ω system, we have

$$P = kTB = 4.00 \times 10^{-21}\ \text{W}$$

$$= -174\ \text{dBm}$$

This represents the best performance that could be obtained in a 1-Hz system at room temperature.

Noise figure, F_s, can be defined as the ratio of the S/N of a hypothetically perfect (noiseless) receiver to that of a real receiver, both measured at the output. Noise figure and temperature can be related by the following formula:

$$F_s\ (\text{dB}) = 10 \log_{10}\left(1 + \frac{T_e}{T_r}\right) \tag{8.21}$$

where

T_e = equivalent noise figure
T_r = reference temperature (290 K)

For example, for a 20-dB noise figure, the equivalent noise temperature of the receiver is

$$T_e = T_r\left[\log^{-1}\left(\frac{F_s}{10}\right) - 1\right]$$

$$= 290\left[\log^{-1}\left(\frac{20}{10}\right) - 1\right] = 28{,}710\ \text{K}$$

Referring to equation (8.20), the factor $10 \log_{10} BW$ represents the change in noise power due to change in bandwidth. Obviously, the wider the bandwidth, the greater the noise power and the higher the noise level. C/N represents the desired carrier-to-noise ratio. This ratio can represent a signal level that is barely detectable, or it can be considered to be the level that will allow the signal modulation to be reproduced with various ratings as to its fidelity. For example a 0-dB ratio C/N is sometimes used. This means that the output signal and noise power are equal. An example illustrates the calculation of receiver sensitivity.

EXAMPLE 8.4

Calculate the system sensitivity for a receiver with a 1-MHz bandwidth and a 20-dB noise figure. Assume an output SNR of 10 dB. What is the sensitivity for 0-dB SNR out?

Solution Using equation (8.20),

$$S = -174 + 20 + 10 + 10 \log_{10}(1{,}000{,}000)$$

$$= -174 + 20 + 10 + 60 = -84 \text{ dBm}$$

This assumes an input noise temperature of 290 K. It may be much smaller (80 K) for a satellite dish or much greater (thousands of degrees Kelvin at HF (High Frequency)). For 0-dB SNR out,

$$S = -174 + 20 + 0 + 10 \log_{10}(1{,}000{,}000)$$

$$= -94 \text{ dBm}$$

The −94-dBm figure is the level at which the signal power would be equal to the noise power in the 1-MHz bandwidth. Note from this example that the lower the acceptable SNR, the better the sensitivity must be.

The output SNR is used with all types of receivers. This ratio can be determined by measuring the output with and without a signal using a true RMS voltmeter. In a DSB-SC or SSB system, the required ratio C/N is equal to the specified output SNR. The specification becomes more complicated in FM or digital systems. In an FM system, some minimum carrier-to-noise ratio is required for proper operation of the discriminator. Improvement factors may make the output SNR much greater than the ratio C/N. The BER (10^{-y}) is used with digital data receivers. It is defined as the signal level required to produce a digital output in which not more than 1 bit in 10^{y} will be incorrect. In order to convert BER to required carrier-to-noise ratio, it is necessary to know the type of RF modulation used. Generally speaking, a graph of BER versus carrier-to-noise ratio is available.

8.5.1 Third-Order Intercept Point

An important receiver parameter is the third-order intercept point. (Refer to Figure 8.17.) Curve 1 represents the actual receiver gain. Note that it is linear for low input levels and becomes nonlinear at high input levels. Curve II represents the third-order intermodulation products generated by the receiver. The output level corresponding to the intersection of the linear extensions of curves I and II is defined to be the third-order intercept point. This is an unattainable level used only as a figure of merit for a particular amplifier or series of amplifiers, including those contained in a receiver.

Intermodulation product distortion (IMD) occurs when two large unwanted signals beat together (intermodulate) in a nonlinear receiver stage to produce a product at the unwanted frequency. Dynamic effects such as intermodulation and cross-modulation effectively limit receiver dynamic range. Second-order intermodulation products result in frequencies of ($f_1 \pm f_2$), where f_1 and f_2 are the two first-order unwanted frequencies. For example, if the desired frequency is 10 MHz, two unwanted frequencies of 11 and 21 MHz results in a beat frequency at 10 MHz. Note that the other frequency (32 MHz) is outside the passband of any octave filter that passes the signal frequency, f_s. Third-order intermodulation products result in frequencies of $f_1 \pm 2f_2$ or $f_2 \pm 2f_1$. For example, for $f_1 = \pm 0.4$ MHz and $f_2 = 10.8$ MHz, this results in two frequencies close to the desired signal frequency, f_s, of 10 MHz. Intermodulation performance is typically specified as the levels of two unwanted signals not less than 20 KHz off the main signal frequency to give a 0-dBμV (1-μV EMF) response. A good receiver is shown in Figure 8.18. Note that interception results because the third-order products increase at a rate greater than the desired signal as the input signal level is increased.

Dynamic range can be described loosely as the range of input signals over which dynamic interference effects produce outputs which are not significant. That is, the effects are below the noise level. Dynamic range is defined as being two-thirds of the difference in level between the noise floor and the intercept point. An example will illustrate the procedure for calculating dynamic range.

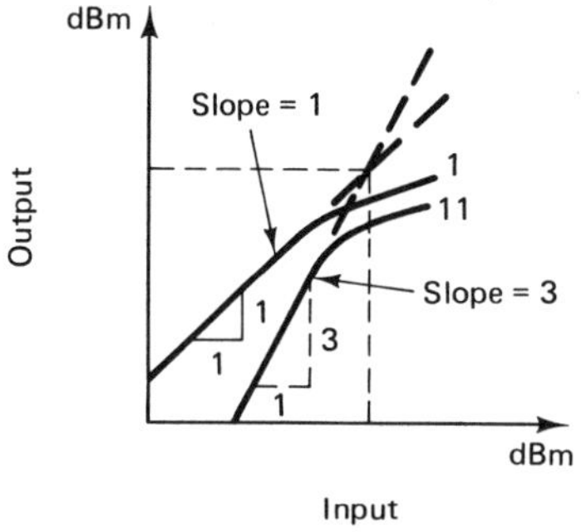

Figure 8.17 Third-order intercept point.

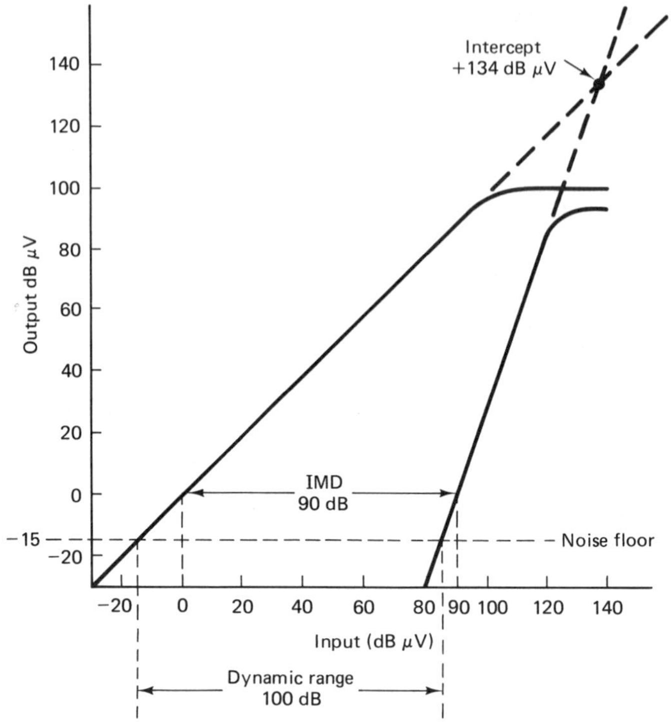

Figure 8.18 Good receiver third-order intermodulation performance is shown as 100 dB spurious-free dynamic range.

EXAMPLE 8.5

Calculate the dynamic range for a 20-dB noise figure receiver, with a 1-MHz bandwidth and a 0-dB output SNR. The third-order intercept occurs at an input signal level of +5 dBm. Assume, also, that any preamplifier has a third-order intercept point sufficiently high that the amplifier is not a limiting factor.

Solution

$$\text{Sensitivity} = -174 + 20 + 0 + 10 \log_{10} 1{,}000{,}000$$

$$= -94 \text{ dBm}$$

$$\text{Dynamic range} = \frac{2}{3}[5 \text{ dBm} - (-94 \text{ dBm})]$$

$$= 66 \text{ dB}$$

EXAMPLE 8.6

A receiver with a 20-dB noise figure is preceded by a preamplifier with a gain of 24 dB and a noise figure of 5 dB. Assuming the output third-order intercept of the amplifier occurs when the input to the receiver is +5 dBm, calculate the dynamic range of the system. The bandwidth is 1 MHz.

Solution

$$F_s = F_{pa} + \frac{F_{r-1}}{G_{pa}}$$

$$F_{pa} = \log^{-1}[\tfrac{5}{10}] = 3.16$$

$$G_1 = \log^{-1}[\tfrac{24}{10}] = 251$$

$$F_r = \log^{-1}[\tfrac{20}{10}] = 100$$

The system noise figure, F_s, is

$$F_s = 3.16 + \frac{100 - 1}{251} = 3.55$$

$$= 5.5 \text{ dB}$$

The system sensitivity, S, referenced to the preamp input, is

$$S = -174 + 5.5 + 0 + 60$$

$$= -108.5 \text{ dBm}$$

$$= \text{noise floor}$$

The system third-order intercept referred to the input of the preamp, is +5 dBm − 24 dB = −19 dBm. The dynamic range is then

$$\text{Dynamic range} = \tfrac{2}{3}[(\text{third-order intercept}) - (\text{noise floor})]$$

$$= \tfrac{2}{3}[-19 \text{ dBm} - (-109.5 \text{ dBm})]$$

$$= 59.7 \text{ dB}$$

Comparing Examples 8.5 and 8.6, we see that the sensitivity increased less than the gain of the preamplifier. That is:

> −108.5 dBm versus −94 dBm without the preamplifier. A net improvement of 14.5 dB is realized.

A consideration of the results of the these two examples reveals also that the greatest sensitivity is realized by using a preamplifier with the lowest noise figure and highest available gain. This serves to mask the higher noise figure of the

TABLE 8.7 SYSTEM TRADE-OFF CONSIDERATIONS

	Receiver only	Receiver + 10-dB gain preamplifier	Receiver + 24-dB gain preamplifier
F (dB)	20	11.2	5.5
Sensitivity (dBm)	−94	−102.8	−108.5
Third-order intercept point (dBm)	+5	−5	−19
Dynamic range (dBm)	66	65.2	59.7

receiver. A little thought should convince the reader that preamplifier gain also reduces the third-order intercept of the receiver in proportion to the preamplifier's gain. Therefore, in order to maintain the best dynamic range, it is best to use only the amplification needed to obtain the desired system noise figure. Excessive gain should not be employed. Some important system trade-offs are itemized in Table 8.7.

8.6 COMMUNICATION LINK ANALYSIS

At this point we understand the importance of received signal strength to proper receiver operation. That is, an adequate SNR must be present at the receiver to realize the expected BER. The general problem, then, lies in determining the required transmitter power to ensure the required received SNR. We must determine the system link-loss budget.

Consider the communication system shown in Figure 8.19. Transmitter power P_T is applied to the transmitting antenna, where it is directed to a receiving antenna located a distance L away. The propagating medium connecting the transmitter and the receiver is called the channel, in this case *free space*. The concept of free space assumes a channel region free of all objects that would affect RF propagation. This includes effects such as absorption, reflection, refraction, and so on. Inherent in this assumption is a perfectly uniform and nonabsorbing channel and that the earth is infinitely far away or its reflection coefficient is negligible. The RF propagation over a microwave link must contend with the factors we have just excluded and the free-space assumption is modified.

The performance of a communication system degrades primarily through the attenuation of signal power relative to undesired waveform power. We refer

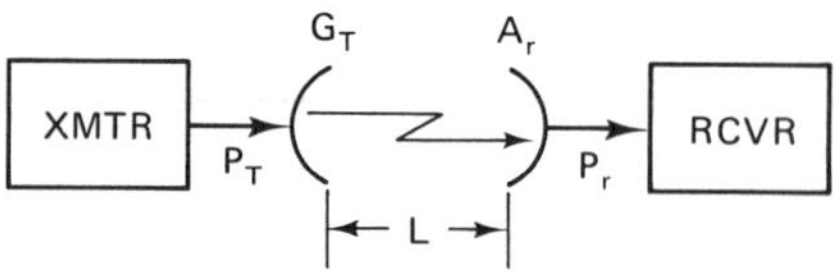

Figure 8.19 Typical free-space communication link.

to these mechanisms as *loss* and *noise*, respectively. Losses originate due to inverse square law RF propagation, reflection, and so on. Noise occurs when unwanted signal energy is injected into the link or when thermal noise is generated within the link. There are two main classes of noise (see Section 8.5):

- Thermal noise
- Intermodulation noise

Intermodulation noise is caused by system nonlinearities and its effects are usually grouped with other system losses as a loss parameter. Consequently, we limit our discussion to thermal noise. We assume the power spectral density of this noise to be flat, up through the gigahertz range. In communication receivers, it is generally accepted to be an AWGN process. Thermal noise generated by a resistance can be calculated using equation (8.19).

For the following development, assume an isotropic antenna transmitting power, P_t, uniformly over 4π steradians. The received power density, P, at a distance L from the antenna may be expressed as

$$P_d = \frac{P_t}{4\pi L^2} \frac{\text{W}}{m^2} \tag{8.22}$$

The power intercepted by the receiving antenna may be expressed as

$$P_r = P_d A_r = \frac{P_t A_r}{4\pi L^2} \text{ W} \tag{8.23}$$

where

A_r = effective aperture area of the receiving antenna

From equation (8.23), the effective aperture area is given by

$$A_r = \frac{P_r}{P_d} \tag{8.24}$$

where for a dish-type or horn-type antenna, the receiving antenna's effective aperture area and physical cross-sectional area are related by

$$A_r = \eta A_{\text{phy}} \tag{8.25}$$

where η is an efficiency parameter. Thus, the total power incident upon the dish is not absorbed. Some is lost through reradiation, scattering, and so on. Typical values for η are 0.55 for a dish and 0.75 for a horn antenna. The gain of an antenna relates to the maximum power density of the radiation pattern. This antenna gain may be defined as

$$G = \frac{\text{maximum power density of the radiation pattern}}{\text{power density of an isotropic antenna}} \tag{8.26}$$

We see that the gain, then, results from concentrating the RF energy into a region less than 4π steradians. The effective power radiated in the maximum-gain direction relative to that which would be radiated by an isotropic source is defined as the effective isotropic radiated power (EIRP),

$$\text{EIRP} = P_t G_t \tag{8.27}$$

where G_t is the maximum transmitting antenna gain. Using this expression, we can rewrite equation (8.23) as

$$\text{Pr} = \frac{\text{EIRP}\cdot A_r}{4\pi L^2}\ \text{W} \tag{8.28}$$

It can be shown that the relationship between antenna gain and antenna effective area is given by

$$A_r = \frac{\lambda^2}{4\pi} G_r \tag{8.29}$$

where λ is the wavelength of the radiation. Using this expression, the received power may be expressed as

$$P_r = \frac{\text{EIRP } G_r \lambda^2}{(4\pi L)^2}\ \text{W} \tag{8.30}$$

or

$$P_r = G_t P_T \left(\frac{\lambda}{4\pi L}\right)^2 \tag{8.31}$$

where $(\lambda/4\pi L)^2$ represents the free space loss. For frequency in megahertz and L in miles, this expression becomes (in decibels)

$$(\text{FSL})_{\text{dB}} = 36.6 + 20 \log L + 20 \log f \tag{8.32}$$

where FSL represents the actual path loss. Actually, the FSL could more accurately be termed a *spreading* loss. If we consider a constantly expanding sphere surrounding the transmitting antenna, the energy density on the surface of the sphere becomes less with increasing radius. Thus the receiving antenna intercepts less energy as its distance from the transmitter increases. The received power, P_r, may be computed as

$$P_r = P_t + G_t + G_r - \text{FSL} \tag{8.33}$$

An example illustrates the use of this equation in computing link margins.

EXAMPLE 8.7

A rocket is to be launched to the moon. Calculate the transmitter power needed to ensure a received power of 10 dBm. Assume the following system parameters:

$$G_t = 10 \text{ dB}$$
$$G_r = 45 \text{ dB}$$
$$f = 960 \text{ MHz}$$
$$\text{Range} = 250{,}000 \text{ mi}$$

Solution

$$FSL_s = 36.6 + 20 \log 250{,}000 + 20 \log 960$$
$$= 204.16 \text{ dB}$$
$$P_t = P_r - G_t - G_r + L_s$$
$$= 10 - 10 - 45 + 204.16$$
$$= 159.16 \text{ dBm}, \qquad \text{or } 8.25 \times 10^{-12} \text{ W}$$

In evaluating the performance of a system, the parameter of greatest interest is not the receiving power, P_r, but is instead the SNR in the receiver prior to the detector. We have observed repeatedly that the BER is a function of the SNR. In a digital communication system, the desired signal is a modulated carrier waveform. We are often interested, then, in the average ratio C/N or P_r/N as the SNR of interest. We can rewrite equation (8.30) as

$$\frac{P_r}{N} = \frac{\text{EIRP}\cdot G_r \lambda^2}{(4\pi L)^2 \cdot N} \tag{8.34}$$

This equation applies to any one-way satellite RF link. In analog systems the noise bandwidth is generally greater than the signal bandwidth. Consequently, the quantity P_r/N is the primary parameter for measuring signal detectability and quality of performance. By way of contrast, correlators or matched filters are used in digital receivers. Thus the signal bandwidth is taken to be equal to the noise bandwidth. The practice in digital communication systems is to replace the input noise power with noise density. Making this change, equation (8.34) becomes

$$\frac{P_r}{N_0} = \frac{\text{EIRP}\cdot G_r \lambda^2}{(4\pi L)^2 N_0} = \frac{\text{EIRP}\cdot \lambda^2 [G_r/(T_e + T_{\text{Nant}})] H_{L_s}}{(4\pi L)^2 k} \tag{8.35}$$

where $N_0 = kT_e$ has been used. The quantity k is Boltzman's constant and T_e is the effective system noise temperature. That is,

$$T_e = (F_o - 1)T_o \tag{8.36}$$

T_{Nant} is the effective noise temperature of the noise on the arriving signal referred to the antenna terminals.

In equation (8.35), a transfer function factor, H_{Ls}, has been introduced. This parameter, which is less than unity, allows for a large assortment of losses not considered in the other terms in the equation, such as intermodulation losses and signal fading due to the weather. This equation summarizes the key parameters of any link analysis:

- Received signal power-to-noise density, P_r/N_0
- Magnitude of transmitted power, EIRP
- Sensitivity of the receiver, G_r/T_e
- Path loss, $(4\pi L/\lambda)^2$
- Other losses, H_{Ls}
- Antenna effective noise temperature, T_{Nant}

In using this equation, if we assume that all the received power results from the modulating signal (suppressed carrier), then we can write

$$\frac{P_r}{N_0} = \frac{S}{N_0} = \left(\frac{E_b}{N_0}\right)R \tag{8.37}$$

where R is the bit rate. If some of the received power is contained in the carrier, equation (8.37) can still be employed. The carrier power is represented as a loss in the parameter, H_{L_s}.

We recognize the parameter E_b/N_0 in equation (8.37) as the bit energy-to-noise ratio associated with the BER. In this context, we have considered the E_b/N_0 ratio required to yield a specified BER. In order to facilitate calculating a margin or safety factor M, we need to differentiate between the required E_b/N_0 and the actual or receiver E_b/N_0. We can introduce a link margin parameter M into equation (8.37) as follows

$$\frac{P_r}{N_0} = \left(\frac{E_b}{N_0}\right)_r R = M\left(\frac{E_b}{N_0}\right)_{\text{reqd}} \cdot R \tag{8.38}$$

where

$$\left(\frac{E_b}{N_0}\right)_r = \text{received bit energy-to-noise ratio}$$

$$\left(\frac{E_b}{N_0}\right)_{\text{reqd}} = \text{required bit energy-to-noise ratio}$$

$$M = \text{link margin}$$

$$(M)_{\text{dB}} = (E_b/N_0)_{r,\text{dB}} - (E_b/N_0)_{\text{reqd,dB}}$$

As an example, consider a system specified to operate at a required $(E_b/N_0)_{\text{dB}} = 10$ dB with BER $= 10^{-4}$. We may, for example, require a link margin of 4 dB to achieve a BER $= 10^{-6}$. We can view this margin in two ways:

- The system has an excess of 4 dB more E_b/N_0 than is actually needed to meet the required BER $= 10^{-4}$ specification.
- The system is operating at an $(E_b/N_0)_{\text{dB}} = 14$ dB with an operating BER $= 10^{-6}$. The performance margin is 100 times greater than that required.

We noted earlier in this chapter (Section 8.1) that the parameter $(E_b/N_0)_{\text{reqd}}$ reflects differences in different communication systems, depending on choice of modulation and/or coding schemes.

From equations (8.35) and (8.38), we can express the link margin, M, as

$$M = \frac{\text{EIRP}\cdot\lambda^2 G_r/(T_e + T_{\text{Nant}})}{(4\pi L)^2 k\,(E_b/N_0)_{\text{reqd}}\cdot R} H_{LS} \tag{8.39}$$

Generally, a link budget analysis is calculated in decibels. Equation (8.39) may be expressed as

$$\begin{aligned} M(\text{dB}) = {} & \text{EIRP(dBW)} + G_r(\text{dB}) - \left(\frac{E_b}{N_0}\right)_{\text{reqd}}(\text{dB}) \\ & - R\left(\text{dB}\cdot\frac{b}{s}\right) - [k(T_e + T_{\text{Nant}})]\text{dBW/Hz} \\ & - L_S(\text{dB}) - \left(\frac{4\pi L}{\lambda}\right)^2(\text{dB}) \end{aligned} \tag{8.40}$$

where (L_S) dB $= -10\log_{10}(H_{LS})$. The values of the parameters in this expression constitute the link margin and trade-offs can be made in a system design. A note of caution is, however, in order. Binary PSK modulation outperforms binary FSK in terms of BER. However, if the system operates in a scintillating environment, we would avoid PSK in favor of FSK. This impacts $(E_b/N_0)_{\text{reqd}}$. As another example, the physical size of the antenna may impose a limit on gain. An example illustrates a typical system calculation.

EXAMPLE 8.8

The receiver in Example 8.4 receives a signal from a 2.11-GHz 1-W transmitter located 40 mi away. Both the transmitter and receiver have antennas with a 25-dB gain. Assume a worst-case fading of 6 dB and a required E_b/N_0 of 16 dB. $T_{\text{Nant}} = 0$.

(a) If an NRZ code is used, what is the maximum data rate for the receiver?

(b) Calculate $(E_b/N_0)_r$.

(c) Calculate the link margin.

Solution

(a) From Figure 3.3

$$\text{BW} \approx \frac{R}{2}$$

$$R = (2)(10^6) = 2 \text{ megabits/s}$$

(b)
$$\left(\frac{E_b}{N_0}\right)_r = \frac{1}{R}\frac{P_r}{N_0} = \frac{\text{EIRP}\cdot\lambda^2 G_r/T_e}{R(4\pi L)^2 k L_S}$$

$$\left(\frac{E_b}{N_0}\right)_r \text{dB} = \text{EIRP(dBW)} + G_r(\text{dB}) - R\left(\text{dB}\cdot\frac{b}{s}\right)$$
$$- kT_e(\text{dBW/Hz}) - L_S(\text{dB}) - \left(\frac{4\pi L}{\lambda}\right)^2(\text{dB})$$

$$F = 20 \text{ dB} \rightarrow 100$$

$$T_e = (F - 1)T_o = (100 - 1)(290) = 2610 \text{ K}$$

$$kT_e = (1.38 \times 10^{-23})(2610) = -194.4 \text{ dBW/Hz}$$

$$\text{EIRP} = G_t(\text{dB}) + 10 \log \frac{P_T}{1 \text{ W}}$$

$$= 25 + 10 \log \frac{1}{1} = 25 \text{ dBW}$$

$$FSL_P = 36.6 + 20 \log L + 20 \log f$$

$$= 36.6 + 20 \log 40 + 20 \log 2110 = 135.1 \text{ dB}$$

$$L_S = 6 \text{ dB}$$

$$\left(\frac{E_b}{N_0}\right)_r = 25 + 25 - 63 - (-194.4) - 6 - 135.1 = 40.3 \text{ dB}$$

(c) $M = (E_b/N_0)_r = (E_b/N_0)_{\text{reqd}} = 40.3 - 16.0 = 25.3 \text{ dB}$

In this example, note that we implicitly assume that a modulation format is chosen that results in no power in the carrier. If there is power in the carrier, this represents a loss as far as the modulation signal is concerned and this loss must be included in L_s. This results in a corresponding reduction in $(E_b/N_0)_r$. The actual amount of margin designed into a system varies. For example, at C-band, where the statistical parameters are well known, a 1-dB link margin for satellite links is sufficient. Some of the Intelsat systems have a 4- to 5-dB margin. Typically, a fading loss of 6 dB is used for satellite links below 10 GHz. For land links, the simplest scheme for assessing fading loss is to assume that the fading is random

TABLE 8.8 FADING LOSS FOR LAND LINKS VERSUS PROPAGATION RELIABILITY

Single-hop propagation reliability (%)	Fade margin (dB)
90	8
99	18
99.9	28
99.99	38
99.999	48

with time and may be described by a Rayleigh distribution. Table 8.8 may be used to determine the appropriate fading loss.

8.7 CHANNEL-SIMULATION TEST FOR DIGITAL RADIOS

A channel-simulation test for digital radios involves establishing a controlled environment in the laboratory that will realistically simulate real-word effects. This implies laboratory devices that will mimic such multiple channel impairments as multipath fading, Doppler shift, thermal noise, adjacent and cochannel interference, bandlimiting, and nonlinearities.

We mentioned earlier that digital radios contain fading compensation circuits. A test setup that includes a *fading simulator* can be used to check these circuits. The effects of fading on the received signal are illustrated in Figure 8.4, and a fading simulator is illustrated in Figure 8.20. The incoming IF input is split

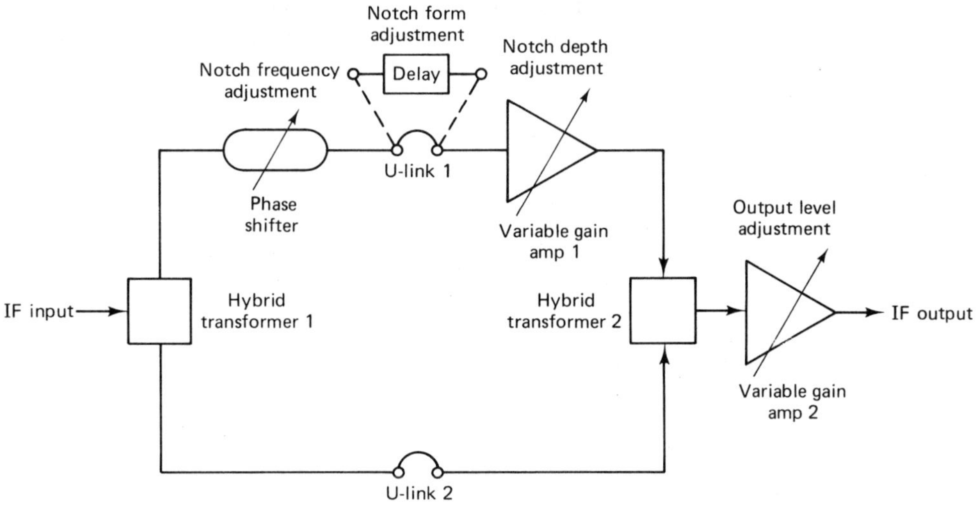

Figure 8.20 The facing simulator can be constructed from a bag full of common IF components.

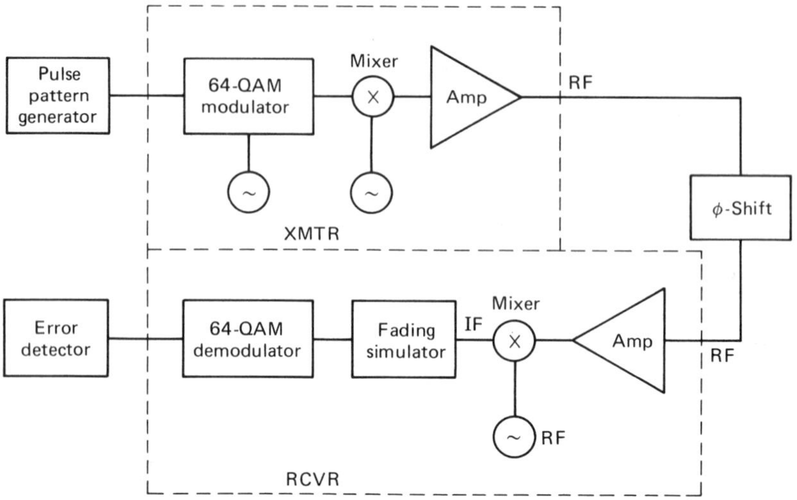

Figure 8.21 System test configuration.

into two signals: (1) a direct wave, and (2) a reflected interference wave. The interference wave is shifted 180° out of phase with respect to the direct wave and is fed to a 360° variable phase shifter, which imposes a notch in the signal's frequency characteristics. This phase shifter is used to set the notch frequencies that are required to simulate the fading characteristics (see Figure 8.4). The form factor of the notch can be adjusted using the variable delay line, and the variable gain amplifier permits notch depth adjustment (~20 dB). Thus adjustment of the notch frequency, form factor, and depth permits a realistic and accurate evaluation of the fading compensation circuit. The second hybrid transformer essentially sums the two signals. A complete system test setup is illustrated in Figure 8.21.

The effects of fading on the system can be assessed by observing the BER of the system under test and the receiver AGC. The AGC circuit compensates for level variations due to fading by varying in proportion to the strength of the received signal. In general, fading is measured with respect to some threshold voltage (see Figure 8.22). The fading time rate is the percentage of time the AGC

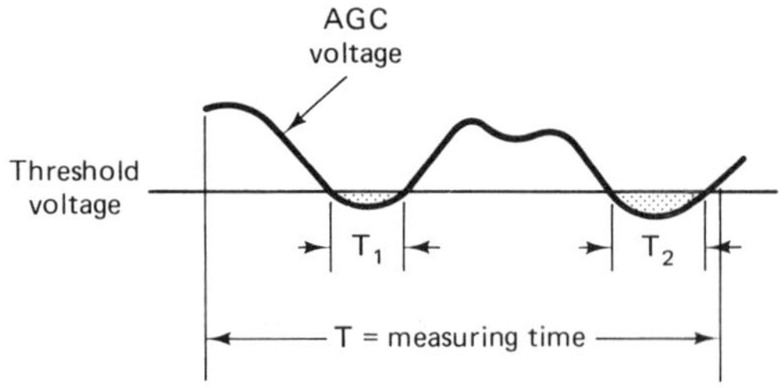

$$\text{Time rate} = \left[\frac{T_1 + T_2}{T}\right](100)\ \%$$

Figure 8.22 Fading time rate.

falls below the threshold. Correspondingly, during this period of time, the BER is unacceptable.

As noted earlier, the fading impairments to a radio link also arise from other factors such as cochannel and adjacent-channel interference. Nonlinearities introduced by power amplifiers can severely affect the performance of a multicarrier link. Also, if the receiver bandwidth is too small, signal energy is lost, further reducing link performance. Adjacent-channel interference is readily checked using a spectrum analyzer.

Coherent detection in fading channels represents a special challenge. A carrier-tracking loop can maintain lock only if the loop noise bandwidth is much larger than the Doppler frequency. However, an increase in noise bandwidth reduces noise immunity. A trade-off exists, then, between the error floor and power performance.

8.7.1 Automatic Adaptive Equalization

Figure 8.21 indicates a 64-QAM digital transmitter and receiver. The recent development in systems such as this rely heavily on the techniques discussed in this chapter. This includes the use of raised cosine filtering and the sharing of both transmitter and receiver filtering to keep adjacent-channel interference within a closed limit. The key to making microwave receivers feasible with this number of signal states lies in the development of an adaptive transversal equalizer by AT&T Bell Laboratories in 1982. This device is the key to controlling multipath fading.

A baseband receiver equipped with an adaptive transversal equalizer is shown in Figure 8.23. In this particular receiver, the QAM IF signal is split into I and Q rails, each 45 megabits/s. After filtering with raised cosine shaping, the four level signals enter the transversal equalizer for removal of the linear ISI previously generated by multipath propagation, imperfect Nyquist filtering, and

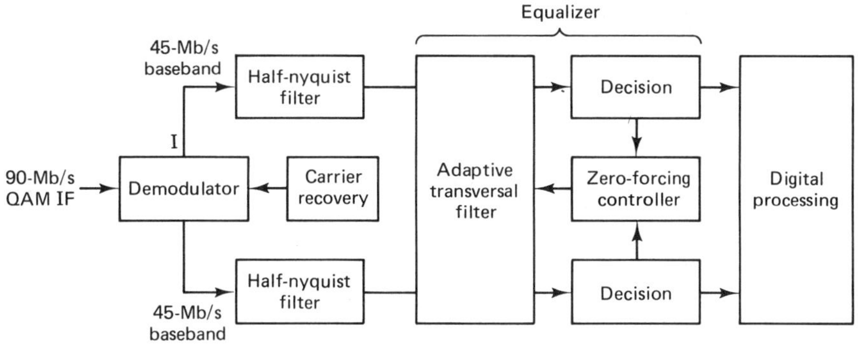

Figure 8.23 Baseband receiver with adaptive transversal equalizer.

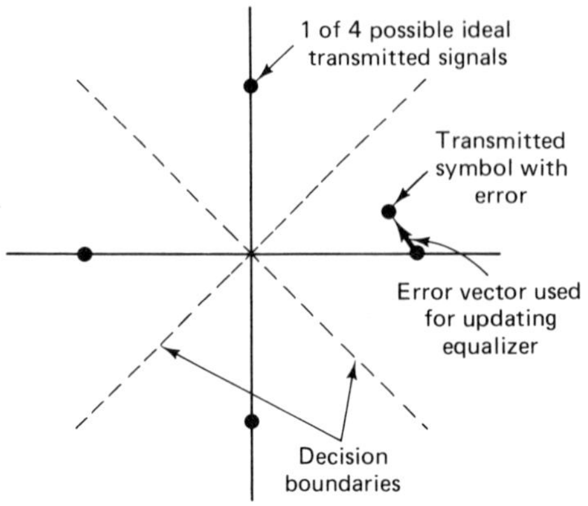

Figure 8.24 Automatic adaptive equalizer operation.
Reprinted with permission from the September 1982 issue of COMPUTER DESIGN, copyright 1982, PennWell Publishing Company, Advanced Technology Group.

so on. The equalized signals are then applied to the decision device. Note that an adaptive transversal equalizer is dedicated to each rail.

Refer to Figure 8.23. Note that a feedback path is provided from the decision device to the adaptive equalizer. The principal of operation for an adaptive equalizer is illustrated in Figure 8.24 for a four-phase (state) signal. The decision device first samples the signal element. Imaginary decision boundaries that bisect the quadrants are set up and then used to determine in which quadrant the signal belongs. If the signal falls in region 1, for example, an error vector is computed by the decision device to be used by the equalizer in minimizing future errors.

The most often used form of an automatic adaptive equalizer is a digital transversal filter (Figure 8.25). The input to the equalizer is the Nyquist filter output. This output is represented by a complex number, $\hat{Y}_i$. Sampled values are stored in the digital delay line, which has the outpu'/ tap shown. The output of the equalizer is the sum of the products of the tap output and a complex equalizer

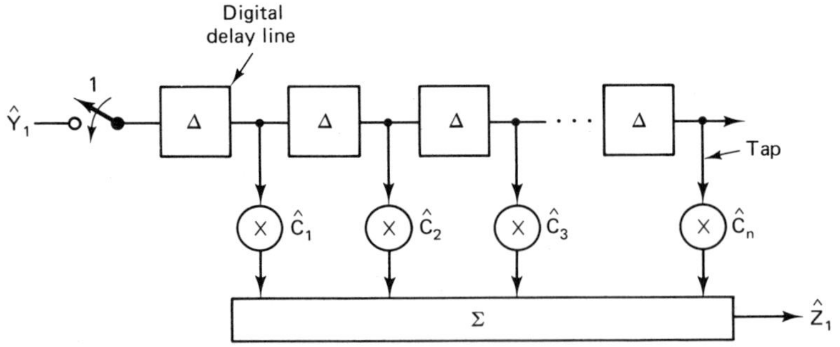

Figure 8.25 Digital transversal filter to accomplish automatic adaptive equalization. The filter derives from values stored in a digital delay line plus complex equalizer coefficients.
Reprinted with permission from the September 1982 issue of COMPUTER DESIGN, copyright 1982, PennWell Publishing Company, Advanced Technology Group.

coefficient. For an n-tap equalizer, the ith output of the equalizer, $\hat{Z}_i$, is

$$\hat{Z}_i = \sum_{j=1}^{n} \hat{C}_j \hat{Y}_{i-j} \tag{8.41}$$

where $\hat{C}_j$ is the sampled value of the impulse response of the adaptive equalizer filter. $\hat{Z}_i$ is a complex number since both $\hat{C}_j$ and $\hat{Y}_{i-j}$ are complex. A simple derivation for the coefficient update is as follows. First assume that all the equalizer coefficients are correct except $\hat{C}_m$, which has error $\Delta\hat{C}_E$ and no other perturbation. The error vector, E_i, may be expressed as

$$\hat{E}_i = \hat{Y}^*_{i-m} \Delta \hat{C}_E \tag{8.42}$$

where $\hat{E}_i$ is the error vector at time i. Thus the error in the equalizer coefficient is

$$\begin{aligned} \Delta \hat{C}_E &= \frac{\hat{E}_i}{\hat{Y}^*_{i-m}} \\ &= \frac{\hat{E}_i \hat{Y}_{i-m}}{|\hat{Y}_i|^2} \end{aligned} \tag{8.43}$$

Equation (8.43) implies that as $\hat{Y}_{i-m}$ becomes smaller, the update amount will be larger. This equation is correct if there is no other error source. Since other sources of error are usually present, however, the probability that an error will be caused by a shift in $\hat{C}_m$ increases as $\hat{Y}_{i-m}$ increases. This yields a correction factor proportional to $\hat{Y}_{i-m}$ as well as $\hat{E}_i$. We can express this as

$$\Delta \hat{C}_m \propto \hat{E}_i \hat{Y}_{i-m} \tag{8.44}$$

For purposes of minimizing self-noise and ensuring stability, the update is also scaled by a factor k of approximately 0.01. Equation (8.44) becomes

$$\Delta \hat{C}_m = k \hat{E}_i \hat{Y}_{i-m} \tag{8.45}$$

At each baud time i, all the equalizer coefficients are updated based on the new error vectors represented by equation (8.45). The new equalizer coefficients are equal to

$$\hat{C}_j = \hat{C}_j - k \hat{E}_i \hat{Y}_{i-j}; \qquad j = 0, \ldots, n \tag{8.46}$$

The factor k is normally based on such factors as ease of generation (i.e., 2^{-n}), self-noise, and filter-tracking ability. With n coefficients, we are in effect solving n simultaneous equations with n unknowns. In operation, the coefficients will slowly migrate to those values that yield the least mean square of the error vectors. Further background information is contained in the *Bell System Journal* (April 1965, pp. 547–588).

PROBLEMS

8.1. **(a)** For the given receiver, what is the output of the LPF? $a(t)$ is the information pulse with maximum frequency, $f_M \approx T_b^{-1}$ and $s(t)$ is the modulated signal described by

$$s(t) = a(t)\exp(j\omega_c t) + a(t)\exp(-j\omega_c t)$$

(b) What form of modulation does $s(t)$ represent?

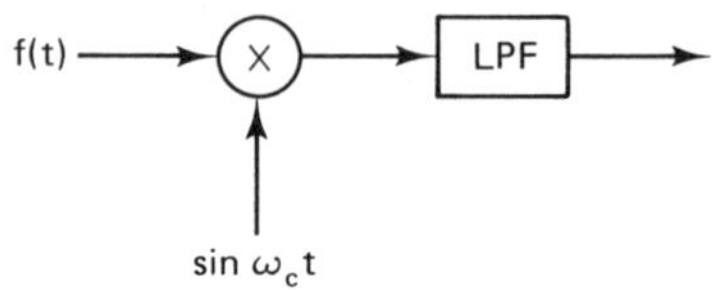

8.2. Are there only odd-order intermodulation products due to two tones in a nonlinear device? Explain.

8.3. Suppose path loss had been defined as $4\pi L^2$ instead of $(4\pi L/\lambda)^2$. What effect would this have on the computation of received power, P_r (equation (8.30))? Discuss the advantage of computing link margins using the real definition.

8.4. **(a)** List the basic design goals for the front end of a receiver.

(b) The front-end noise figure is improved by X decibels. Does this result in an output SNR improvement of X decibels? Explain.

8.5. A data rate of 60 megabits/s is to be transmitted through a transponder that has a 20-MHz bandwidth.

(a) Which modulation technique would you choose?

(b) What E_b/N_0 ratio is required for BER $= 10^{-6}$?

(c) For the E_b/N_0 ratio of part (b), what is the carrier-to-noise ratio?

8.6. Explain the difference in phase transitions in offset QPSK, conventional QPSK, and MSK. What is the reason for no abrupt phase transitions in MSK?

8.7. Given that the Nyquist frequency is 1 MHz, calculate the frequency at which the attenuation of the raised cosine filter is 30 dB. Assume (a) $\alpha = 0.3$; (b) $\alpha = 0.5$.

8.8. The cutoff frequency of a brick wall channel is 1.5 MHz. At what data rate in bits per second, can we transmit data in each case?

(a) A binary code is used.

(b) A four-level code is used.

8.9. NRZ data is to be transmitted through a raised cosine filter. The data rate is 32 kilobits/s. We wish a minimum of 30 dB attenuation at a frequency of 1.5 times the Nyquist bandwidth. Calculate the value of α.

8.10. The impulse response of raised cosine filter having $\alpha = 0$, $\alpha = 0.5$, and $\alpha = 1.0$ is shown in Figure 8.8.

(a) Explain why a filter with a steeper roll-off has a larger peak-to-peak jitter.

(b) Why is the ISI not affected by the roll-off factor?

8.11. The signal-state space for 16 QAM is shown in Figure 3.34.
(a) Sketch a modulator that will produce the QAM signal.
(b) How many threshold comparators are required to detect the I and Q channels in the demodulator?
(c) Assuming a data rate of 10 megabits/s, what is the theoretical minimum bandwidth?
(d) If premodulation low-pass filters ($\alpha = 0$) are used in the I and Q arms of the modulator, what is the minimum (Nyquist) bandwidth?

8.12. A QPSK system is illustrated in Figure 8.6.
(a) Explain why a four-phase ambiguity is introduced by conventional carrier recovery circuits in the QPSK demodulator. Comment on the possibility of using a Costas loop as the carrier-recovery loop.
(b) What coding scheme can be utilized to resolve the ambiguity?

8.13. Prove that QAM modulation is as efficient in required power and bandwith as SSB modulation.

8.14. Show that the given circuit can be used as an FSK demodulator.

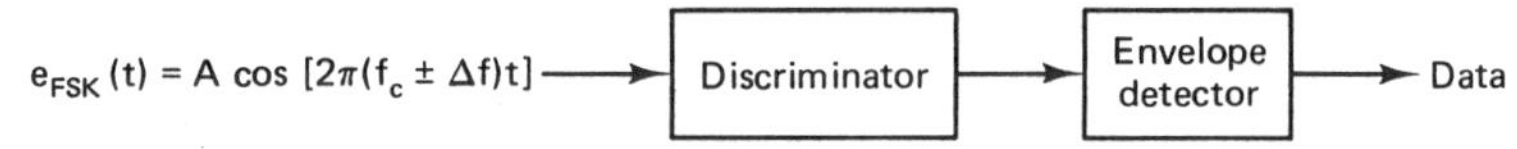

8.15. **(a)** Plot the receiver and transmitter filter transfer functions given by equations (8.13) and (8.15), respectively.
(b) Combine the plots in part (a) to get the total system response.

8.16. For a 11-GHz communication system, calculate the required attenuation for each value.
(a) 10.970 GHz
(b) 11.040 GHz

8.17. Verify the first two columns in Table 8.7.

8.18. A receiver has a sensitivity of 0.5 μV. Convert this to:
(a) dBμV
(b) dBm (50 Ω)

8.19. A receiver has a noise bandwidth of 3 KHZ and a noise figure of 10 dB. The manufacturer claims a sensitivity of 0.15 μV for an SNR of 10 dB. Comment on the accuracy of this specification.

8.20. The bandwidth of a receiver is reduced from 3 KHz to 300 Hz. By what factor in decibels does the noise voltage change?

8.21. A 3-KHz bandwidth receiver has a dynamic range of 100 dB with an intercept point of 140 dBμV (+27 dBm). What is the noise-floor level?

8.22. Repeat Example 8.8 for an RZ code and a modulation format with 6-dB carrier suppression.

8.23. An X4 circuit is used in the carrier-recovery circuit for QPSK demodulation. Explain why four-phase ambiguity is present in the recovered carrier signal.

8.24. A fourth-order Butterworth filter is shown in the accompanying figure.

(a) Find the transfer function and determine the -3-dB cutoff frequency.

(b) Plot the magnitude of the transfer function (Bode plot) and compare this with a raised cosine filter ($\alpha = 0.5$). How closely does this filter approximate the theoretical curve?

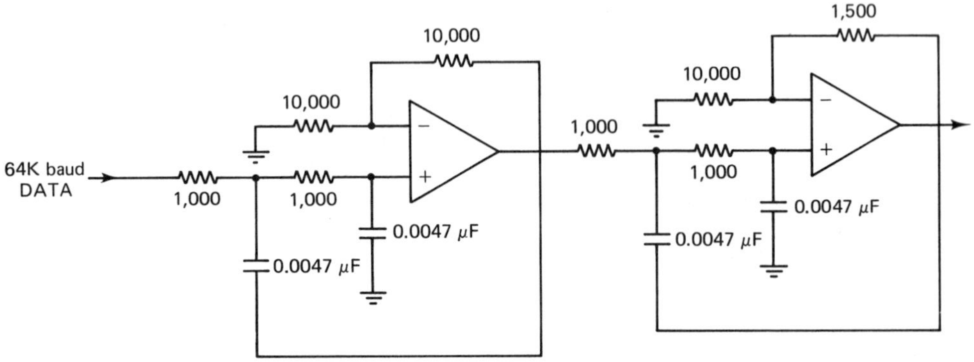

Prob. 8.24

8.25. If we wished to approximate a brick wall filter, which of the following filters would you choose: Butterworth, Chebyshev, or Bessel? Give reasons for your choice.

9

SPREAD-SPECTRUM COMMUNICATION SYSTEMS

9.0 INTRODUCTION

The principles of spread-spectrum communications were introduced in Chapter 5 in conjunction with techniques for PN code generation. We noted that spread-spectrum communication spreads the transmitted (rf) signal over a bandwidth that is as large as is practical. This can result in an rf bandwidth that is 10 to 100 times the information bandwidth. Frequency-hopping and direct-sequence modulation are the two principal spread spectrum techniques. A third technique, time hopping, often combines with frequency hopping. An important trend in secure communications is to mix these techniques in order to gain the advantages of each. A second important trend involves faster frequency hopping and code sequencing.

The technique for spread-spectrum modulation is illustrated in Figure 9.1. Digital data is added modulo-2 to a PN bit stream. The combined data is then used to modulate an RF carrier in either an M-ary FSK or PSK fashion. Three primary modulation techniques are used for spread spectrum systems: PSK, FSK, and amplitude (pulse), or OOK. At this point we are well acquainted with these digital modulation techniques.

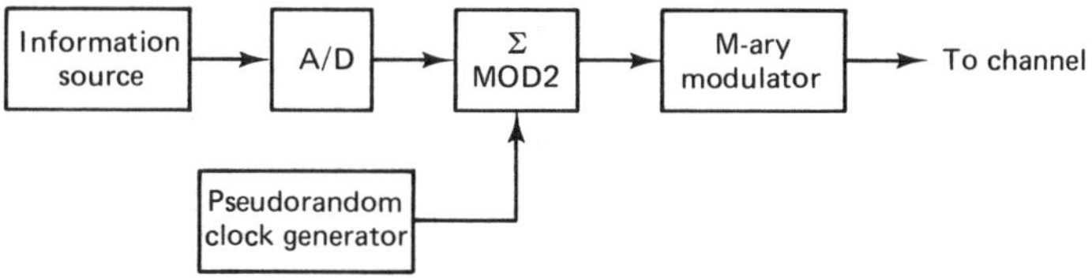

Figure 9.1 Spread-spectrum modulation system.

The type of spread-spectrum system is determined by the form of modulation and the modulation rate. In the direct-sequence type, the carrier phase is usually shifted in phase between 0° and 180° (biphase PSK). This shift is in response to a single bit of the modulo-2 sum. Quadraphase and higher-order direct-sequence signals can also be generated. In frequency-hopping systems, the data and the PN code are assembled into an N-bit PN word, which represents an instantaneous value of a variable between 0 and 2^N. This instantaneous value designates one of the 2^N frequency cells within the carrier frequency range. Time hopping implemented by OOK can be used with PSK or FSK to achieve hybrid spread-spectrum modulation.

9.1 PROCESSING GAIN AND JAMMING MARGIN

The difference between the output SNR and input SNR of any processor is defined to be its processing gain. For example, $SNR_o = 15$ dB and $SNR_i = 8$ dB results in a processing gain of 7 dB. Spread-spectrum techniques develop their processing gain through the process of spreading and despreading. The processing gain of both direct-sequence or frequency-hop type of spread spectrum is the ratio of the spread or transmitter bandwidth, W_t, to the baseband signal bandwidth. That is,

$$\begin{aligned} G_p = \text{processing gain (dB)} &= 10 \log_{10} \frac{(S/N)_0}{(S/N)_i} \\ &= 10 \log_{10} \frac{W_t}{W} \end{aligned} \tag{9.1}$$

If there is a jamming (interference) signal at the receiver, the despread operation affects it just as the PN spreading signal affected the original information signal at the transmitter. Thus even a narrow-band jamming signal in the center of the information band will be spread to the bandwidth of the spreading signal. Assume that the spreading signal has bandwidth W_t. If the jamming signal has power J watts, its average density can be treated as wideband noise $J_o = J/W_t$ watts per hertz. A received signal with S watts at data rate R bits per second has an energy per bit $E_b = S/R$ watts. The parameter E_b/J_o that determines BER performance in the presence of wideband jamming noise can be expressed as

$$\frac{E_b}{J_o} = \left(\frac{S}{J}\right)\left(\frac{W_t}{R}\right) \tag{9.2}$$

Neglecting thermal noise, the ratio of jamming power to signal power is

$$\begin{aligned} \frac{J}{S} &= \left(\frac{W_t}{R}\right)\left(\frac{E_b}{J_o}\right) \\ &= G_p\left(\frac{E_b}{J_o}\right) \end{aligned} \tag{9.3}$$

This equation illustrates that E_b/J_o is the minimum ratio of bit energy to jamming noise density needed to support a given BER; also, if W_t/R is the ratio of spread bandwidth to the original data rate, then J/S is the maximum ratio of jamming power to signal power that can be tolerated.

We can express the jamming margin, M_j, as

$$(M_j)_{\text{dB}} = (G_p)_{\text{dB}} - \left[(L_{\text{sys}})_{\text{dB}} + \left(\frac{S}{N}\right)_{0,\text{dB}}\right] \tag{9.4}$$

where L_{sys} is the system-implemented losses. For example, a system with a processing gain of 25 dB, system losses of 1.5 dB, and minimum $(S/N)_o = 10$ dB would have a jamming margin of 13.5 dB. That is, it is not reasonable to expect the system to operate with interference greater than 13.5 dB above the desired signal.

9.2 SPREAD-SPECTRUM SIGNAL RECEPTION

Spread-spectrum systems illustrating direct sequence and frequency hopping are shown in Figure 9.2. The transmitter transfers baseband information to the spread bandwidth domain at RF frequencies. The receiver must detect, synchronize, and demodulate these signals after they have propagated through a corruptive medium. At the receiver, the signal is similar to and far below the power level of the background noise. In fact, signal-to-noise levels on the order of −15 to −30 dB are common.

In order for the receiver to detect and demodulate the received signal, certain information must be known to the receiver. This information includes the following:

- Existence of the signal buried in the noise
- Carrier frequency of the signal
- Modulation type
- PN code rate
- The code phase (propagation time delay)

The designated receiver has access to all these except the last. That is, it does not have access to the starting point of the code. The receiver must have this information in order to collapse the expanded signal coherently. The receiver must be able to derive this information from the signal. Obviously, then, this synchronization function becomes one of the most important functions performed in a spread-spectrum receiver. In general, this process becomes one of searching through the time and frequency uncertainty regions to achieve and maintain synchronization. Code sync is the key to despreading the desired signal and spreading the undesired signals.

Refer to Figure 9.2. Each system possesses certain advantages and disadvantages. One advantage of frequency hopping is that the receiver need not be phase coherent with the transmitter over the entire spread-spectrum bandwidth. This translates into faster receiver acquisition, since the receiver is not required to spend time determining the correct phase of the transmitted signal. The noncoherent approach, however, is more prone to noise than a coherent scheme such as direct-sequence modulation. In terms of noise, direct-sequence systems enjoy a major advantage because of its phase-coherent nature. Unfortunately, phase coherence requires that the receiver be rather complex, since it must acquire and track a signal within a single chip time.

A block diagram for a serial search spread-spectrum receiver is shown in Figure 9.3. The local code clock and carrier-frequency generator are adjusted (local code races forward or backward in time relative to incoming code) until the correlator output is a maximum (see Figure 5.8). After acquisition, the carrier-tracking loop tracks the carrier in frequency and phase while the code loop remains

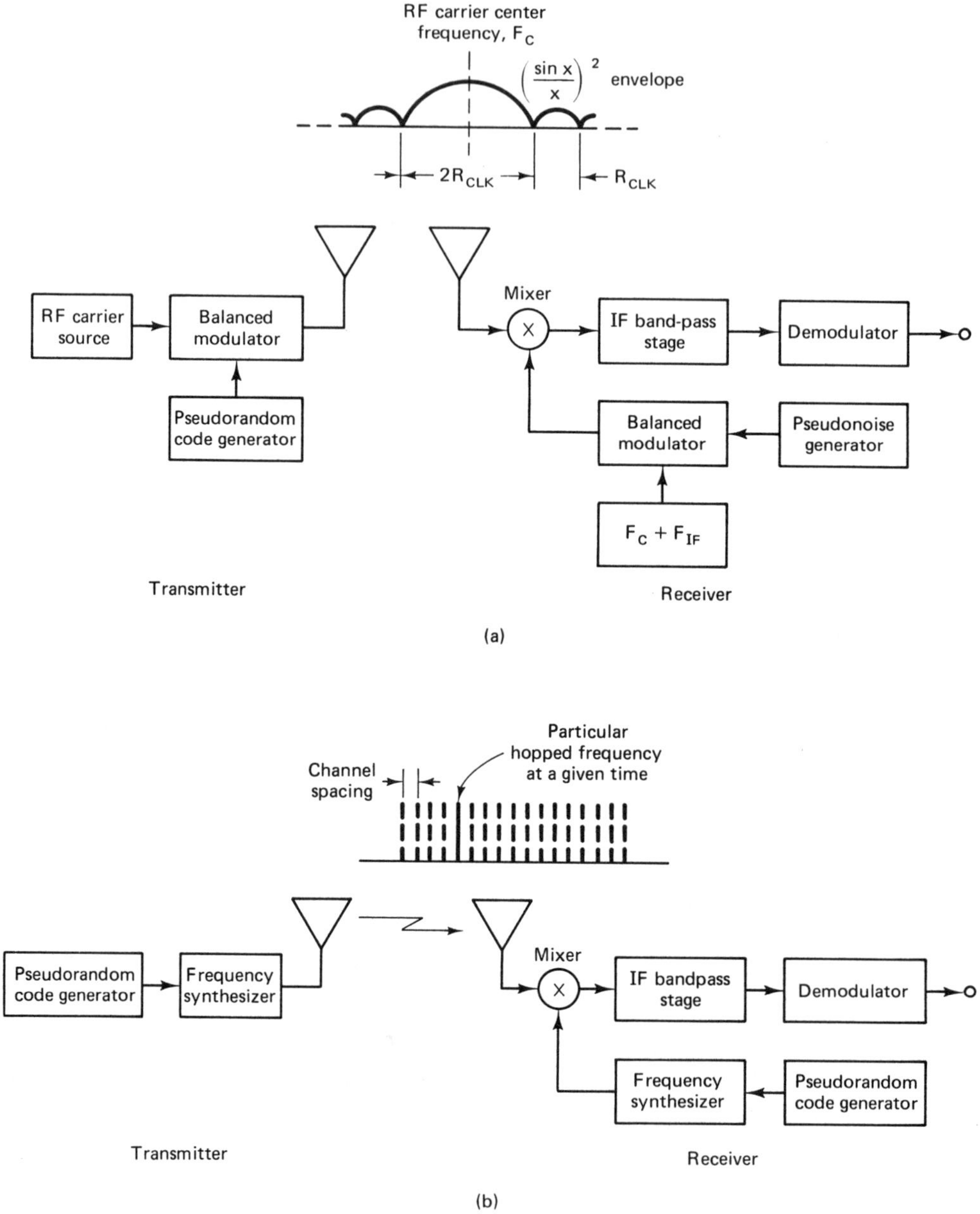

Figure 9.2 (a) Direct sequence. (b) Frequency hopping.

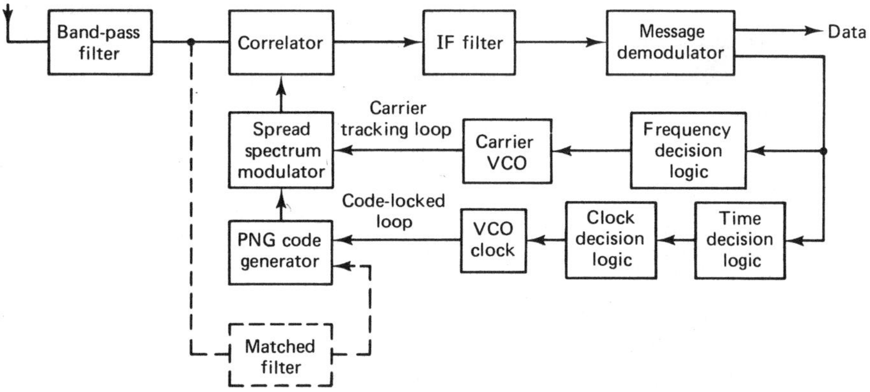

Figure 9.3 Spread-spectrum receiver.

in step (locked) to the transmitted PRN code. The input band-pass filter must be wide enough to allow the entire spread spectrum to pass. The IF filter must be wide enough to allow all expected Doppler shifts.

In receivers of this type, tracking is simplified if in the transmitter, the code sequence clock rate is designed to be a submultiple of the carrier frequency. That is, the carrier and code clock are derived from the same source and are coherently related. Thus both the code and carrier are affected the same by propagation delay and Doppler shift. Thus once synchronization is achieved, tracking the carrier automatically keeps the codes synchronized.

9.2.1 Synchronization

We noted earlier that code synchronization is one of the most important functions the spread-spectrum receiver must perform. With respect to the receiver of Figure 9.3, both code-phase and carrier-frequency uncertainties needed to be resolved before the receiver could operate. If accurate frequency sources are contained in both the transmitter and receiver much of the uncertainty is removed. Doppler frequency shift, however, cannot be necessarily predicted, since it is a function of velocity. Considering a transmitted frequency f_t, the received frequency is

$$f_r = f_t \pm \frac{f_t V}{C} \tag{9.5}$$

where

V = relative velocity between transmitter and receiver
C = speed of light, 3×10^8 m/s

The plus sign is applicable for increasing frequency when the transmitter and receiver are approaching each other and the minus sign applies to the transmitter and receiver moving away from each other. This is the nonrelativistic equation. For very large velocities, the relativistic equation

$$f_r = f_t \sqrt{\frac{C - V}{C + V}}$$

must be used.

EXAMPLE 9.1

A direct-sequence system with a 5-megabit/s code clock is traveling at a relative velocity of 1000 mi/h away from the receiver. The carrier frequency is 400 MHz. What is the received code clock rate and carrier frequency?

Solution

$$f_{r(\text{code})} = f_t - f_t \frac{V}{C} = 5 \times 10^6 - 5 \times 10^6 \frac{1000}{186{,}000}$$

$$= 4.973657 \text{ Hz}$$

$$f_{r(\text{carrier})} = f_t - f_t \frac{V}{C} = 4 \times 10^8 \frac{1000}{186{,}000}$$

$$= 397.849463 \text{ MHz}$$

As might be expected, initial synchronization is the most difficult. The simplest correlation technique is a sliding correlator. For example, if the code-sequence generator is operated at a rate different from the transmitter's code generator, the two code sequences effectively slip in phase with respect to each other. Figuratively, the two codes slide past each other and stop when the point of coincidence is reached. A version of a sliding correlator is shown in Figure 9.4. Initially, the output phase k of the local PN generator is set to $k = 0$ and a partial correlation is performed by examining λ chips. If the integration (low-pass filter) output is below the threshold $V_T(\lambda)$, k is set to $k = 1$ and the procedure is repeated. When V_I exceeds $V_T(\lambda)$, this indicates that acquisition has taken place.

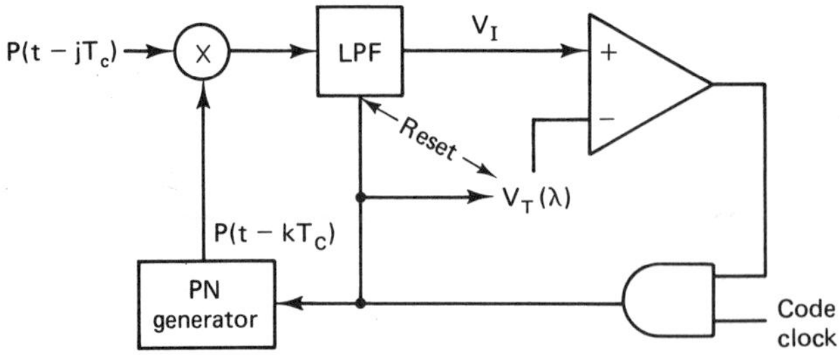

Figure 9.4 Sliding correlator.

In the worst case, it may be necessary to set $k = 0, 1, 2, \ldots, 2N_c - 1$ before finding the correct value of k. Assuming during each correlation that λ chips are examined, the worst-case correlation time is given by

$$T_{\text{ACQ,MAX}} = 2\lambda N_c T_c \tag{9.6}$$

where

T_c = chip time
N_c = number of bits in the PN code

EXAMPLE 9.2

For the sliding correlator of Figure 9.4, determine the maximum time to acquire for the following parameters:

$$f_{\text{clock}} = 1 \text{ megabits/s}$$

$$N_c = 1024 \text{ bits}$$

$$\lambda = 100 \text{ bits}$$

Solution

$$T_{\text{ACQ,MAX}} = 2\lambda N_c T_c$$

$$= (2)(100)(1024)(10^{-6}) = 204.8 \text{ ms}$$

The computed synchronization time of 204.8 ms would be intolerable for a dynamic voice communication channel. The sync time for a channel of this type should be less than 50 ms. Hybrid solutions are available for solving problems of this type. Refer to Figure 9.3. A matched filter is shown by the dashed lines. The initial part of the data transmission would involve sending a spectral waveform (pattern) that could be easily detected by this filter. This filter utilizes a principal that produces a maximum-output SNR when excited by a special waveform. When the receiver sees the matched filter precede its threshold, it sets the local code generator to a known phase. A correlator such as the sliding type then takes over and performs fine synchronization and tracking. Using this technique, total acquisition time can be reduced to acceptable values.

Matched filter synchronizers generally make use of delay elements and thus are quite different from integrate-and-dump filters such as the low-pass filter in Figure 9.4. In fact, a delay line can be utilized to form a matched filter (Figure 9.5). Suppose that a 7-bit preamble (1100011) is transmitted as biphase modulation. The signal propagates into the delay line, where each delay corresponds to one chip time, T_c. The signal sets $T_1T_2T_6T_7$ and $T_3T_4T_5$ are summed separately. After inverting $T_3T_4T_5$ the two signal sets are summed to form the filter output. Since the signal energy in all seven delay segments now add in phase, the output is

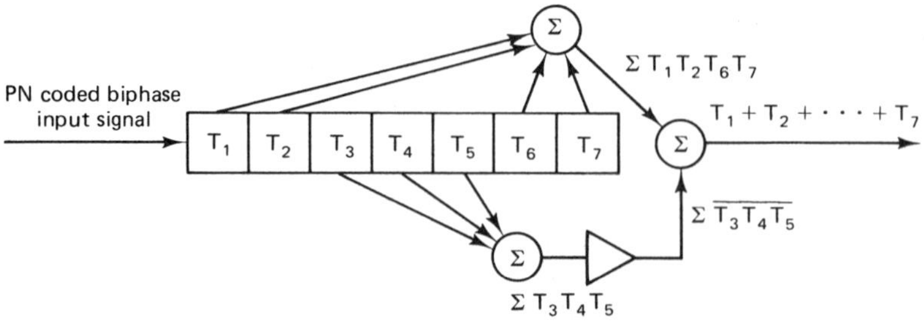

Figure 9.5 Delay line matched filter.

seven times greater than a single input. Greater signal enhancement is available simply by adding additional delay elements. The filter gain may be expressed by

$$G_f(\text{db}) = 10 \log_{10} N \tag{9.7}$$

where N = number of delay elements. Obviously, maximum gain is realized if the delay exactly matches the chip time, T_c; otherwise only partial correlation is achieved.

Refer to Figure 9.5. We see that the matched filter compresses seven inputs into a single output. Since this occurs in the time domain, we have pulse compression. A simplified model of the sequence of events is illustrated in Figure 9.6. In this example, the RF bandwidth at the receiver input must be $2/\tau$, or $10/T$.

A large number of matched-filter configurations are based on the transversal filter (see Figure 8.25). A matched-filter version is shown in Figure 9.7. This particular filter uses a form of detection in which the signal is correlated with a

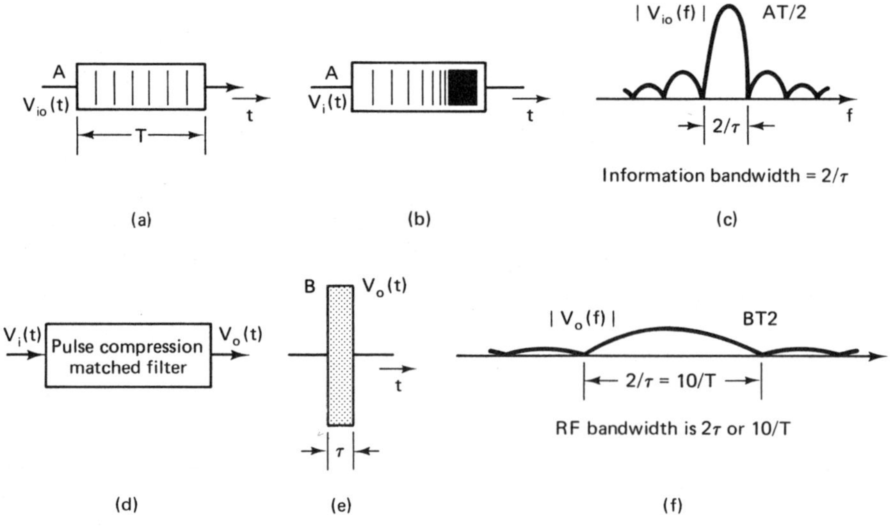

Figure 9.6 Simplified model of pulse compression.

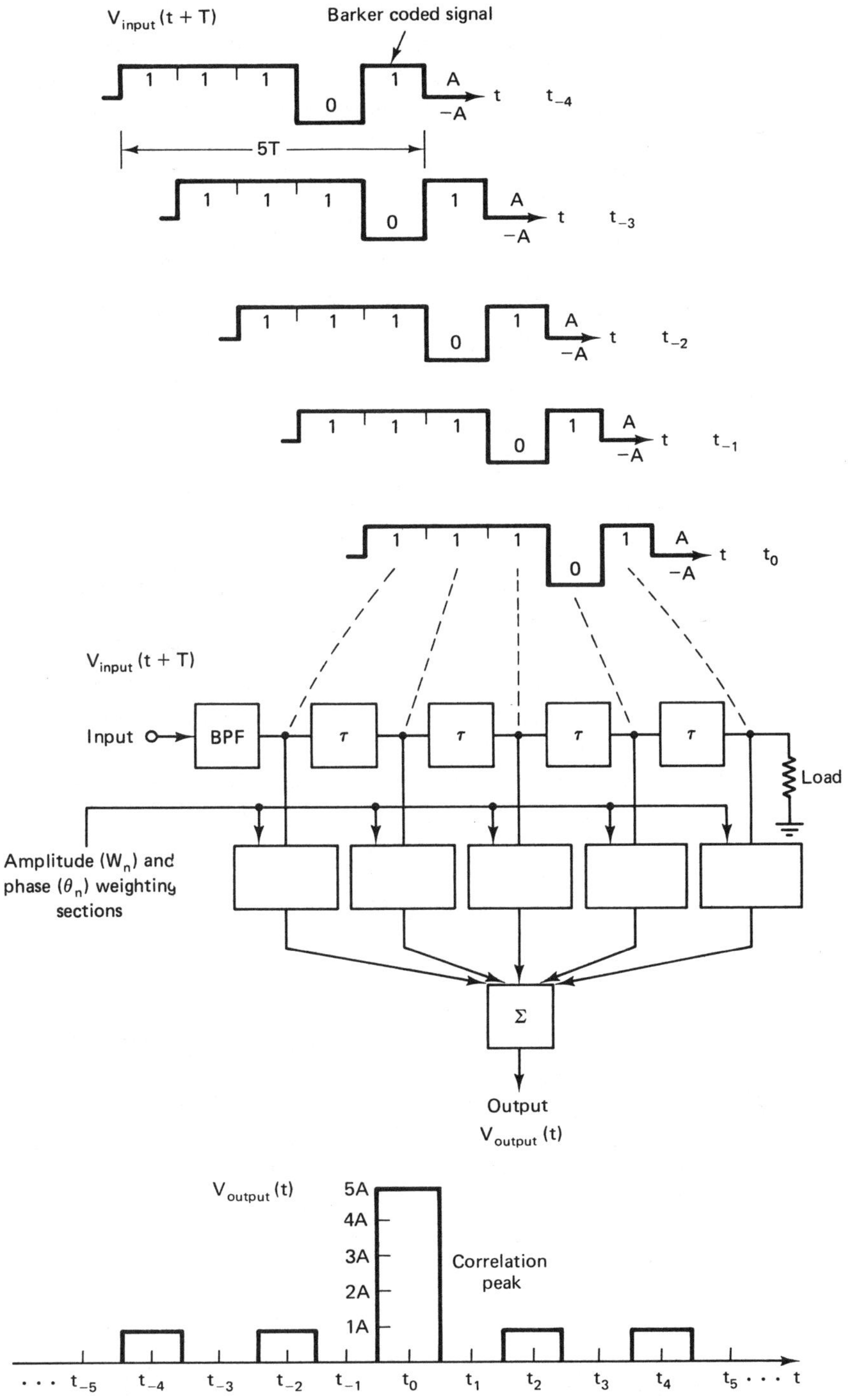

Figure 9.7 Matched filter version.

stored replica of itself. When the signal and the reference are timed coincidently, the resulting output SNR is a maximum. Consider an input pulse with a relatively low amplitude A and long duration T. It is converted to an output of relatively high amplitude, B, and short duration, τ. Neglecting loss, the output energy is equal to the input energy. Thus

$$E_i = P_i T = \frac{A^2}{4R} T = E_o = \frac{B^2}{4R} T \tag{9.8}$$

The processing gain is then

$$G_P = \frac{B^2}{A^2} = \frac{T}{\tau} = \frac{\text{BW}_{\text{rf}}}{\text{BW}_{\text{info}}} \tag{9.9}$$

As noted previously, the processing gain is also a function of the number of taps, n. That is,

$$G_P = \frac{T}{\tau} = \frac{n\tau}{\tau} = N \tag{9.10}$$

which corresponds to equation (9.7). Note that equations (9.9) and (9.10) implicitly assume that the output and input noise is equal.

Solid-state technologies are helping make certain types of matched filters easier to construct. One is the surface acoustic wave, or SAW device. The other involves the transfer of charge in a metal-oxide semiconductor (MOS) structure and thus is called a *charge-transfer device*, or CTD. Both are sampled analog devices whose versatility, small size, and low power requirements make them especially attractive for matched-filter applications. SAW technology can handle analog signal processing and filtering over a frequency range from 3 MHz to 2 GHz. The device uses phonon propagation and requires no clocking. An SAW device is illustrated in Figure 9.8. Acoustic energy travels along the substrate

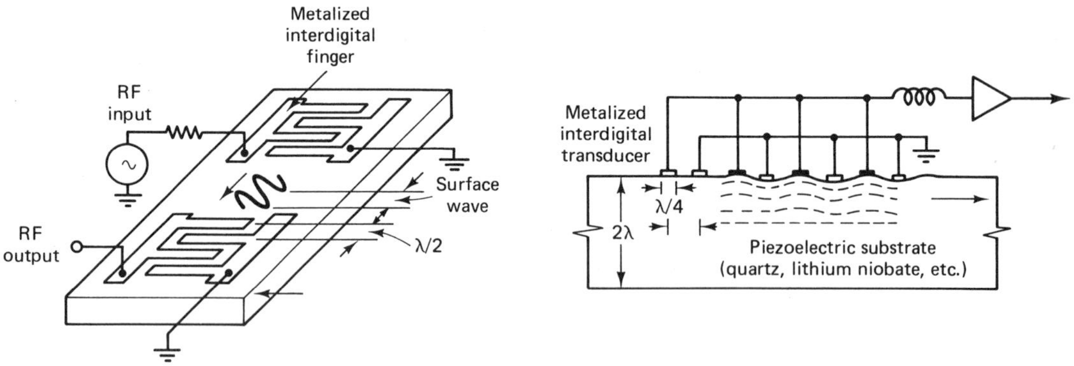

Figure 9.8 The metalized interdigital fingers of SAW device convert electricity into acoustic energy using the piezoelectric properties of the substrate material, using quartz or lithium niobate. Most of the energy is confined to an area within one wavelength of the surface of the substrate.

surface and reacts with metallized interdigital transducer elements deposited on the piezoelectric material.

9.2.2 Tracking

After sync acquisition has been achieved (to within one chip), tracking or fine synchronization takes place. This includes chip synchronization and, for coherent systems, carrier-phase tracking. The two most common techniques for tracking the incoming code rate for direct-sequence systems are similar and are called *tau-dither* and *delay-lock*.

As an illustration, the basic tracking loop for a direct-sequence spread-spectrum system using PSK modulation is shown in Figure 9.9. The received carrier of frequency, ω_o, is amplitude modulated by the product of the data $d(t)$ and the PN sequency $P(t)$. Referring to the figure, the tracking loop contains a local PN generator, which is offset in phase from the incoming sequence $P(t)$ by a time that is less than one-half the chip time. The tracking loop PN generator generates two PN sequences, delayed from each other by one chip. At the outputs of multipliers 1 and 2, we have the following:

$$e_{01} = [P(t)d(t)\cos(\omega_o t + \theta)]\, P\left(t + \frac{T_c}{2} + \tau\right) \tag{9.11}$$

and

$$e_{02} = [P(t)d(t)\cos(\omega_o t + \theta)]\, P\left(t - \frac{T_c}{2} + \tau\right) \tag{9.12}$$

The two band-pass filters are designed to have a two-sided bandwidth, B, equal

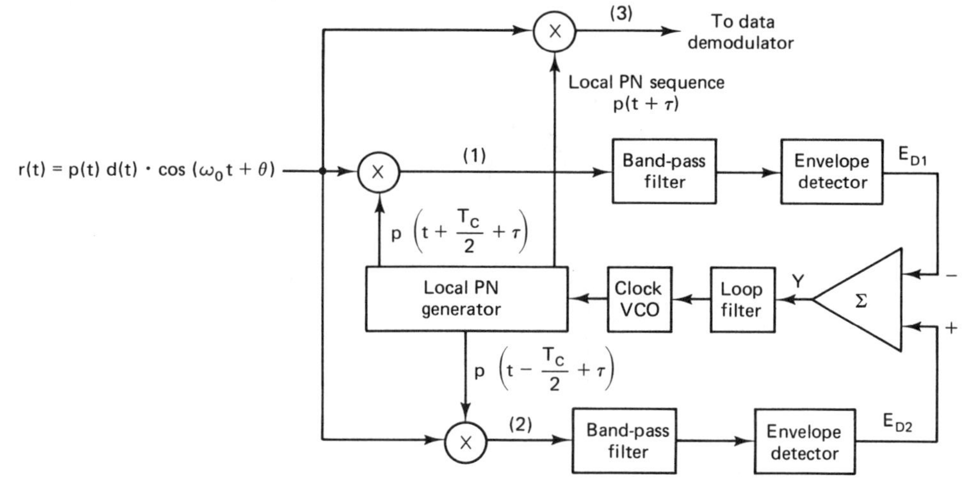

Figure 9.9 Delay-locked loop for tracking direct-sequence PN signals.

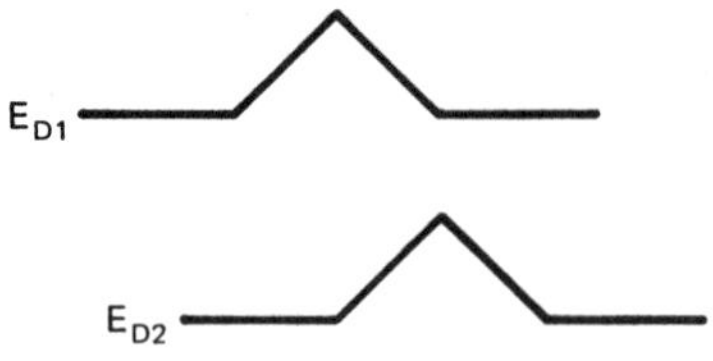

Figure 9.10 Correlation waveforms in a delay-locked loop.

to twice the bit rate. That is,

$$B = 2R = \frac{2}{T_b} \tag{9.13}$$

Thus the data $d(t)$ is passed, but the product of the two PN sequences $P(t)$ and $P(t \pm T_c/2 + \tau)$ is averaged. The square-law action of the envelope detector eliminates the data, since $|d(t)| = 1$. At the envelope detector output, we are left with the correlation function between the encoded incoming signal and the local reference. That is,

$$e_{D1,2} = \left| \overline{P(t)P\left(t \pm \frac{T_c}{2} + \tau\right)} \right| = \left| R_P\left(\tau + \frac{T_c}{2}\right) \right| \tag{9.14}$$

where $R_p(\tau + T_c/2)$ is the autocorrelation functions of the PN waveform, as shown in Figure 5.7. The triangular correlation functions are each 2 bits wide with their peaks offset by one chip time, that is, the amount of delay between the local reference signals (see Figure 9.10).

The correlation waveform E_{D1} is next summed out of phase with E_{D2} and the output $Y(t)$ of the adder is illustrated in Figure 9.11. For τ positive, a positive voltage proportional to Y, results in an increase of the VCO frequency. This in turn, forces τ to decrease. A negative τ reduces VCO frequency, thereby forcing

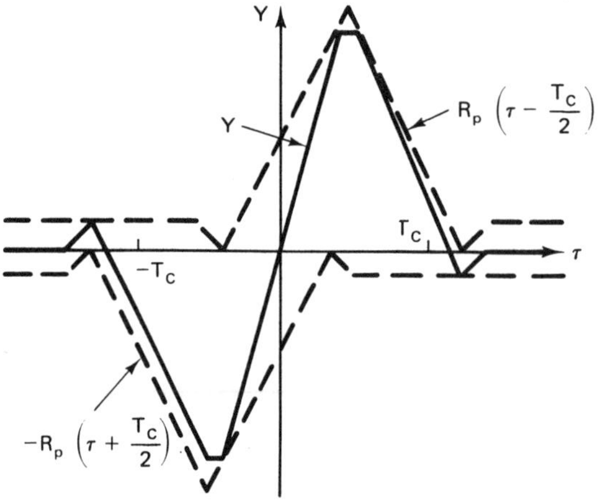

Figure 9.11 Variation of Y with τ.

τ to increase toward 0. Thus the receiver's code will track the incoming code at a point halfway between the correlator's output peaks.

When the tracking error, τ, is zero, the output of multiplier 3 is

$$\begin{aligned} e_{03} = e_{\text{PSK}} &= P^2(t)d(t)\cos(\omega_o t + \theta) \\ &= d(t)\cos(\omega_o t + \theta) \end{aligned} \tag{9.15}$$

The despread signal, e_{PSK}, is applied to the data demodulator where the data, $d(t)$, is recovered.

The second tracking technique for a direct-sequence system is the tau-dither loop illustrated in Figure 9.12 along with control waveforms. This loop is a delay-lock loop with only a single arm. The control waveforms (Figure 23(b)) are used effectively to generate both arms of the delay-lock loop. The simplicity of this loop often makes it the designer's choice.

In operation, note that the control waveforms generate the signal

$$V_P(t) = g(t)P\left(t + \tau - \frac{T_c}{2}\right) + \bar{g}(t)P\left(t + \tau + \frac{T_c}{2}\right) \tag{9.16}$$

Note that either one or the other, but not both, of the waveforms occurs at each instant of time. The multiplier output is

$$\begin{aligned} e_m &= e_r(t) \cdot V_P(t) \\ &= d(t)P(t)\cos(\omega_o t + \theta)\left[g(t)P\left(t + \tau - \frac{T_c}{2}\right)\right. \\ &\quad \left. + \bar{g}(t)P\left(t + \tau + \frac{T_c}{2}\right)\right] \end{aligned} \tag{9.17}$$

As in the delay-lock loop, the band-pass filter performs averaging. That is, it is designed to pass only the data and control signals. Data is eliminated by the envelope detector; thus the envelope detector output, $E_{d(t)}$ is

$$E_d(t) = g(t)\left|R_P\left(\tau + \frac{T_c}{2}\right)\right| + g(t)\left|R_P\left(\tau - \frac{T_c}{2}\right)\right| \tag{9.18}$$

The input, $Y(t)$, to the loop filter is

$$\begin{aligned} Y(t) &= E_d(t)g'(t) \\ &= g(t)\left|R_P\left(\tau - \frac{T_c}{2}\right)\right| - \bar{g}(t)\left|R_P\left(\tau - \frac{T_c}{2}\right)\right| \end{aligned} \tag{9.19}$$

where the minus sign results from the inversion caused by $g'(t)$. Averaging of $Y(t)$ occurs in the narrowband loop filter. As a result of each term being zero half

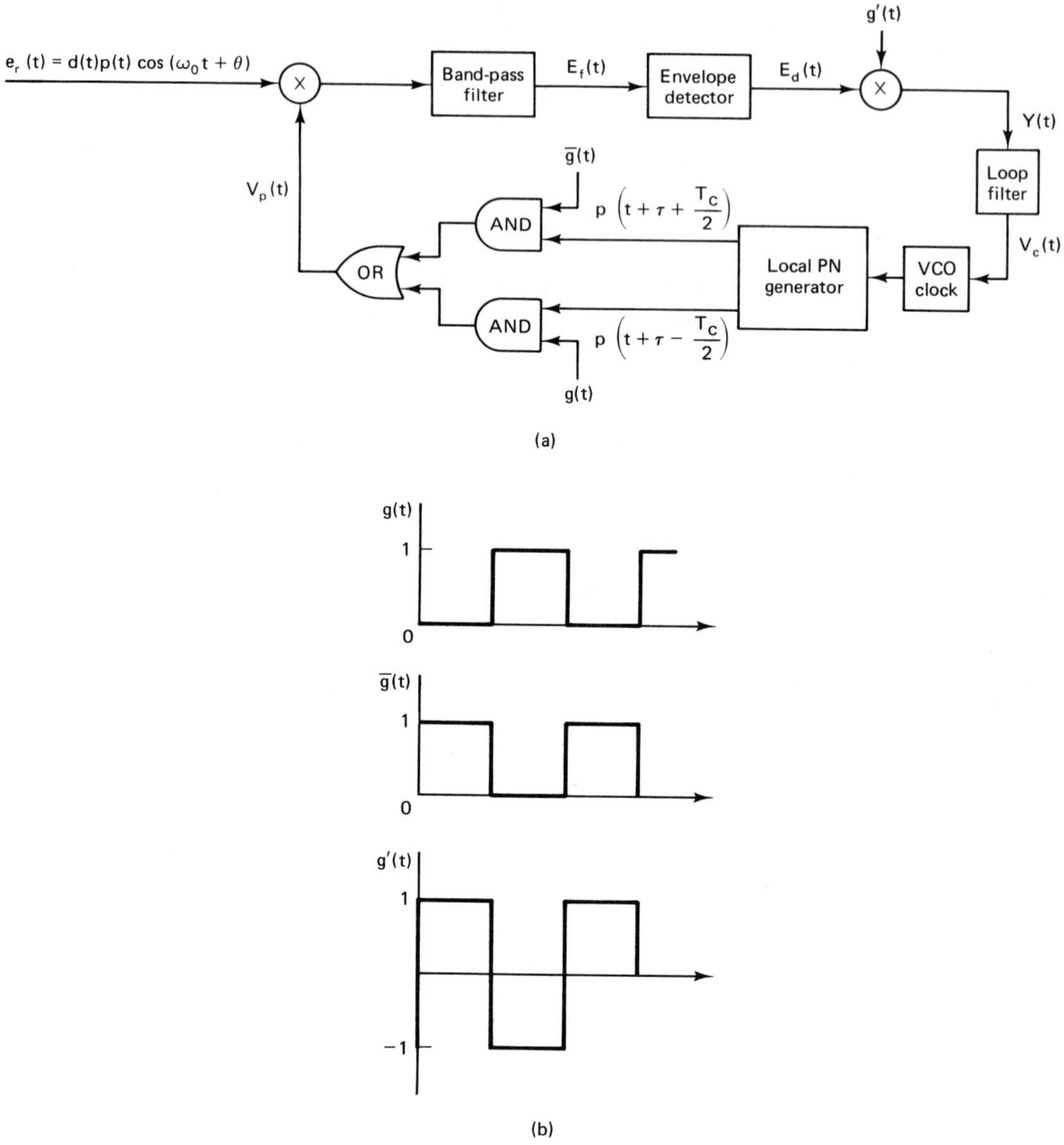

Figure 9.12 The tau-dither loop. (a) Block diagram. (b) Control waveforms.

the time, the VCO control signal, $V_c(t)$, is

$$V_c(t) = \left| R_P\left(\tau - \frac{T_c}{2}\right) - R_P\left(\tau + \frac{T_c}{2}\right) \right| \tag{9.20}$$

which yields the waveform of Figure (9.11).

A typical tracking system for an FSK/FH spread spectrum system is shown in Figure 9.13 along with waveforms. In this example, we assume that acquisition has occurred, but there remains an error of τ seconds between transitions of the incoming signal frequencies and the locally generated frequencies. The band-pass filter is made sufficiently wide to pass the product of the incoming signal $V_1(t)$,

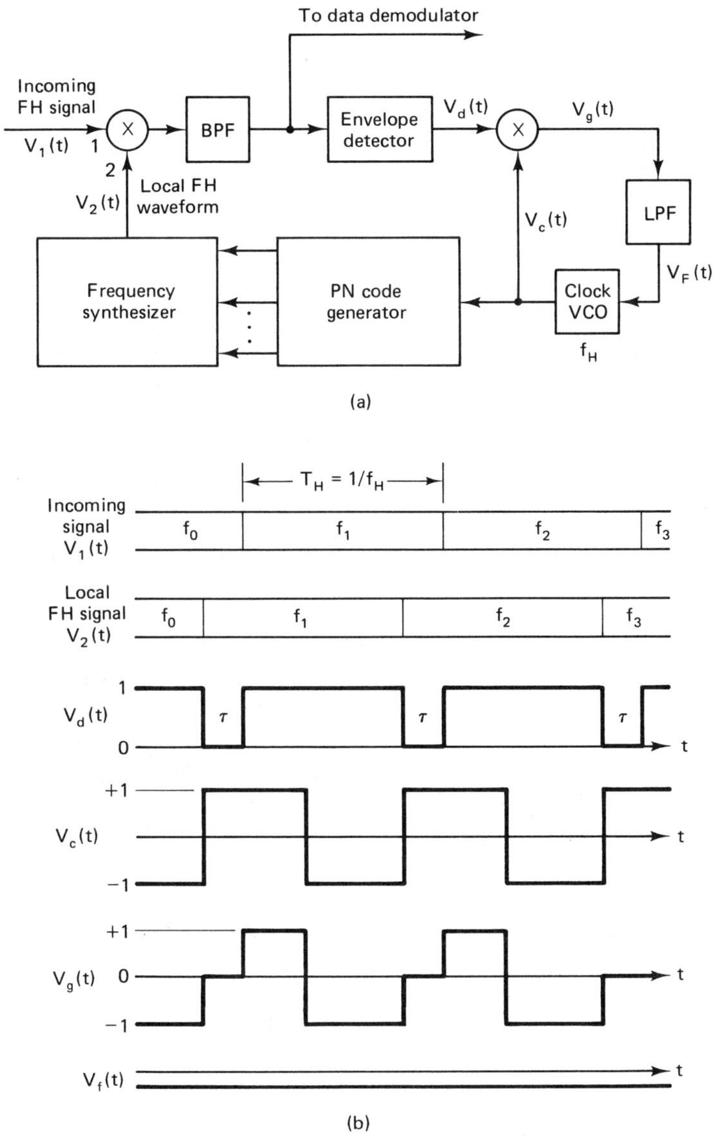

Figure 9.13 (a) Tracking loop for FH signals. (b) Waveforms for tracking an FH signal.

and the local FH signal, $V_2(t)$, where both are at the same frequency, f_i. When $V_1(t)$ and $V_2(t)$ are at different frequencies, f_i and f_{i+1}, the band-pass filter rejects the product. The envelope-detector output, $V_d(t)$, is then unity when the frequencies of $V_1(t)$ and $V_2(t)$ are the same and is zero otherwise. From Figure 9.13(b), we see that $V_g(t) = V_d(t) \cdot V_c(t)$ is a three-level signal. The LPF filters this signal to form a dc voltage. In this case, a negative voltage is presented to the VCO. In fact, if $V_2(t)$ has frequency transitions which precede $V_1(t)$, the voltage to the VCO, $V_F(t)$, is negative, thus delaying the transition. If $V_1(t)$ precedes $V_2(t)$, the voltage is positive, thus speeding up the transition. The total effect of the tracking circuit is to keep the offset time small. Even a small offset can have a major impact on the probability of error of the received data.

9.3 COMMUNICATION SYSTEM PERFORMANCE

In Chapter 8, we developed equations relative to the performance of digital communication systems. The design and evaluation of any communication system makes use of these so-called communication equations. We can write the communications equation as

$$\frac{P_r}{N_o} = P_t G_t \left(\frac{\lambda}{4\pi R}\right)^2 L_\theta L_P L_J L_S G_r \frac{1}{k(T_e + T_N)} \tag{9.21}$$

where the L factors are actually power-transfer (gain) functions less than unity.

L_θ = pointing loss of either the transmitting or receiving antenna

L_P = polarization loss factor; this loss occurs when the polarization characterisitics of the received radiation does not match the polarization characteristics of the receiver antenna

$L_J = \dfrac{J}{W_t k}$ = jamming loss. J is the total jamming power

$L_S = L_A L_C L_M$

where

L_A = losses caused by atmospheric absorption and other propagation losses in the transmission medium

L_C = cabling losses between the receiver antenna terminals and the receiver

L_M = modulation loss due to allocating power for reference generation, or losses due to imperfect detection. Detection losses may be caused by phase errors in receiver carrier tracking, code tracking, or fluctuating signal fading

In terms of the energy per bit we can rewrite equation (9.21) as

$$\frac{E_b}{N_0} = \frac{P_S T_b}{N_0} = P_t G_t \left(\frac{\lambda}{4\pi R}\right)^2 L_\theta L_P L_J L_S \left(\frac{G_r}{T_e + T_N}\right) \frac{T_b}{k} \tag{9.22}$$

The term $G_r/(T_e + T_N)$ is particularly important in terms of the receiver effectiveness. That is, the receiver effectiveness is described in terms of variables that the designer has control over. The $G_r/(T_e + T_N)$ ratio can be determined from the received E_b/N_0 ratio necessary to achieve a specified performance (see equation (8.38)).

In using equation (9.22), the primary addition is the jamming loss factor, L_j. Otherwise, considering perfect spreading and despreading, the discussions in previous chapters regarding digital modulation and coding techniques apply. Consider a system that transmits a data rate of R bits per second spread over a bandwidth of W_t hertz in white noise of power spectral density N_0 watts per hertz. For a data rate R

$$\left(\frac{E_b}{N_0}\right)_{\text{actual}} = \frac{P_S}{N_0 R} = \frac{P_S}{P_N}\frac{W_t}{R} \tag{9.23}$$

where

$$P_S = E_b R \quad = \text{signal power}$$

$$P_N = N_0 W_t = \text{noise power}$$

For the $(E_b/N_0)_{\text{min}}$ necessary to achieve minimum acceptable performance,

$$R \le \frac{P_S}{P_N} \frac{W_t}{(E_b/N_0)_{\text{min}}} \triangleq R_o \tag{9.24}$$

Consider now that the transmitter is transmitting at the maximum rate R_o and a jamming signal with power P_j appears. Equation (9.23) becomes

$$\left(\frac{E_b}{N_0}\right)_{\text{actual}} = \frac{P_S}{P_N + P_j}\frac{W_t}{R_o} = \left(\frac{E_b}{N_0}\right)_{\text{min}} \frac{P_N}{P_N + P_J} \tag{9.25}$$

or

$$\left(\frac{E_b}{N_0}\right)_{\text{actual}} = \left(\frac{E_b}{N_0}\right)_{\text{min}} \frac{N_0}{N_0 + P_J/W_t} \tag{9.26}$$

In the receiver, the spread jamming signal is a noiselike signal, and its interference is added to the noise, N_0. Note that equation (9.26) indicates that if we wish to recover from the effects of the jammer, it is necessary to increase W_t so that the term $N_0/(N_0 + P_J/W_t)$ approaches unity.

As indicated by equation (9.4), in the design of a practical system, the processing gain, G_p, is not by itself a measure of how well a system will perform in a jamming environment. The jamming margin

$$M_{j,\text{dB}} = G_{p,\text{dB}} - \left(\frac{E_b}{N_{0j}}\right)_{\text{min,dB}} - L_{\text{dB}} \tag{9.27}$$

is a measure of the residual advantage that the system has against a jammer after both the minimum required energy/bit-to-jamming noise power spectral density ratio $(E_b/N_{0j})_{\text{min,dB}}$ and implementation and other losses, L, are subtracted. The jamming margin can be increased by reducing the $(E_b/N_{0j})_{\text{min}}$ through the use of coding gain. For example, the block codes and convolutional codes discussed in Chapter 5 can be used to preclude the possibility of a small number of frequency slots being jammed, thus causing an unacceptable error rate. Even if the jammer wipes out a few code symbols, depending upon the error-correcting capability of the code, the data may still be recovered. An analogous situation occurs in direct sequence spreading when a pulse jammer is present.

9.3.1 Antijam Considerations

The single most important application of spread-spectrum techniques is that of resistance to intentional jamming or interference. In the literature, the two most common types of jamming signals analyzed are single-frequency sine waves and broadband noise. Of these two, the simplest case to analyze is that of broadband noise jamming. For a direct-sequence system, if it is assumed that the jamming signal occupies the total RF bandwidth (two times the chip rate), then the despread jammer occupies an even greater bandwidth and appears to the final data detector as approximately white noise. For example, if binary PSK is the modulation format, the average probability of error is approximately given by

$$P_e = Q\left(\sqrt{\frac{2E_b}{N_0 + J_o}}\right) \tag{9.28}$$

This is just the classical result for the performance of a coherent binary communication system in additive white noise. It differs from the conventional result because of the addition of the jamming term, J_o. Equation (9.28) is plotted in Figure 9.14 for a value of $P_J/P_S = 23$ dB, where P_S is the average signal power. The interesting thing to note about Figure 9.14 is that for a given J_o, the curve bottoms out as E_b/N_0 continues to get larger. The presence of the jammer will result in an irreducible error for a fixed value of P_j and W_t. For a fixed P_J, the only way to reduce the error rate is to increase W_t, that is, increase the spreading. This is in keeping with the previous discussion on interference effects.

EXAMPLE 9.3

A spread spectrum receiver is operating in an environment with a jamming signal of 20 dB above the desired signal. An additional jamming signal 10 dB above the desired signal appears. What is the effect of the second jamming signal?

Solution

$$\frac{J}{S} = 20 \text{ dB} \rightarrow 100$$

$$\frac{J}{S} = 10 \text{ dB} \rightarrow 10$$

Total worst case:

$$\frac{J}{S} = 100 + 10 = 110$$

$$= 20.4 \text{ dB}$$

Interference due the second jamming signal is an increase of 0.4 dB.

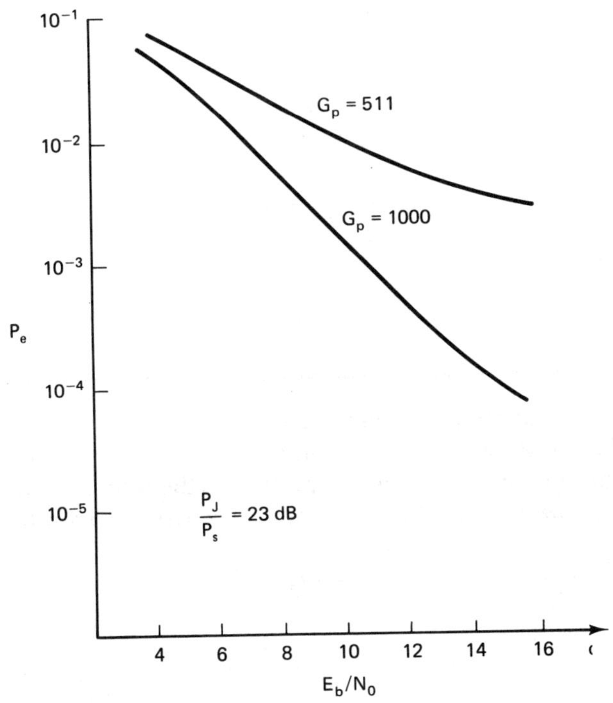

Figure 9.14 Probability of error versus E_b/N_0.

EXAMPLE 9.4

A direct-sequence spread-spectrum receiver with a 10-MHz bandwidth has a 26-dB processing gain. The receiver noise figure is 6 dB and a 2-dB direct-sequence system implementation loss is experienced. An output SNR of 10 dB is desired. Assuming standard temperature (290°K), what must the sensitivity of the receiver be?

Solution Neglecting losses and receiver noise figures,

$$G_p = \frac{(S/N)_o}{(S/N)_i}$$

$$\left(\frac{S}{N}\right)_i = \left(\frac{S}{N}\right)_o - G_p = 10 - 26 = -16 \text{ dB}$$

$$N_i = kTB = (1.38 \times 10^{-23})(290)(10^{-7}) = 4.002 \times 10^{-14} \text{ W}$$

with losses

$$\left(\frac{S}{N}\right)_o = K \cdot G_p \left(\frac{S}{N}\right)_i$$

$$10 = 6 + 2 + 26 + \left(\frac{S}{N}\right)_i$$

$$(S/N)_i = -24 \text{ dB}$$

The sensitivity of the receiver is 24 − 16 = 8 dB below N_i.

$$8 = 10 \log_{10} \frac{4.002 \times 10^{-14}}{S_i} \Rightarrow S_i = 0.6343 \times 10^{-14} \text{ W}$$

For a 50-Ω input impedance

$$E_i = [S_i \cdot R]^{1/2} = [(0.6343 \times 10^{-14})(50)]^{1/2} = 0.563 \ \mu\text{V}$$

It is not always advantageous for a noise jammer to jam the entire RF bandwidth of an FH system. That is, for a given jamming power P_J, the effectiveness of the jammer can be increased by jamming only a fraction of the total bandwidth. This is called *partial band jamming*, and the probability or error is given by

$$P_e = \frac{(N-K)(N-K-1)}{N(N-1)} \frac{1}{2} \exp\left(-\frac{1}{2}\text{SNR}\right) + \frac{K(N-K)}{N(N-1)} \exp\left[-\frac{1}{2/\text{SNR} + 1/\text{SJR}}\right] + \frac{K(K-1)}{N(N-1)} \frac{1}{2} \cdot \exp\left[-\frac{1}{2} \frac{1}{1/\text{SNR} + 1/\text{SJR}}\right] \tag{9.29}$$

Equation (9.29) assumes that the jammer divides its power uniformly among K slots, where a slot is the region in frequency that the FH signal occupies during one of its hops and the modulation format is FSK with noncoherent detection at the receiver. There is a total of N slots, over which the signal can hop. In this expression, SNR is the ratio of signal power to thermal noise power at the output of the receiver MARK band-pass filter in the receiver (assuming that a MARK was transmitted). The same remark applies to the transmission of a SPACE tone. SJR is the ratio of signal power to jammer power per slot at the output of a MARK band-pass filter. Equation (9.29) is illustrated in Figure 9.15 for $K = 1$ and $K = 100$ for a P_J/P_S of 10 dB. The situation for $K = 1$ represents single-tone jamming. That is, all of the power is put into a single slot. An error is made on an average of one out of every N hops, so that the error in the system is approximately $1/N$. Note that a marked improvement is effected by the use of error-correcting coding. For this particular figure, a Golay code was used. A Golay code is a (23, 12) cyclic code with minimum distance 7. Many other types of jamming signals are employed. The reader is referred to the literature for additional reading.

Refer to Figure 9.14. Note that the probability of error, with a fixed jamming power P_j, increases with increasing processing gain. One of the problems with any spread-spectrum system is that the desired wide RF bandwidth trades off interference rejection against system process gain. We noted that the processing gain is approximately equal to the ratio of the spread bandwidth to the information bandwidth (equation (9.1)). As the spread bandwidth gets larger, the receiver's ability to reject interfering signals decreases. For FH systems, the process gain

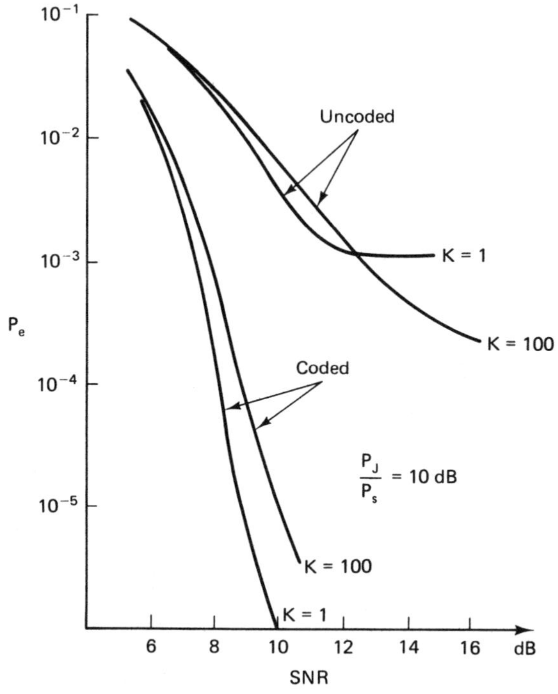

Figure 9.15 Probability of error versus SNR.

is roughly equal to the number of channels the system hops over. For TH systems, it is roughly the inverse of the system's duty factor.

Another problem encountered in spread-spectrum systems is difficulty in achieving very wide RF bandwidth from a single spread-spectrum technique. An example illustrates.

EXAMPLE 9.5

A spread-spectrum system with an RF bandwidth of 1 GHz is desired. An information rate of 10 kilobits is to be transmitted. (a) What is the required code rate for a direct sequence system? (b) What is the required hopping rate for a frequency-hopping system? (c) What is the process gain for a system that hops over 100 frequencies that is modulated by a direct sequence system with a code rate of 10 Mbps? (d) What is the process gain for (a) and (b)?

Solution

(a) $\mathrm{BW}_{\mathrm{RF}} \approx 2R_{\mathrm{CLK}}$

$$R_{\mathrm{CLK}} = \frac{10^9}{2} = 500 \text{ megabits}$$

(b) $$\text{Frequency hops (channels)} = \frac{\mathrm{BW}_{\mathrm{RF}}}{R_{\mathrm{info}}} = \frac{10^9}{10^4} = 100{,}000$$

(c) $$G_p\left(\frac{\mathrm{FH}}{\mathrm{DS}}\right) = G_p(\mathrm{FH}) + G_p(\mathrm{DS})$$

$$= 10 \log(\text{number of channels}) + 10 \log \frac{\mathrm{BW}_{(\mathrm{DS})}}{R_{\mathrm{info}}}$$

$$= 10 \log 100 + 10 \log \frac{2 \times 10^7}{104} = 53 \text{ dB}$$

(d) $$G_p(\mathrm{DS}) = 10 \log \frac{10^9}{10^4} = 10 \log 100{,}000 = 50 \text{ dB}$$

$$G_p(\mathrm{FH}) = 10 \log 100{,}000 = 50 \text{ dB}$$

Note that the process gain is 3 dB higher for the FH/DS system. This occurs because of overlap in the 100 frequency channels.

9.4 CDMA

Our study of conventional communication techniques has acquainted us with TDM and FDM. CDMA provides a relatively new type of channel multiplexing in which many information channels can be established in the same frequency channel simultaneously without interference between channels. This technique allows multiple access to a channel by assigning a unique orthogonal code to each user. If the assigned codes cannot be mutually correlated, each user will not interfere with or even be aware of other channel activity.

The design problem is much more complicated, however, than having a number of spreading sequences with uniformly low cross-correlations, such as the Gold codes discussed in Chapter 5. Typically, the CDMA systems are asynchronous. That is, the data symbols of the different users do not have to coincide. In a DS CDMA system, for example, the key parameters are both the cross-correlation and the partial-correlation properties of the code. In addition to optimization of the user codes, a problem of potentially greater importance is the "near-far" problem. Since the users are likely to be geographically separated, a receiver trying to detect the kth signal may be much closer to the ith transmitter. Thus if each user transmits equal power, the signal from the ith transmitter will arrive at the receiver with a larger power. This problem can be so severe that a DS CDMA system cannot be used.

An alternative to DS CDMA is FH CDMA. If each user is given a different hopping pattern and if all of the hopping patterns are orthogonal, theoretically the near-far problem is solved. An exception to this is spectral spillover from one slot to adjacent slots. Orthogonality implies that no two users will ever use the same frequency simultaneously. This is never completely true, and interference does occur. These events are sometimes referred to as *hits*, and the frequency of hits increases with the number of users.

Variation of FH CDMA systems may be considered based on system requirements. For example, we may use one hop per bit, multiple hops per bit (referred to as *fast hopping*), or multiple bits per hop (referred to as *slow hopping*). Often the characteristics of the channel may be the deciding factor. Frequency-hopping systems achieve an improvement in performance through a different mechanism than direct-sequence systems. That is, FH systems achieve their processing gain through interference avoidance, not interference attenuation as in a DS system. This mechanism also applies if the interference is multipath. As long as the signal is hopping fast relative to the differential time delay between the desired signal and the multipath signal, most of the multipath energy will fall in slots that are orthogonal to the slot that the desired signal currently occupies.

In considering the implementation of a CDMA system, it is clear that the designer needs to compare its relative capacity to a TDMA or FDMA system. In a perfectly linear, synchronous system, the number of orthogonal users for all three systems is the same. This is true, since the number depends only on the dimensionality of the signal space. One attractive feature of CDMA systems is that network synchronization is not required, whereas in TDMA, it is. Another advantage of CDMA is that it is relatively easy to add additional users to the system. For an asynchronous system, ignoring the near-far problem, a rule of thumb for the number of system users is $G_p/10$, where G_p is the processing gain.

In the future, we are likely to encounter increasing uses of spread-spectrum techniques in such areas as mobile telephone and microwave communications in congested areas.

PROBLEMS

9.1. In FH with M-ary FSK, the jammer spaces a series of jamming tones across the bandwidth. Show that

$$P(\text{hop-decision error}) = (M - 1)\frac{N_j}{N_f}$$

where

$$N_j = \text{number of jamming tones}$$
$$N_f = \text{number of orthogonal frequencies}$$

9.2. A frequency-hopping system uses 16-ary FSK modulation and a data rate of 1000 bits/s. After each symbol the system hops to a new frequency.

(a) What is the symbol rate?
(b) What is the hopping rate?
(c) What is the hopping duration?

9.3. **(a)** Draw the block diagram of a hybrid FH/DS transmitter/receiver.
(b) Show that

$$\begin{aligned} G_p\left(\frac{\text{FH}}{\text{DS}}\right) &= G_p(\text{FH}) + G_p(\text{DS})\ \text{dB} \\ &= 10\log_{10}(\text{number of channels}) \\ &\quad + 10\log_{10}\left(\frac{\text{BW}_{\text{DS}}}{R_{\text{info}}}\right)\ \text{dB} \end{aligned}$$

(c) What would be the processing gain of an FH/DS system with a chip rate of 30 megachips/s, a data rate of 4 kilobits/s and 150 frequency channels?
(d) Why does FH/DS offer simpler implementation possibilities?

9.4. Consider a frequency-hopping communication system with a transmission data rate of 6400 bits/s. The modulation is 16-ary FSK.

(a) If we want the system to hop at the rate of once per symbol, what will be the time duration of each hop?
(b) What is the time duration of a chip in part (a)?
(c) If we increase the hopping rate so that the system hops at a rate of 4 times per symbol, what will be the time duration of a chip?
(d) If we decrease the hopping rate so that there are 10 symbols per hop, what will be the time duration of a chip?
(e) If, in part (d), we change to a binary modulation format, what will be the time duration of a chip?

9.5. In Problem 9.2, the system is modified such that each symbol transmission is repeated four times and the bit transmission rate remains fixed. The system is now made to hop faster; that is, it hops at each symbol repetition.
(a) What is the new hopping duration?
(b) What is the new hopping rate?

9.6. Two terminals (XMTR and RCVR) located 150 mi apart are synchronized in real time to within 10 s. For a frequency-hopping system with hop rate 200 KHz, how many bits must the receiver search to gain initial synchronization?

9.7. A maximal-length sequence is generated with a 10-stage register. A delay-line matched filter is to be constructed to match the code sequence. For a code rate of 10 megabits/s what are the major characteristics of the delay line?

9.8. What jamming margin is required if the interfering signal is 300 times greater than the desired signal?

9.9. For the communication system in Problem 9.8, what minimum process gain is needed for a 15-dB output SNR? Assume system losses of 2 dB.

9.10. An FH/DS system operates with a 30-megabit/s code rate and a data rate of 6 KHz. The system hops over 200 frequencies. What is the expected process gain?

9.11. A direct-sequence spread-spectrum system is used to communicate between stations separated by a distance that yields a propagation delay of 0.92 μs. The spreading code is a maximal-length code specified by the generator $P(x) = x^{10} + x^3 + 1$, and the code-generator clock rate is 100 MHz. Determine a set of initial conditions for the transmitter and receiver code generators such that if the generators are started simultaneously, the received signal will be despread immediately.

9.12. The PLL frequency synthesizer shown in the accompanying figure is to be used in an FH system. An overall bandwidth of 25.6 MHz is desired with 128 channels. The lowest output frequency is 43.4 MHz and the coded message bit rate is 3180 bits/s.
(a) Find the reference oscillator frequency, f_1.
(b) Find the VCO center frequency.
(c) Find the range of the divider, n.
(d) Assuming that the divider, n_i, is the output of a PN generator, what is the required PN code generator rate if there are three hops per message bit?

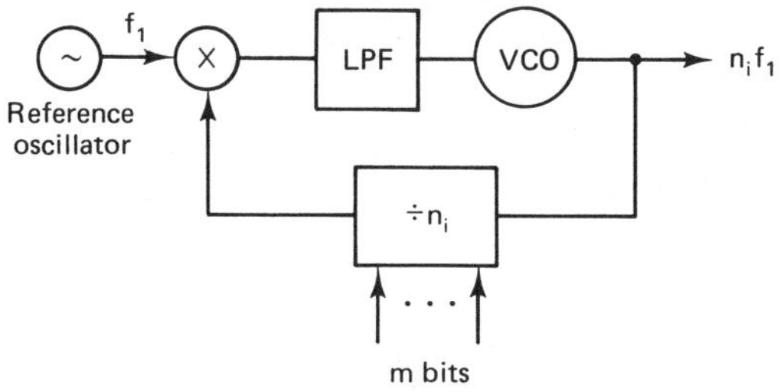

9.13. In Problem 9.12, if 1 bit of the coded message is used to form n_i, find:
(a) The range of n_i.
(b) The PN code rate necessary to obtain the same degree of spreading in a PN sequence system.

9.14. A frequency-hopping spread-spectrum system is to be constructed using the frequency synthesizer of problem 9.12, with the following given parameters:

$$\text{Number of channels} = 2048$$

$$B_s = 61.44 \text{ MHz}$$

$$\text{Center frequency} = 375.015 \text{ MHz}$$

$$\text{Final multiplication} = N = 15$$

(a) Find the reference oscillator frequency f_1.
(b) Calculate the divider values necessary.
(c) If one divider stage requires 4 bits, how many cascaded divider stages will be required?

9.15. The design shown in the accompanying figure can be used to reduce the number of required divisions in the frequency synthesizer of Problem 9.14. If we assume the lowest value of n_i is 27, find:
(a) The uppermost value for n_i
(b) f_{LO}
(c) Minimum required -3 dB frequency for LPF(2)

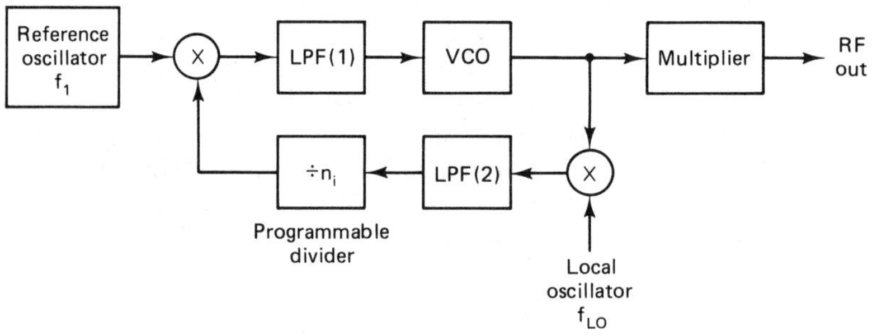

9.16. The results of multipath propagation can cause wide variations in signal intensity at the point of reception (see Section 8.1.3.2). Explain why frequency-hopping systems are able to combat this.

9.17. A direct-sequence system must operate with interference from two jammers whose center frequencies are near. The following parameters apply:

$$\text{PN code rate} = 125 \text{ megabits/s}$$

$$\text{Message bit rate} = 2500 \text{ bits/s}$$

Received jamming power $= 1 \times 10^{-6}$ W and 5×10^{-7} W in a combined bandwidth of 50 KHz

(a) Calculate the approximate interference power at the correlator output.
(b) Under worst-case conditions, what is the effect of the smaller jammer relative to the larger?

9.18. An FH/PN hybrid spread-spectrum system is to be designed. The following design parameters apply:

$$\text{Processing gain} \geq 41 \text{ dB}$$

$$\text{Code bits/message bit} = 20$$

$$\text{Coded message bit rate} = 4.8 \text{ kilobits/s}$$

(a) How many frequencies must the synthesizer produce? What is the switching speed?
(b) What is the bandwidth of the final RF amplifier if all the frequencies are used?

9.19. An FH system is designed with the following parameters:

20 hops/message bit

2048 frequencies

Message bit rate = 5000 bits/s

Required:
(a) Processing gain in decibels.
(b) PN code generator clock rate (see Problem 9.12).
(c) Frequency separation in kilohertz.
(d) RF signal bandwidth.
Would this be considered a fast-hopping or a slow-hopping system?

9.20. We wish to transmit a 4-KHz signal over a PN spread-spectrum system. The signal is digitized to 128 quantization levels.
(a) Find the chip rate to yield a processing gain of 20 dB.
(b) If the sequence length is to be greater than 5 h, find the length N of the shift register.

10

OPTICAL COMMUNICATIONS

10.0 INTRODUCTION

Optical communications is the oldest and also one of the most modern methods of communication. The most powerful aspect of optical communication is the tremendous bandwidth available. For example, a wavelength of 1 μm corresponds to 3×10^{14} Hz (300,000 GHz). Thus a single 1-GHz channel occupies a mere 3.3×10^{-6} μm of wavelength spread. Optical channels with a 1-GHz bandwidth are in use today with the prospect of continued improvement.

The transmission of information by optical means requires an optical source, a modulation technique, a method of transmitting the energy, and a detector to extract the information. One of the simplest methods of transmitting optical energy is propagation through free space. The extremely short wavelengths involved allow propagation with very low divergence. For example, at a wavelength of 1 μm, a 1-m antenna (lens) has a beam divergence on the order of 1 μrad. The use of free space propagation in the earth's atmosphere is limited by such effects as scattering and optical absorption. Significant advantages exist, however, in the vacuum of free space. Among these are

- Small antenna (lens) size
- Low side lobes
- Low power requirements

Optical communications between two satellites in geosynchronous orbit and a satellite in low Earth orbit is shown in Figure 10.1.

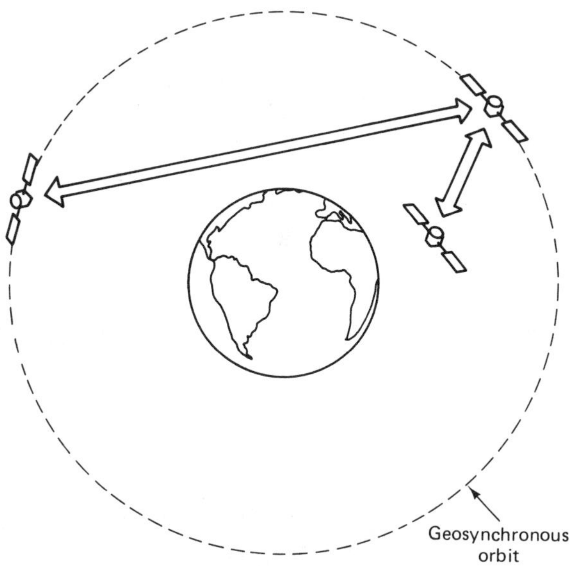

Figure 10.1 The large bandwidth of the optical channel can speed communications between satellites in space. This diagram shows communication between a low-flying satellite and one in geosynchronous orbit and between two satellites, both of which are in geosynchronous orbit.

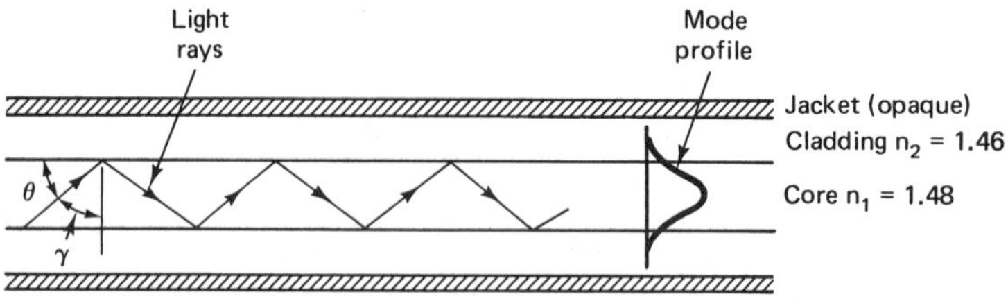

Figure 10.2 Schematic diagram of an optical fiber showing light rays bouncing at an angle less than the critical angle and a field profile for a bound mode.

The development of low-loss optical fibers has resulted in an explosion of the use of optical communication. An optical fiber is a dielectric waveguide (see Figure 10.2). The fiber consists of a core with index n_1 surrounded by a cladding with lower refractive index, n_2. The light will propagate as shown with total internal reflection as long as the angle Θ is less than the 90° complement of the critical angle, γ_{crit}. The minimum angle that supports total internal reflection is given by

$$\sin \Theta_{min} = \frac{n_2}{n_1} \tag{10.1}$$

The advantage of optical fiber is that it is low loss, very wideband, flexible, and lightweight. Core diameters are on the order of 50 to 100 μm. Semiconductor sources and detectors are available with these same small dimensions.

A fundamental physical limitation to optical detection is the effect of signal shot noise at optical frequencies. At microwave frequencies, the quantum nature of an electromagnetic field is not readily apparent. For example, at a 10-cm wavelength (30 GHz), the energy of a single photon is only 0.1 eV. At room temperature thermal energy, kT, is 26 meV. Thus, at a 10-cm wavelength, 260 photons are required just to equal the thermal energy. Considering that a single watt of power carries with it 10^{24} photons, the granular nature of the required electromagnetic energy is completely masked. At an optical wavelength of 1 μm, the photon has an energy of 1.2 eV. This energy is considerably in excess of the thermal energy and also the number of photons per watt is much less. At optical frequencies, then, the randomness of arrival of these energetic photons generates shot noise in the detected signal.

Most high-speed photo detectors operate on the principal of arriving photons generating free carriers. We can represent the electric field of the incident optical signal by

$$\begin{aligned} e_s(t) &= E_s \cos \omega_s t \\ &= \text{Re}[E_s e^{j\omega_s t}] \end{aligned} \tag{10.2}$$

This incident signal results in a current, $i_s(t)$, through the external circuit loads on the photodetector, which is proportional to the incident optical power, P_s. The power in turn is proportional to E_s^2. Shot noise is generated by the signal current and is described by

$$i_N^2 = 2qi_sB \tag{10.3}$$

where

q = electron charge, 1.6×10^{-19} C
B = bandwidth in hertz

For a detector with a quantum efficiency, η, the signal current, i_s, is

$$i_s = \frac{\eta P_s q}{h_f} \tag{10.4}$$

where

h = Plank's constant
f = frequency in hertz

Using equations (10.3) and (10.4), the SNR at the output of the photodetector is

$$\text{SNR} = \frac{i_s^2}{i_N^2} = \frac{\eta P_s}{2hfB} \tag{10.5}$$

Note that the SNR is proportional to received signal power. Theoretically speaking, this is the best SNR that can be expected. This limit is difficult to reach for several reasons. In the case of lasers, for example, intensity fluctuations exist that are greater than those associated with shot noise. The fluctuations are a fixed percentage of the laser power; thus the effect is referred to as *relative intensity noise*. Considering the relative intensity noise alone, this places an upper limit on the SNR. Attenuation of the signal as it propagates causes the effect of shot noise on the SNR to increase relative to the intensity fluctuations. At the receiver, the first noise source to overtake the effect of relative intensity noise is the thermal noise of the receiver. In order to achieve theoretical shot noise limited noise performance from an attenuated signal, heterodyne detection can be employed.

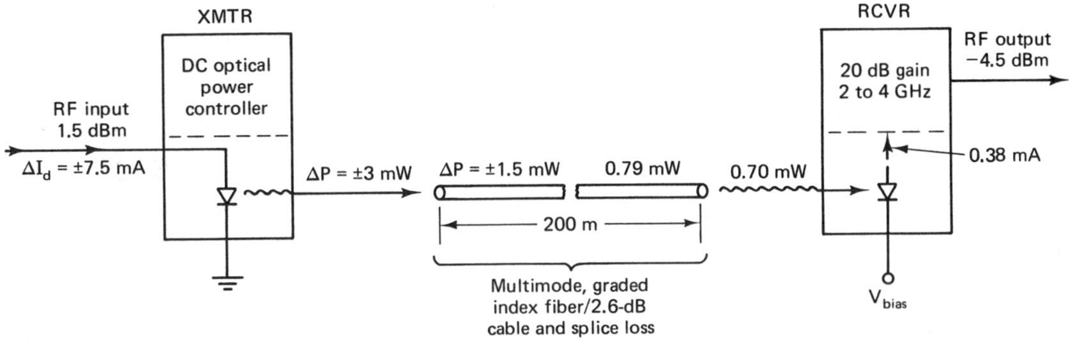

Figure 10.3 A wideband fiber optic link in the 2–4-GHz band.

Heterodyne detection is a familiar practice in radio engineering. At optical frequencies the technique is still in the laboratory, primarily because of difficulties in stabilizing the source and local oscillator frequencies. If the required frequency stability can be achieved, it will be possible to achieve essentially quantum limited performance in an optical channel. In a heterodyne detection scheme, we have not only the incident electric field, $e_s(t)$ but also an electric field, $e_{LO}(t)$, from the local oscillator. Just as in equation (10.2), we can describe this field by

$$e_{LO}(t) = \text{Re}[E_{LO}e^{j\omega_{LO}t}] \tag{10.6}$$

where ω_{LO} is the frequency of the local oscillator. The photodetector current, $i(t)$, is now proportional to

$$\begin{aligned} i(t) &= | E_{s}e^{j\omega_s t} + E_{LO}e^{j\omega_{LO}t} |^2 \\ &= E_s^2 + E_{LO}^2 + 2E_sE_{LO}\cos(\omega_s - \omega_{LO})t \end{aligned} \tag{10.7}$$

If we now make the local oscillator power much greater than the signal we are trying to detect, the shot noise is generated by the average detected current. Thus it is proportional to the local oscillator power, P_{LO}.

Note from equation (10.7) that the signal current, $i_s(t)$, can be identified with the difference frequency term and is proportional to the product of the electric fields, $[P_SP_{LO}]^{1/2}$. Therefore, the SNR is proportional to signal power, P_S, as it is for ordinary shot noise. In addition, the signal is increased by a factor equal to $[P_{LO}/P_S]^{1/2}$, thus allowing it to rise above the thermal noise of the amplifier.

Some of the parameters in a wideband fiber optic link are illustrated in Figure 10.3. This link operates in the 2- to 4-GHz band over 200 m of multimode fiber. The fiber itself is a 50-μm core graded index fiber. As indicated in the figure, the maximum excursion of the output modulated laser power is ±3 mW. For a quantum efficiency of 27 percent, this translates to 3/7.5 = 0.4 mW per milliamp of drive current. At the fiber a 3-dB coupling loss occurs, and an additional 2.8-dB cable loss occurs. An additional output coupling loss to the photodiode of 0.5 dB occurs. The quantum efficiency of the photodiode itself is 66 percent, which at a wavelength of 840 μm translates to 0.54 mA/mW. Current through the photodiode exhibits a DC component of 0.89 mA with a modulated AC component of ±0.38 mA. Considering the RF power (into 50 Ω), the end-to-end signal loss is 26 dB. The 20-dB gain amplifer yields an RF output power level of −4.5 dBm. In the following section we take a more detailed look at the design principles involved in Figure 10.3.

10.1 OPTICAL FIBER CHARACTERISTICS

We saw in the previous section that optical waveguide fibers have unique characteristics that make them attractive for communications. The large bandwidths and low transmission losses are particularly applicable to long-haul, high-speed telecommunications. To these advantages we can add

- Immunity to electromagnetic interference
- Small size
- Light weight
- Security against unauthorized tapping

In designing fiber optic systems, it is essential to have an understanding of the characteristics of optical fibers and how these characteristics effect signal transmission.

There are several types of fibers available or in development. The characteristics of these fiber types exhibit important differences. The first communication-grade fibers had core diameters on the order of the wavelength of light. Light propagated in a single mode down the fiber. The difficulty of coupling light into such a small core led to the development of multimode fibers. Core diameters were on the order of 50 to 100 μm. These fibers were used with semiconductor light sources emitting at 0.8- to 0.9-μm wavelengths. Later research showed that lower losses along with wider bandwidths could be achieved at a wavelength of 1.3 μm. Meanwhile, improvements have been made at coupling light into small-diameter fibers. Single-mode fibers with very high bandwidths at 1.55 μm are becoming available. One of the most important results of the new technology lies in making a wider variety of fiber types and components available for design applications.

When choosing an optical waveguide fiber, three major performance criteria must be considered:

- Signal losses
- Ease of light coupling and interconnection
- Bandwidth

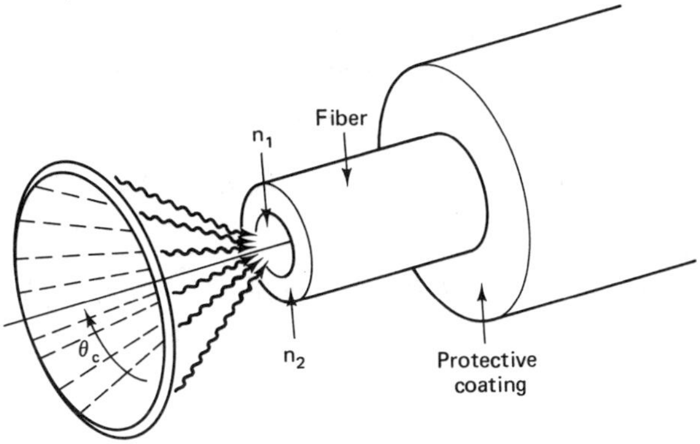

$$NA = \sin\theta_c = \sqrt{n_1^2 - n_2^2}$$
$$= n_1\sqrt{2\Delta}$$

Figure 10.4 Schematic drawing of the angular light acceptance of a fiber. Light impinging on the core within a critical angle, θ_c, is coupled into the fiber and will propagate. The numerical aperature (NA) is related to θ_c, and the core and cladding refractive indicies n_1 and n_2 as shown.

Signal losses depend on attenuation within the fiber, whereas ease of coupling and interconnection depend on the fiber's light-carrying core diameter and its numerical aperture. The numerical aperture is a measure of the angle over which light is accepted (Figure 10.4). The bandwidth depends on dispersion. The information-carrying capacity (bandwidth) of a fiber is inversely related to its total dispersion. The total dispersion is a combination of three components: mode dispersion, material dispersion, and waveguide dispersion. Modal dispersion occurs in multimode fibers because different modes travel different effective distances through the fiber. An example illustrates.

EXAMPLE 10.1

(a) For the figure, find the minimum angle that supports total internal reflection.
(b) Calculate the total path length of a ray entering at this angle.
(c) Compare the arrival-time difference with an axial ray.

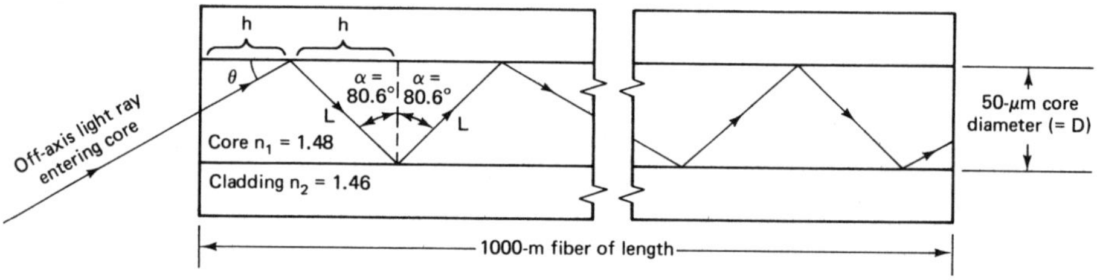

Solution

(a) Using Snell's law, we have

$$\sin \gamma_{min} = \frac{n_2}{n_1} = \frac{1.46}{1.48}$$

which gives $\Theta_{max} = (90° - \gamma_{min}) = 90° - 80.6° = 9.4°$

(b) From this figure,

$$h = D \tan \gamma_{min} = (50\ \mu m) \tan 80.6° = 302.02\ \mu m$$

$$L = \frac{h}{\sin \gamma_{min}} = \frac{h}{\sin 80.6°} = 306.13\ \mu m$$

$$\text{Number of } h\text{'s} = \text{number of } L\text{'s} = \frac{1000 \text{ m}}{h} = 3.311 \times 10^6 \text{ m}$$

$$\text{Zigzag path distance} = (\text{number of } L\text{'s})\,(L)$$
$$= (3.311 \times 10^6)(306.13 \times 10^{-6})$$
$$= 1013.6 \text{ m}$$

(c) $$\Delta_{\text{distance}} = 1014 - 1000 - 14 \text{ m}$$

$$\Delta t_{\text{diff}} = \frac{\Delta_{\text{distance}}}{v} = \frac{\Delta_{\text{distance}}}{c/n}$$

$$= \frac{14}{(3 \times 10^8)/1.48} = 69 \times 10^{-9} = 69 \text{ ns}$$

Note from this example that if the bouncing ray and the axial ray both enter the fiber at the same time, the bouncing ray arrives 69 ns after the axial ray. Obviously, there will be rays intermediate between these two. The end result as far as digital transmission is concerned is bit smearing or ISI. As estimate of the dispersion-limited bandwidth is

$$\text{BW} \approx \frac{1}{\Delta t} = \frac{1}{69 \times 10^{-9}} = 14.5 \text{ MHz}$$

The fiber in Example 10.1 supports numerous modes; thus it can be referred to as a *multimode step-index* type of fiber (see Figure 10.5(a)). From electromagnetic field analysis, there exists an interesting parameter called the normalized frequency, V. It is defined as

$$V = \frac{\pi d}{\lambda}\sqrt{n_1^2 - n_2^2} \tag{10.8}$$

where d is the fiber core diameter and λ is the optical wavelength. The number of propagating modes, N, in a step-index fiber can be estimated by (for V and N both $\gg 1$)

$$N \approx \frac{V^2}{2} \tag{10.9}$$

Note from equation (10.8) that there are three ways to reduce the number of modes in a step-index fiber:

1. Reduce the fiber diameter.
2. Decrease the difference between the cladding and core indices.
3. Increase the wavelength.

Multimode dispersion can also be reduced by making the refractive index of the fiber core graded in the radial direction, hence the name *graded-index* fibers

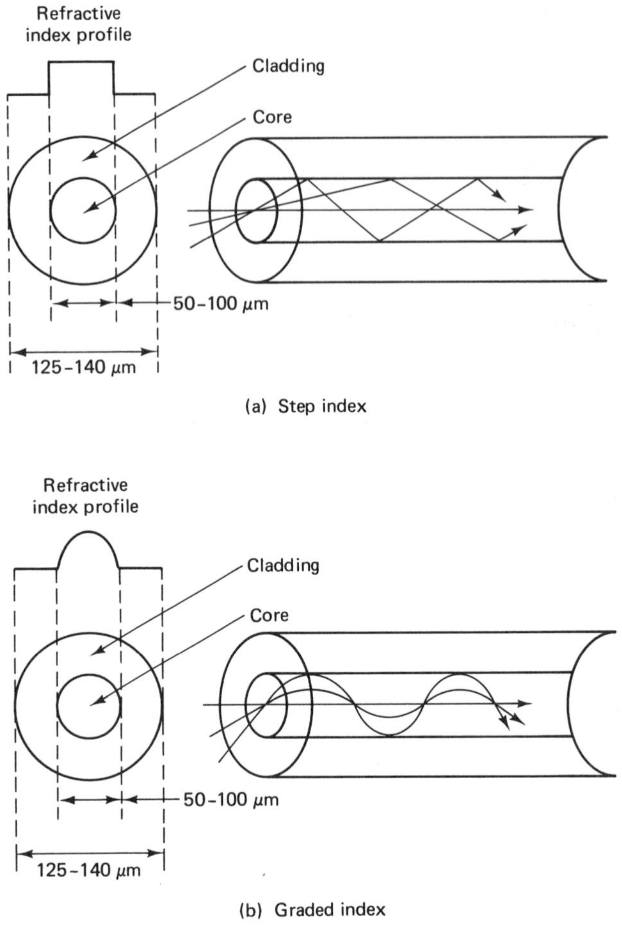

Figure 10.5 Schematic illustration of the propagation of light rays in multimode. (a) Step-index fibers. (b) Graded-index fibers. In step-index fibers the rays may be thought to follow a zig-zag path, whereas in graded-index fibers, the paths are insusoidal.

(see Figure 10.5(b)). The rays in a graded-index fiber follow sinusoidal paths. Thus much of the travel time is in a low refractive index region where it travels more rapidly. Careful design of the index profile results in a design wherein nearly equal transit times for all rays is possible. Bandwidths on the order of 1 GHz/km are available in commercial graded-index fibers. This is in marked contrast to 20 MHz/km for multimode step-index fiber.

Improvements in technology have resulted in a returning interest in single-mode fibers. Refer to equation (10.9). Studies indicate that if V is less than 2.405, only a single mode (an axial ray) can propagate. For example, with $\lambda = 0.82$ nm, $n_1 = 1.48$, and $n_2 = 1.46$, use of equation (10.8) predicts a maximum-core diameter of 2.6 μm. With only a single mode present, there is no modal dispersion present, and very high bandwidths are possible. Material and waveguide dispersion, however, prevents the bandwidth from being infinite. Material and waveguide dispersion are subtly different effects that originate because light propagation through materials is a function of wavelengths. Both are measured in picoseconds (of pulse spreading) per nanometer (of source spectral width) per kilometer (of fiber length). This measurement reflects both the increase in magnitude with source linewidth and the accumulation of dispersion over the length of the fiber. The total dispersion for a single mode fiber is the sum of the two.

Physically, material dispersion is caused by the variation in the refractive index of glass. It is also sometimes referred to as *spectral* or *chromatic* dispersion because this same effect causes a prism to separate the spectrum. This wavelength dependence results in pulse spreading even when different wavelengths follow the same path. Mathematically the index of refraction may be expressed as

$$n = \frac{c}{v} \tag{10.10}$$

where c is the speed of light in a vacuum (3×10^8 m/s) and v is the speed of light in the fiber. By way of contrast, waveguide dispersion occurs because light in a single mode fiber is not completely confined to the core. Roughly 20 percent of the light travels in the cladding surrounding a step-index core. Referring to Figure 10.2, we see that the index of refraction of the cladding is lower than that of the core, so light wants to travel faster in the core. However, with a single mode all the light must travel at the same speed. An effective velocity is established based on the core and the cladding. Dispersion results because this effective velocity is wavelength dependent.

The material and waveguide dispersion can have opposite signs, and the net effect is a complete cancellation. This *zero-dispersion* wavelength is on the order of 1.3 μm for conventional germanium-doped silica fibers. The bandwidth does not become infinite at this wavelength but as a practical matter does exceed most requirements. This wavelength is also important for graded index fibers because it removes the increase in dispersion with increase in spectral bandwidth of the source. This makes possible the use of spectrally broadband LEDs at 1.3 μm.

The size of the fiber core is related to the ease of coupling light from the source of the fiber and for interconnecting fibers. Generally speaking, the core area should be no smaller than the emitting area of the source. For single-mode fibers, the light sources should be lasers emitting in a single transverse mode. An optical fiber with a core diameter $2a$, a fractional index difference between core and cladding $\Delta = \Delta n/n$, and a homogeneous core refractive index will transmit a single mode for all wavelengths longer than the cutoff wavelength λ_c, where

$$\lambda_c = \frac{2\pi a n\sqrt{2\Delta}}{2.405} \tag{10.11}$$

and $\Delta = (n_1 - n_2)/n_1$. At shorter wavelengths the fiber will support two or more modes. For operation at 1.3 or 1.55 μm, the diameter is on the order of several micrometers. This leads to tight mechanical tolerances in coupling and to a small numerical aperture. The payoff is very broad bandwidth.

EXAMPLE 10.2

Calculate the numerical aperture for the fiber of figure 10.2. What is the acceptance cone half-angle?

Solution

$$NA = \sin \Theta_c = \sqrt{n_1^2 - n_2^2} = \sqrt{1.48^2 - 1.46^2} = 0.242$$

$$\Theta_c = \sin^{-1} 0.242 = 14.03°$$

10.2 OPTICAL TRANSMITTERS AND RECEIVERS

The link shown in Figure 10.3 will operate satisfactorily only as long as the received optical power exceeds some minimum level, P_R, called the *receiver sensitivity*. As an optical pulse propagates down the fiber, it is attenuated exponentially. Thus, at a distance l km from the transmitter, the signal power is given by

$$P(l) = P_T\, 10^{-Al/10} \tag{10.12}$$

where P_T is the transmitted power and A is the fiber attenuation expressed in decibels per kilometer. This implies a maximum-loss limited-transmission distance given by

$$L_{\text{MAX}} = \frac{10}{A} \log_{10} \frac{P_T}{P_R} \tag{10.13}$$

where P_T and P_R are the transmitted and received power, respectively. Note that L_{MAX} is very sensitive to fiber attenuation and only weakly (logarithmic scale) dependent on transmitted power. This is in contrast to RF, where signal attenuation obeys an inverse square law. The desire to reduce fiber attenuation has resulted in the gradual evolution from an operating wavelength of 0.8 μm to 1.55 μm. This in turn has forced the change from gallium arsenide–based optical sources to new longer-wavelength technology based on indium phosphide.

The transmitted power in equation (10.13) is optical power actually coupled into the fiber. Typically, manufacturers plot the source-beam pattern on a polar diagram (see Figure 10.6). These curves are derived by measuring the relative radiant intensity on a small detector that swings through a 180° arc. As shown in Figure 10.6(a), the intensity of a uniform surface emitter varies with the cosine of the angle between the line perpendicular to it and another line to the observation point. This is called a *Lambertian source*, and the emission power profile can be described by

$$P = P_o \cos \phi \tag{10.14}$$

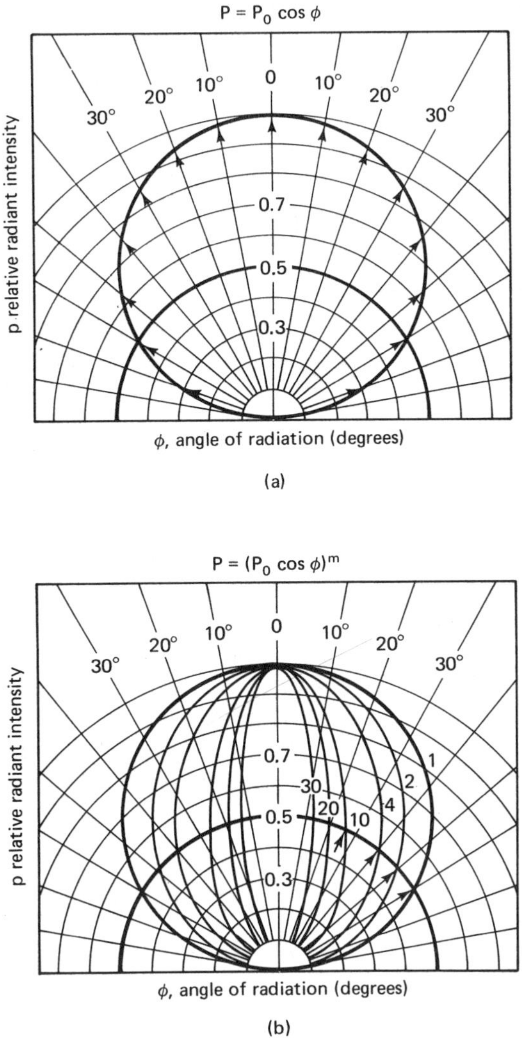

Figure 10.6 A wide variety of emission-beam patterns appears in sources available today. (a) Uniform surface emitters typically have Lambertian-type profiles. (b) Some sources exhibit much narrower beam profiles.

where P_o is the radiant intensity along the line $\phi = 0$. Other sources exhibit narrower beam profiles, which can be approximately described by

$$P = P_o(\cos \phi)^M \tag{10.15}$$

A set of curves for various values of M are shown in Figure 10.6(b). We can calculate the amount of power coupled into a fiber by the expression

$$P_C = P_T[1 - (\cos \theta)^{M+1}] \tag{10.16}$$

where P_T is the total source power and θ is the fiber's acceptance cone half-angle.

EXAMPLE 10.3

Calculate the aperture loss for Example 10.2 if a Lambertian source is used.

Solution

$$P_C = P_T[1 - (\cos\theta)^{M+1}] = P_T[1 - \cos^2\theta]$$

$$= P_T \sin^2\theta = P_T(NA)^2$$

$$\frac{P_C}{P_T} = (NA)^2 = 0.242^2$$

$$NA_{\text{loss}} = 10 \log \frac{P_C}{P_T} = 10 \log 0.242^2 = -13.15 \text{ dB}$$

Solid-state laser diodes are exceptionally well suited for use in digital fiber optic systems. Injection laser diodes (ILDs) exhibit superior performance over light emitting diodes (LEDs). The laser diode is preferable to the LED because of its superior modulation speed capability (>4 gigabits/s), and better optical coupling efficiency into single-mode fibers. Stripe-contact versions of the ILD have active emitter regions measuring 1 × 20 μm. While emission through the narrow slit produces a rather broad beam in the plane perpendicular to the junction (20° to 40° half-angle), the beam in the plane parallel to the junction is much narrower (Figure 10.7). The spectral width is considerably purer than that of an LED. The beam is far more directional than that of an LED and typical *NA* losses into flat-ended fibers are on the order of 6 dB.

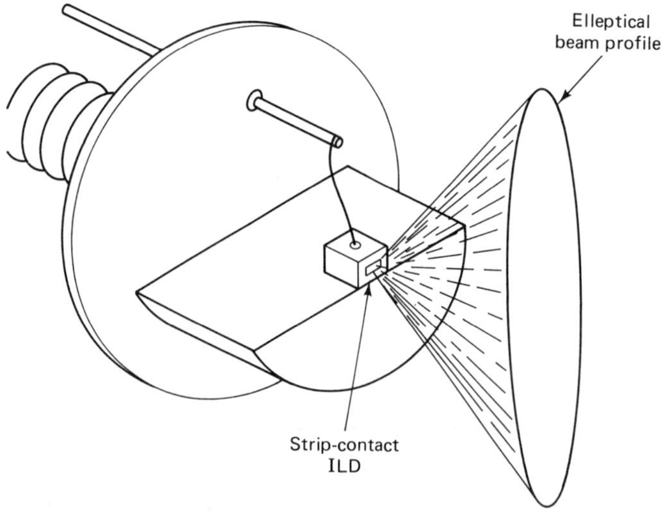

Figure 10.7 Significant NA-loss reduction results from the narrow-beam profile (in the plane parallel to the junction) of stripe-contact-configured laser diodes.

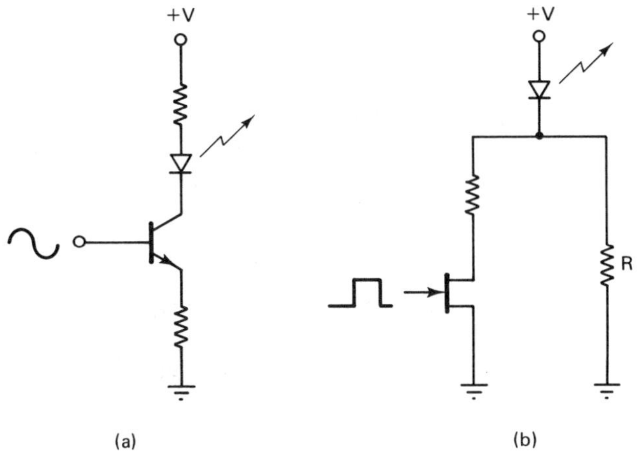

Figure 10.8 Straightforward basics suffice when designing source drive circuits. (a) A simple transistor circuit works well for analog-intensity modulation. (b) An FET circuit works well for pulsed-modulation schemes.

Basic drive circuits for diode light sources are shown in Figure 10.8. The circuit of Figure 10.8(a) is preferred for linear analog intensity modulation, whereas the circuit in Figure 10.8(b) is preferred for digital (pulse) modulation techniques. For PPM or digital OOK, we can use an FET to switch an LED or ILD. For example, with an ILD with a sharp "knee" at the lasing threshold current, we can adjust R to bias the ILD just below threshold. This action increases the switching speed, since the circuit state change does not have to begin with zero current. Increased optical power can be obtained by paralleling or connecting diodes in series. The best switching speed is maintained with the diodes in parallel. More current, however, is required.

Current light wave receiver design is markedly different from the digital radios discussed in Chapter 8. These receivers employ direct detection of the incoming signal rather than first performing a heterodyne conversion. As noted in the introduction, heterodyne receivers at optical frequencies are still in the laboratory stage. We basically have two choices for photodetectors: PIN photodiodes and avalanche photodiodes (APDs). In general, PIN diodes are preferred at bit rates below 100 Mb/s. APDs are the choice at high-bit-rate applications (≥ 1 Gb/s). For bit rates between 10 Mb/s and 2 Gb/s and over the wavelength range 0.8–1.6 m, the best light wave receivers require a flux on the order of 300–1000 photons per bit. This corresponds to a sensitivity of

$$P_R \sim 2 \times 10^{-13}\, B \text{ mW} \tag{10.17}$$

where B is the bit rate.

Before considering system parameters further, it is of interest to look at receiver designs first. The PIN photodiode mentioned previously consists of a large intrinsic (very lightly doped) region sandwiched between p- and n-doped semiconducting regions (Figure 10.9). Photons of light energy absorbed in this region create electron-hole pairs, which are then separated by an electric field.

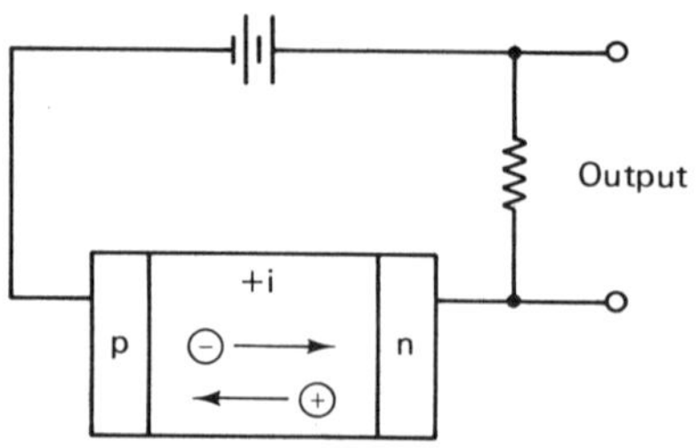

Figure 10.9 Electron-hole pairs, created by photons absorbed in the intrinsic region of a PIN diode, are separated by the diode's electric field to produce a current in the load circuit.

This generates an electric current in the load circuit. The efficiency of the optical photo–electron-hole conversion process is specified by the photodiode's quantum efficiency. This is a measure of the average number of electrons released by each incident photon. A number near 1 is a highly efficient diode detector.

The response of the diode to a light stimulus is a more meaningful design parameter. It is a ratio of the output current to the incident optical power and may be expressed as

$$r = \frac{\lambda^2}{1.24} \text{ A/W} \tag{10.18}$$

The signal current generated when P watts of optical power are incident on the diode is

$$i_s = rp = \frac{r\lambda P}{1.24} \text{ A} \tag{10.19}$$

For a typical PIN diode, the current is generally less than 1 A/W.

Some basic photodiode circuits are shown in Figure 10.10. The transimpedance design suits current sources such as PIN or APD diodes. The op-amp effec-

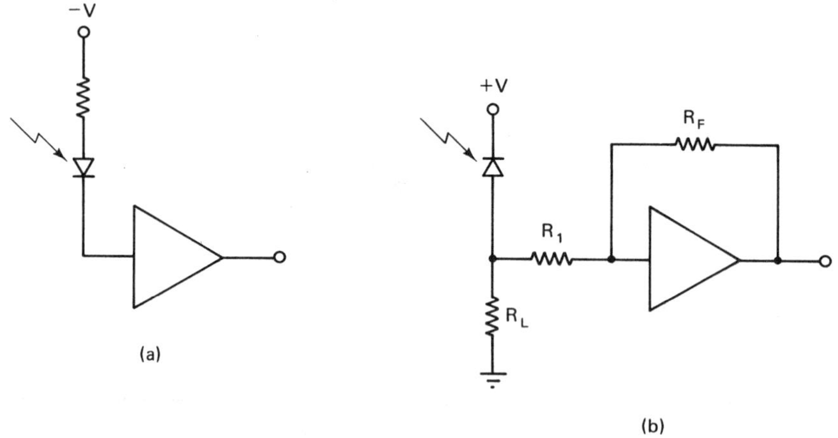

Figure 10.10 (a) Transimpedance receiver. (b) Voltage gain receiver.

tively converts current into a voltage out. The second design amplifies the signal voltage developed across R_L. Optical receivers are available commercially with the photodiode and amplifier as an integrated package.

As noted previously, with an optical source, receiver, method of modulation, and a fiber, we have a light wave system. A measure of performance for a light wave system is the product of the bit rate (B) and the length (L). Curves of L_{max} based on equations (10.13) and (10.17) are shown in Figure 10.11. The nearly horizontal shape of the curves illustrates that it is possible to increase $B \cdot L$ by making large increases in the bit rate while suffering only a small increase in length. Note, however, at the breakpoints, dispersive effects limit this technique.

Remember that dispersion arrives because different components of a pulse arrive at the receiver at different times (see Example 10.1). The typical spread in propagation times of the different components is called the *delay spread*, S. For many systems the degradation is small provided that S is

$$S < \frac{1}{2B} \tag{10.20}$$

That is, S is less than half a symbol period. Delay spread in high-quality graded-index multimode fibers increases with propagation distance at a typical rate of 0.25 ns/km. From equation (10.20) we have

$$B \cdot L < 2 \text{ Gb/s} \cdot \text{km} \tag{10.21}$$

This relation is indicated in Figure 10.11 in the leftmost dashed line.

Note from Figure (10.11) that the dispersive effects for single-mode fibers is much less. In fact, with a single mode there is no modal dispersion, leaving

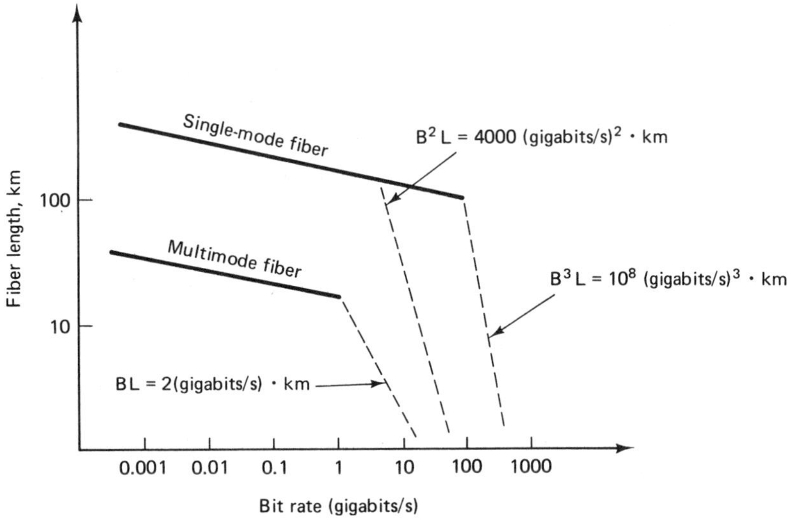

Figure 10.11

material dispersion that results from the frequency dependence of the refractive index of the fiber material. The Fourier components of an optical pulse propagate at different speeds, and the resulting delay spread is given by

$$S \approx D \cdot L \cdot W \tag{10.22}$$

where

D = absolute value of the linear dispersion of the medium
L = fiber length
W = pulse bandwidth

The dispersion, D, for typical silica fibers is shown in Figure 10.12. Note the zero dispersion at 1.3 m. At 1.55 μm, fibers exhibit minimum loss, but the dispersion is still small. In terms of $B \cdot L$ performance, best results are obtained at 1.5 m. The pulse bandwidth for an ideal transmitter is approximately equal to the bit rate. Using this fact, equations (10.20) and (10.22) yield the following dispersion constraint:

$$B^2 L < 4000 \text{ (gigabits/s)}^2 \cdot \text{km} \tag{10.23}$$

This constraint is shown in Figure 10.11.

A promising technique for achieving even higher $B \cdot L$ performance involves the use of a dispersion-shifted fiber. The dispersion $D = 0$ is shifted up to coincide with the wavelength of minimum loss. At this wavelength, the dispersive effects are very small (second order). Instead of being linear in B, the delay spread becomes quadratic. This leads to a dispersion constraint given by

$$B^3 L < 10^8 \text{ (gigabits/s)}^3 \cdot \text{km} \tag{10.24}$$

Referring to Figure 10.11, it appears that bit rates above 10 gigabits/s over spans of 100 km should be achievable. The primary technical challenge is not the fiber but rather the optical and electronic components at the fiber ends. As of this writing, research activity aimed at achieving these goals is intense. Data rates have been doubling every two years; thus it appears that it is only a matter of time until the transmission rate of 10 gigabit/s over long distance is achievable.

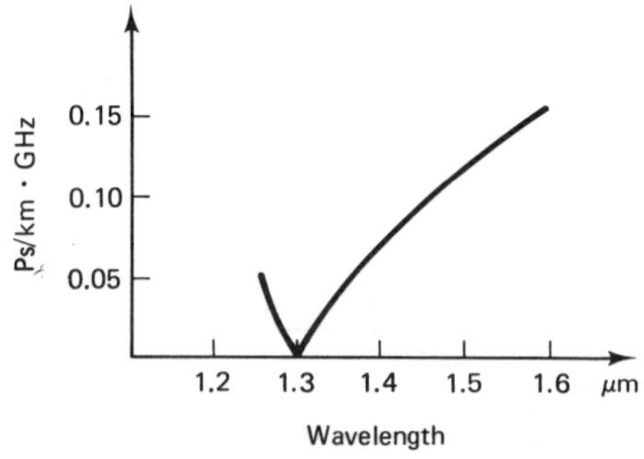

Figure 10.12 Linear material dispersion of silica fiber. The linear dispersioon is zero near 1.3 μm wavelength, but at 1.55 μm it is large enough to cause significant impairments.

10.3 DIGITAL MODULATION AND DETECTION TECHNIQUES

Despite the rapid progress made in light wave technology in recent years, direct modulation of the source (OOK) and direct detection at the receiver using a PIN diode or APD remains the mainstay of present systems. Recent advances promise improvements in this time-honored approach. For example, external modulation of the laser with an electro-optic device demonstrated modulation bit rates greater than 4 gigabits/s. As noted in the introduction to this chapter, coherent lightwave communications using single-mode optical fibers and the optical analog of superhetrodyne radio for reception promise the best improvements. All the modulation formats that have been discussed for digital radios can, in principle, be used in a coherent optical communication system. Specifically, this refers to amplitude, phase, and frequency-modulation schemes. The modulation is carried out using external electro-optic acoustopic modulators.

10.3.1 Intensity Modulated Systems

Intensity modulation (IM) systems are the easiest to manufacture and are also the least expensive. These operate directly on the input signal to drive a laser or LED (see Figure 10.8). Most digital optical-fiber communication systems utilize

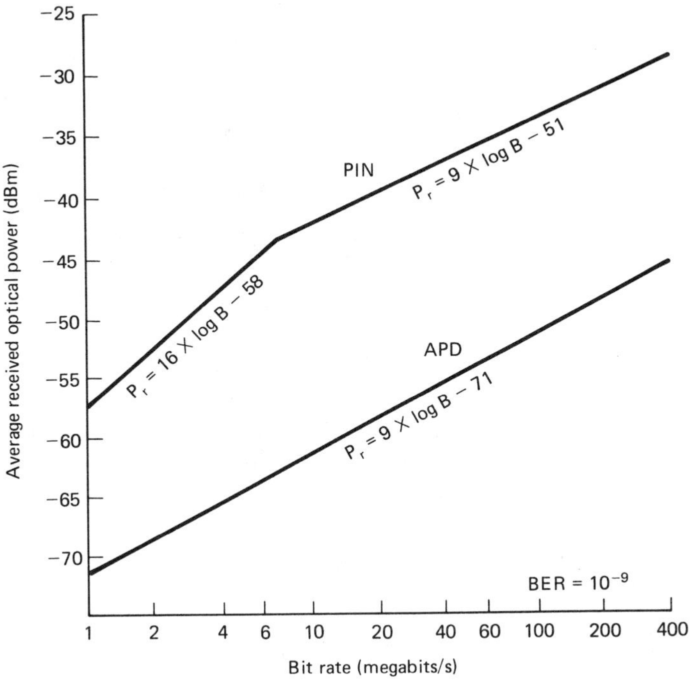

Figure 10.13 Required optical power versus bit rate.

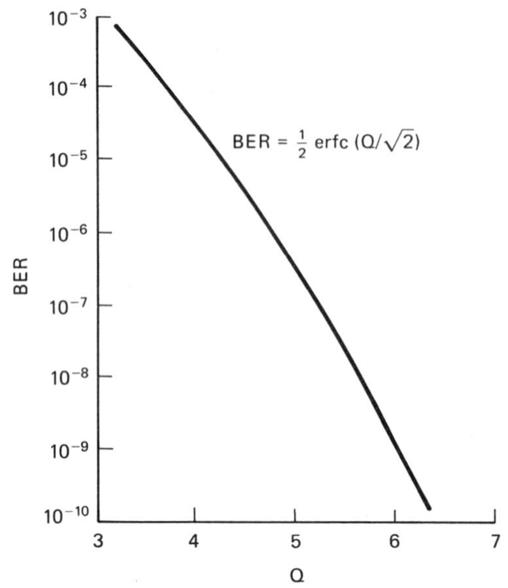

Figure 10.14 BER versus Q for binary PCM reception.

binary intensity modulation of the optical source. Baseband PCM transmission of this sort is designated as PCM-IM. The required received optical power versus BER $= 10^{-9}$ is shown in Figure 10.13 for both PIN and APD diodes. For a bipolar preamplifier and a 500-kHz bit rate, the required received power is

$$P_r = 9 \times \log B - 51$$

$$= 9 \times \log 0.5 - 51 = -53.7 \text{ dBm } (4.3 \times 10^{-6} \text{ mW})$$

The minimum average optical power, $P_{\min}$, is given by

$$P_{\min} = \frac{h\nu}{\eta e} Q(i_{\text{eff}}^2)^{1/2} \tag{10.25}$$

where Q is a function of the error probability, P_e.

$$Q = \sqrt{2}/\text{erfc}^{-1}(2P_e) \tag{10.26}$$

and i_{eff} is the effective value of the noise current. The quantity $h\nu$ is equal to $(2 \times 10^{-19})/(\lambda/\mu\text{m})$. For a given BER, we can obtain $P_{\min}$ if we know Q and i_{eff}. Equation (10.26) is plotted in Figure 10.14. From Figure 10.14, $Q = 6$ for BER $= 10^{-9}$. Thus, using equation (10.25), we have

$$(i_{\text{eff}}^2)^{1/2} = \frac{P_{\min}}{Q(h\nu/\eta e)}$$

$$= \frac{4.3 \times 10^{-9}}{6[92 \times 10^{-19}/0.8)(1.0)(1.6 \times 10^{-19})]}$$

$$= 0.458 \times 10^{-9} = 0.458 \text{ nA}$$

For $P_e = 10^{-4}$, $Q = 3.6$. Thus, using equation (10.25), $P_{\min}$ is

$$P_{\min} = \frac{(2 \times 10^{-19}/0.8)(3.6)(0.458 \times 10^{-9})}{(1)(1.6 \times 10^{-19})}$$

$$= 2.58 \times 10^{-6} \text{ mW } (-56.02 \text{ dBm})$$

Once the required received optical power for a given BER is obtained, it remains to determine the required transmitter power. This can readily be determined from a link analysis. For satisfactory system performance, it is necessary to budget the optical losses. We can express the allowable link losses, L, as

$$L = P_s - P_r \tag{10.27}$$

where P_s is the source power and P_r is the required power at the receiver. Based on the preceding discussion, the link-loss budget can also be expressed in terms of its components as

$$L = \text{ICL} + \text{CSL} + \text{AL}(D) + \text{OCL} \tag{10.28}$$

where

ICL = input-coupling loss
CSL = connector and/or splice loss
AL(D) = attenuation loss in the fiber
OCL = output-coupling loss

In this expression, D is the power-limited distance of the fiber. Beyond this distance, the received power falls below the required received power for a particular BER. Neglecting dispersion, then, D determines the maximum repeater spacing. A design margin is usually added by allowing an extra 2 or 3 dB of loss. Suppose that we have a system with the following loss parameters:

$$L = \text{ICL} + \text{CSL} + \text{AL}(D) + \text{OCL} + \text{safety}$$

$$= 12.2 + 3 + \text{AL}(D) + 1 + 3$$

$$P_s = 2 \text{ mW} + 3 \text{ dBm}$$

$$P_r = -35 \text{ dB}$$

Then

$$L = P_s - P_r = 3 - (-35) = 38 \text{ dB}$$

Therefore, AL(D) = 18.8 dB.

Typical system calculations for a typical fiber optic link are as follows:

System Component Specifications

Source:

P_o = 125 W at 100 mA
NA = 0.7 (for 10 dB effective)
Core diameter = 200 μm
Wavelength = 900 nm

Detector:

Responsivity (R) = 0.4 A/W at 900 nm
NA = 0.7 (for 10 dB effective)
Core diameter = 200 μm
I(dark) = 200 nA at 25°C

Fiber:

Length = 50 m
Attenuation = 25 dB/km at 900 nm
NA = 0.32 (at 50 m)
Core diameter = 200 μm

Connector (2):

Gap = 0.15 mm typical
Misalignment = 0.05 mm typical
(Gap and alignment attentuation from manufacturer's data; for NA 0.7 = 2.7 dB; for NA 0.32 = 1.5 dB

System Calculations

	dB
LED-to-fiber connector	2.7
LED-to-fiber fresnel (typical)	0.2
LED-to-fiber NA (20 log 0.7/0.32)	6.79
Fiber attenuation (25 dB/km × 0.05 km)	1.25
Fiber-to-detector connector	1.5
Fiber exit fresnel	0.2
Detector entry fresnel	0.2
Fiber-to-detector NA	0.2
Total system loss	0
	12.84

Power input to the detector can now be calculated from the relationship

$$\text{System loss} = 10 \log[P(\text{in})/P(\text{out})]$$

$$12.84 = 10 \log 125\ \mu\text{W}/P(\text{out})$$

$$P(\text{out}) = 6.5\ \mu\text{W}$$

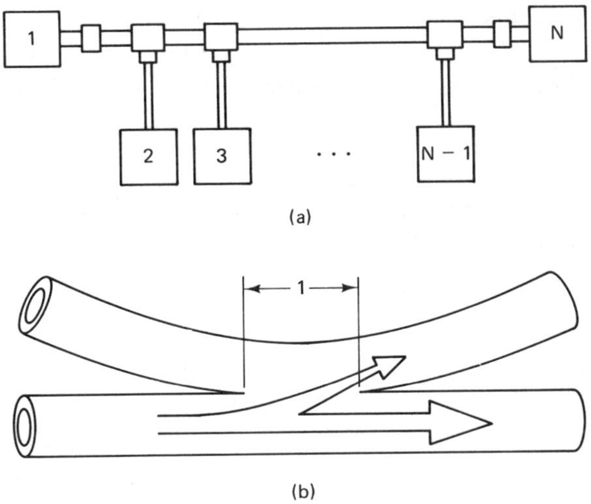

Figure 10.15 (a) Terminals tap or inject optical energy onto the main trunk in the in-line bus configuration. (b) One approach to the T-coupler implementation of this configuration is to fuse the fibers together. The amount of light coupled varies with interaction length, l, and core-to-core proximity.

From the detector responsivity, we can also calculate the signal current output from

$$\begin{aligned} I_o &= P(\text{in}) \times R \\ &= 6.5\ \mu\text{W} \times 0.4\ \mu\text{A/W} \\ &= 2.6\ \mu\text{A} \end{aligned}$$

From the detector noise rating (dark current, I_d), we can also determine the SNR: SNR = 10 log(2.6/0.002) = 3.1 dB.

For multiterminal systems, loss budgeting is somewhat more complex. In this situation we must account for such factors as coupler tap, ratio losses, insertion losses, varying distance to terminals, and dynamic range considerations. A typical configuration is illustrated in Figure 10.15. This configuration requires use of a *T-coupler*. This coupler is often referred to as an *in-line* data bus. From a consideration of the architecture (Figure 10.15), we see that the optical signal suffers a loss at each T-coupler on the bus. The loss increases linearily with the number of terminals on the bus.

10.4 PULSE-MODULATION TECHNIQUES

In terms of intensity modulation, other techniques readily come to mind. For example, as in RF systems, subcarriers can be modulated and the summed outputs used to intensity modulate the light. Thus, for example, we have AM-IM, FM-IM, PM-IM, and so on. In addition, the familiar pulse-modulation techniques are available. Strictly speaking, pulse modulation is an analog technique. The technique, however, can be so intermarried with digital transmission that it warrants inclusion in this text. In this context, we have available to us such schemes as pulse-amplitude modulation (PAM), pulse-width modulation (PWM), pulse-position modulation (PPM), and pulse-frequency modulation (PFM), among others. In particular we wish to discuss PFM-IM, in which the effects of wide band FM gain may be obtained. In terms of SNR, this technique provides a yardstick against which the other pulse-modulation techniques may be measured. As in other pulse-type systems, the rise time plays a prominent part. Prior to discussing PFM-IM systems, it is of interest to first consider the end to end-system rise time of a fiber optic system.

10.4.1 System Rise Time

A consideration for system bandwidth must be the overall system rise time. The source and detector rise times, for example, can distort the shape of a signal (pulse). Typically, these range from 5 to 15 ns for LEDs and 0.1 to 2 ns for ILDs. At the receiving end, PINs and APDs have rise times between 1 and 4 ns. The total system rise time may be expressed as

$$T_{\text{system}} = 1.1\sqrt{T_{\text{source}}^2 + T_{\text{modal}}^2 + T_{\text{material}}^2 + T_{\text{detector}}^2} \qquad (10.29)$$

where the T's are the 10 percent and 90 percent rise times. The weighting factor of 1.1 results from a comparison of the rise time edge with the overall pulse dispersion. The -3-dB bandwidth, and therefore maximum bit rate, is usually defined in terms of T_{system} and the rise time of an RC circuit (Figure 10.16).

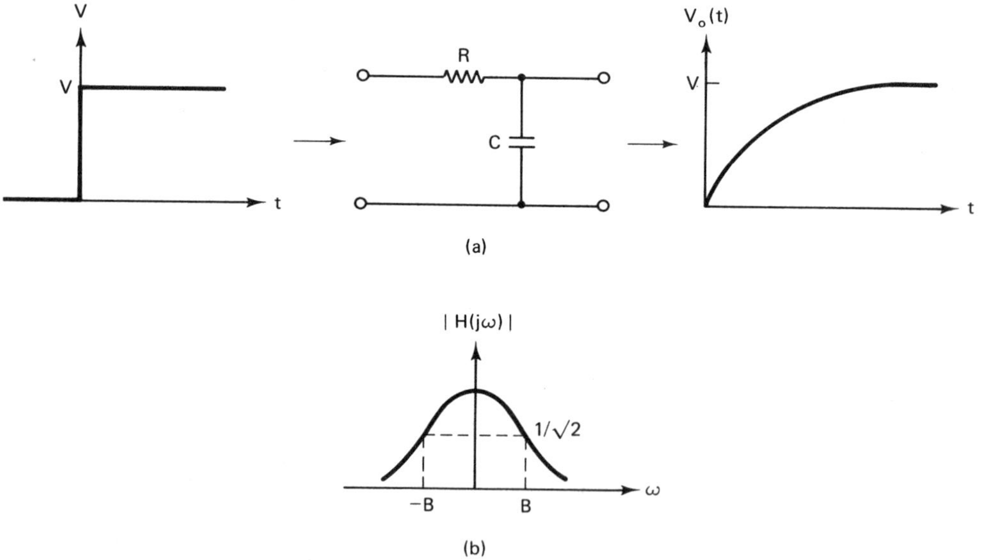

Figure 10.16 (a) The response of a low-pass RC filter circuit to a voltage step input. (b) The transfer function $H(\omega)$ for the circuit in (a).

With a step input voltage of amplitude V, the output voltage, $V_{out}(t)$, is given by

$$V_{out}(t) = V(1 - e^{-t/RC}) \tag{10.30}$$

and the 10 percent to 90 percent rise time, t_r, is

$$t_r = 2.2RC \tag{10.31}$$

The transfer function (Figure 10.16(b)) is readily determined to be

$$H(j\omega) = \frac{1}{1 + j\omega RC} \tag{10.32}$$

From this expression we have the familiar result for the -3-dB bandwidth; that is

$$\text{BW} = \frac{1}{2\pi RC} \tag{10.33}$$

Combining equations (10.31) and (10.33) results in

$$t_r = \frac{2.2}{2\pi} = \frac{0.35}{BW} \tag{10.34}$$

where $t_r = T_{system}$ for a fiber optic link.

Obviously, different results are obtained for different filters. For fiber optic systems, the constant 0.35 is a conservative number and is often used. In order to retain the shape of the pulse at the filter output with reasonable fidelity, the -3-dB bandwidth must be at least large enough to satisfy the condition $\text{BW}\cdot\tau = 1$, where τ is the pulse duration. Thus from equation (10.34) we have

$$T_{sys} = t_r = 0.35\tau \tag{10.35}$$

For an RZ code, the bit rate is $R = \text{BW} = 1/\tau$. Using equation (10.35), this results in

$$R_{max} = \frac{0.35}{T_{sys}} \tag{10.36}$$

On the other hand, for an NRZ code, $R = \text{BW}/2 = 1/2\tau$ and R_{max} is given by

$$R_{max} = \frac{0.7}{T_{sys}} \tag{10.37}$$

We begin to see the bounds on bit rate associated with different codes.

EXAMPLE 10.4

An optical fiber system is to be designed to operate with repeater spacing at 8 km. Components available have the following specifications:

Fiber: intermodal	5 ns/km
Pulse spreading: intramodal	1 ns/km
Pin diode	5 ns
LED	7 ns

Estimate the maximum bit rate the link will support for (a) RZ and (b) NRZ.

Solution

(a) The total system rise time is

$$T_{\text{system}} = 1.1[7^2 + (8 \times 5)^2 + (8 \times 1)^2 + 7^2]^{1/2}$$

$$= 46.17 \text{ ns}$$

$$R_{\max} = \frac{0.35}{T_{\text{sys}}} = \frac{0.35}{46.17 \times 10^{-9}} = 7.58 \text{ MHz}$$

(b) $$R_{\max} = \frac{0.7}{46.17 \times 10^{-9}} = 15.16 \text{ MHz}$$

The required optical power for the link can be determined as discussed in Section 10.3.1.

10.4.2 Data Encoding

Encoding data often permits a long string of 1s or 0s to be transmitted. This can result in problems in a high-speed fiber optic link. Since light pulses are unipolar (either on or off), the receiver must be ac-coupled to eliminate dc-drift errors. Thus the data must contain a transition during each bit period. NRZ can prove to be unsuitable for high data rates. Manchester coding (biphase) can be used to transmit data at better than 1 megabit/s. The disadvantage of biphase, however, is that the required bandwidth is twice that for NRZ. This effectively halves the bit rate.

More recently developed codes are better suited than Manchester and in some cases improve bandwidth. The most important codes suitable for fiber optics use are:

Frequency-shift code (FSC)
Frequency-shift keying (FSK)
Modified frequency-shift keying (MFSK)
Phase-shift keying (PSK)
Modified phase-shift keying (MPSK)
Delayed modified phase-shift keying (DMPSK)
Phase-delayed keying (PDK)
Modified phase-delay code (MPDC)

TABLE 10.1 CHARACTERISTICS OF DATA TRANSMISSION CODES

Code	Bit duration* Min.	Bit duration* Max.	Patterns to achieve high speed** Worst	Patterns to achieve high speed** Best
NRZ	t		Alternate 1s and 0s	All 1s and 0s
Manchester	$\frac{t}{2}$	t	All 1s or all 0s	Alternate 1s and 0s
FSC	$\frac{t}{2}$	t	All 1s	All 0s
FSK	$\frac{t}{2}$	$\frac{3t}{2}$	All 1s	Alternate 1s and 0s
PSK	$\frac{t}{2}$	t	All 1s or all 0s	Alternate 1s and 0s
MPSK	t		All 1s or all 0s	All 1s or all 0s
DMPSK	$\frac{t}{2}$	$\frac{3t}{2}$	The first 1 after 0	The first 0 after 1
MPDC	t	2t	All 1s or all 0s	The first 0 after 1

* For easy comparison, the tabulated codes each use clocks with periods t.

** Data-coding schemes that feature wide spacing between transitions R decrease the chance of decoding errors.

These codes are based on digital frequency- and phase-modulation schemes and have already been discussed. The characteristics of these data transmission codes are shown in Table 10.1

10.4.3 Pulse-Frequency Modulation

In the past decade, pulse-frequency modulation (PFM) has received considerable attention for video transmission over fiber links. Both PFM and digitally encoded pulse video signals are better suited for optical transmission than are linear IM modulation methods. Since PFM modulation permits frequency multiplexing, digital/PFM fiber interfaces exist in today's systems.

The modulator for PFM may be implemented using a commercially available voltage controlled oscillator (VCO). Demodulation is accomplished using a pulse regenerator followed by a low-pass filter. A block diagram for a PFM-IM optical fiber system is shown in Figure 10.17. The received pulses are detected in a wideband preamplifier and then regenerated with a limiter and a "one-shot." The original modulating signal exists as a baseband component, and recovery is accomplished using a low-pass filter. Detection can also be accomplished without the pulse regeneration; however, the SNR is reduced. The SNR in terms of the

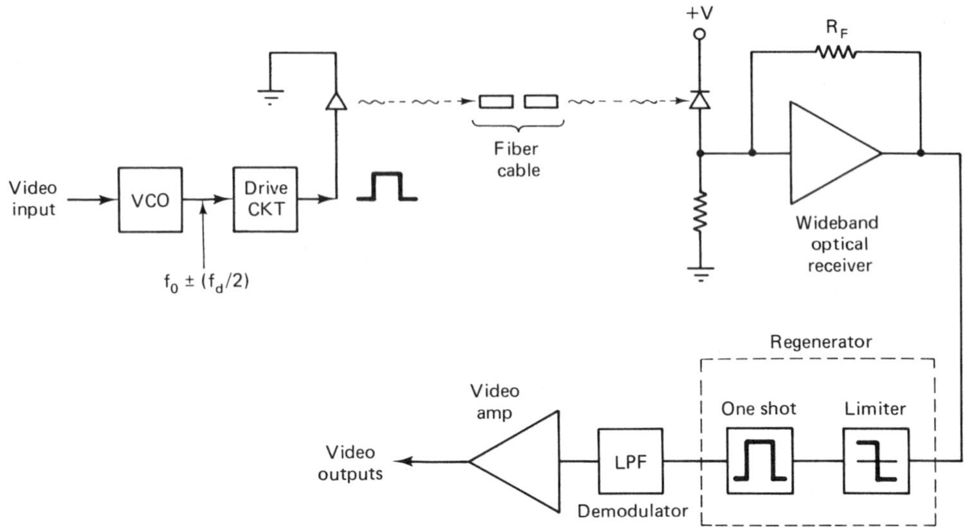

Figure 10.17 A PFM-IM optical fiber system employing regenerative baseband recovery.

peak-to-peak signal power to rms noise power is given by

$$\left(\frac{S}{N}\right)_{P-P} = \frac{12(T_D \Delta f M r P_{so})^2}{(2\pi t_r B)^2 \, \overline{i_{\text{eff}^2}}} \tag{10.38}$$

where

$T_D = \dfrac{1}{fo}$ = normal pulse period
Δf = frequency deviation
M = photodiode (avalanche) multiplication factor
P_{so} = peak receiver optical power
t_r = pulse rise time at the regenerator circuit input
B = postdetection or equivalent baseband noise bandwidth
$\overline{i_{\text{eff}^2}}$ = receiver mean square noise current
r = responsivity

EXAMPLE 10.5

For the receiver in Figure 10.17 and the optical fiber system described by Example 10.4, calculate the peak-to-peak signal-to-rms-noise ratio. Assume an APD photodiode is used in the receiver and the received optical power is -40 dBm. The following system parameters apply:

$$\begin{aligned} \text{Rest pulse rate} &= 10 \text{ MHz} \\ \text{Peak frequency deviation} &= 1.5 \text{ MHz} \\ \text{APD responsivity} &= 0.7 \\ \text{APD multiplication factor} &= 60 \\ \text{Baseband noise bandwidth} &= 5 \text{ MHz} \\ \text{Receiver mean square noise current} &= (0.45 \times 10^{-9})^2 \text{ A}^2 \end{aligned}$$

Solution The pulse rise time at the regenerator circuit should equal the system rise time for an optimum system. Thus

$$tr = T_{\text{system}} = 46.17 \text{ ns}; \quad -40 \text{ dBm} \rightarrow 10^{-7} \text{ W}$$

$$T_D = \frac{1}{fo} = \frac{1}{15} \times 10^{-6} = 6.67 \times 10^{-8} \text{ s}$$

$$\left(\frac{S}{N}\right)_{p-p} = \frac{12[(6.67 \times 10^{-8})(1.5 \times 10^6)(60)(0.7)(10^{-7})]^2}{[(2)(46.17 \times 10^{-9})(5 \times 10^6)]^2[0.45 \times 10^{-9}]^2}$$

$$= 4.98 \times 10^{-8} \rightarrow 73 \text{ dB}$$

We see from this example the possibility of acquiring high SNRs at the output of a regenerative PFM-IM system. Approximately 27 dB of this can be credited to the use of an avalanche photodiode. An additional advantage of PFM is that the power spectrum increases with increasing frequency deviation. This results in preemphasis of the transmitted signal.

A block diagram of a video-data hybrid fiber system is shown in Figure 10.18. As of this writing, systems such as this have been demonstrated in the laboratory. This system supports three full-motion network quality NTSC video channels with stereo audio channels, four ISDN H1 channels, and graphics over a single-mode fiber. The optional replacement of an audio subchannel with 19.6 Kb/s (RS-232) data is also shown. Additional data transmission is possible by wavelength division multiplexing (WDM). We see from this example the enormous impact fiber optics will have on communications.

A marriage of fiber systems and satellites is a natural evolution (see Figure 10.19). Using a Ku-band satellite, a long-distance video-teleconferencing network is established via the connectivity afforded by a shared satellite transponder. Satellite communications is the subject of Chapter 11.

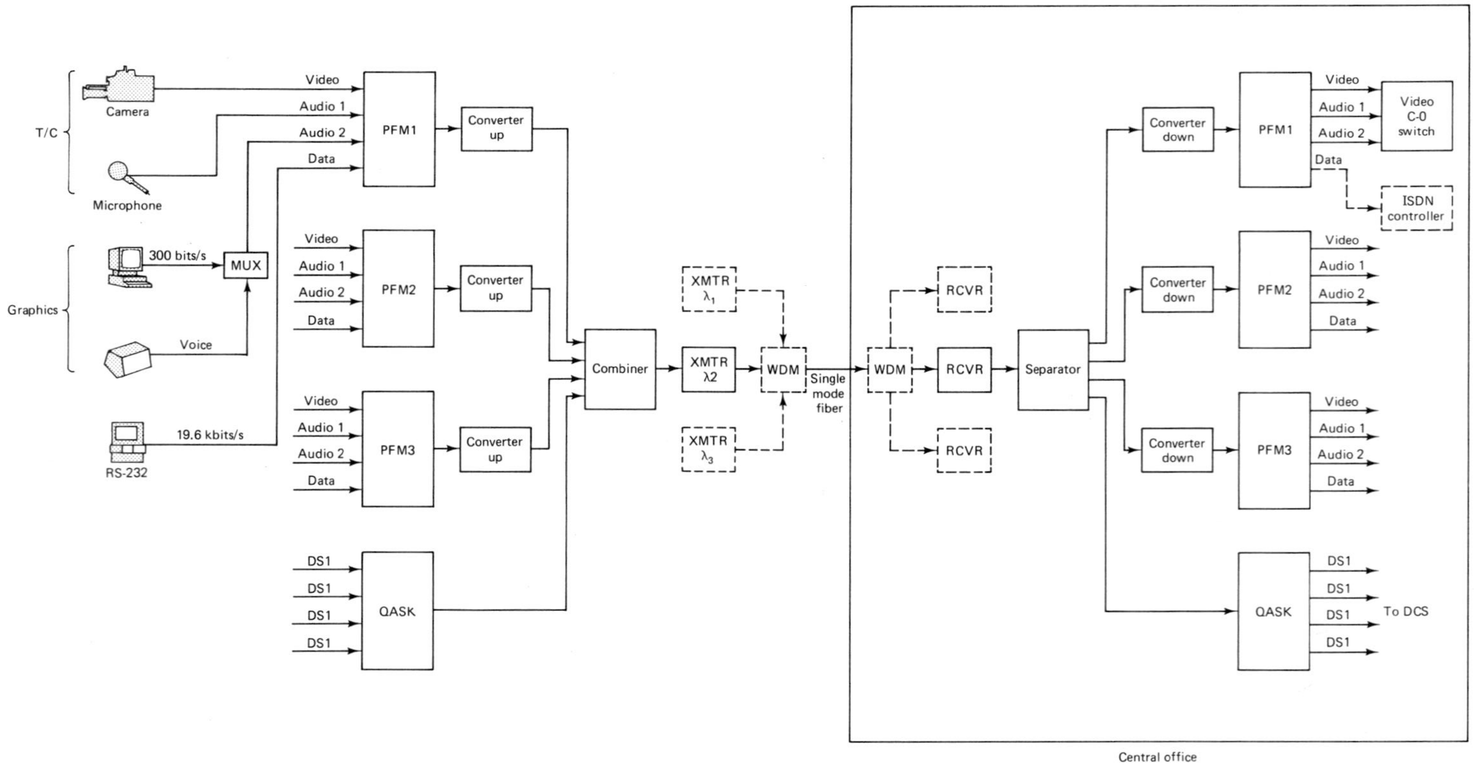

Figure 10.18 Multiplexed fiber transmission system.

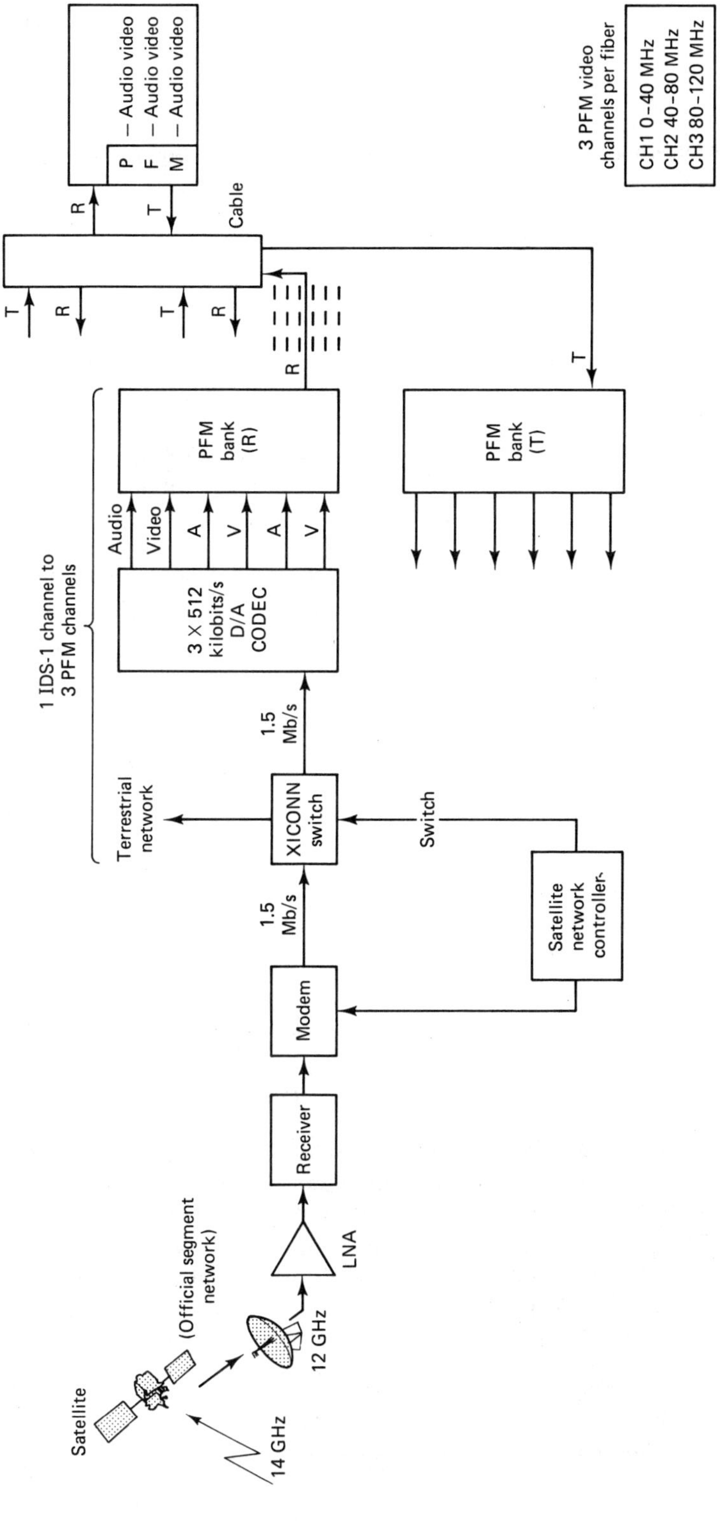
Satellite
14 GHz
(Official segment network)
12 GHz
LNA
Receiver
Modem
1.5 Mb/s
XICONN switch
Terrestrial network
Switch
Satellite network controller
1.5 Mb/s
3 × 512 kilobits/s D/A CODEC
Audio
Video
A
V
A
V
PFM bank (R)
PFM bank (T)
1 IDS-1 channel to 3 PFM channels
R
T
Cable
P – Audio video
F – Audio video
M – Audio video
3 PFM video channels per fiber
CH1 0–40 MHz
CH2 40–80 MHz
CH3 80–120 MHz

Figure 10.19

PROBLEMS

10.1. Twenty-four 4-kHz speech channels are sampled, quantitized, encoded and then time-division multiplexed for transmission as binary PCM on a digital optical fiber link. The quantizer is linear with 0.5-mV steps over a dynamic range of 2.048 V.

(a) Calculate the frame length of the PCM transmission, assuming an additional channel time slot is used for signaling and synchronization.

(b) Find the required channel bandwidth assuming NRZ pulses.

10.2. An analog optical fiber system is to be designed with an SNR of 40 dB at the detector. The postdetection bandwidth is 25 MHz and the operating wavelength is 0.9 μm. If the quantum efficiency of the photodiode is 60%, what is the minimum required optical power?

10.3. A digital optical fiber link employing ideal binary signaling at a rate of 30 megabits operates at a wavelength of 1.3 μm. The detector is a germanium photodiode, which has a quantum efficiency of 60% at this wavelength. An alarm is activated at the receiver when the BER drops below 10^{-5}. Calculate the theoretical minimum optical power required at the photodiode in order to keep the alarm inactivated.

10.4. For the given transimpedance amplifier, show that the following are true:

(a) $V_o \approx I_p R_f$

(b) $f_{-3\text{dB}} \approx A/(2\pi R_F C_F)$

(c) $Z_{\text{in}} \approx \dfrac{R_{\text{in}}}{A}$

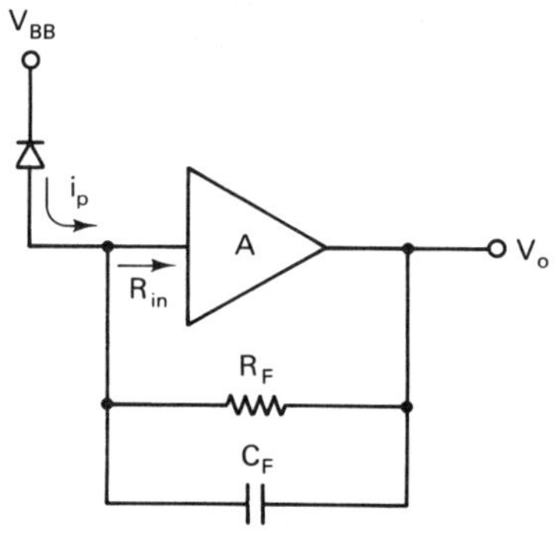

10.5. Equation (10.26) relates the probability of error for binary PCM to the function Q. Develop a relationship between Q and the analog SNR at the output of the photodetector. For an analog SNR of 20 dB, what is the BER?

10.6. Equalization in an optical receiver may be accomplished with the circuit shown. The equalized frequency response is shown. In this design $Rin = 5\ \text{k}\Omega$ and the combined capacitance of the photodiode and amplifier is 5 pF. We desire an upper 3-dB corner frequency for the amplifier-equalizer combination of 30 MHz.

(a) Obtain expressions for ω_1 and ω_2.

(b) Assuming $R_2 = 100\ \Omega$ and $R_0 = 5\ \Omega$, determine values for R_1 and C_1 such that the 3-dB corner frequency is 30 MHz.

(c) What is the corner frequency without the equalizer?

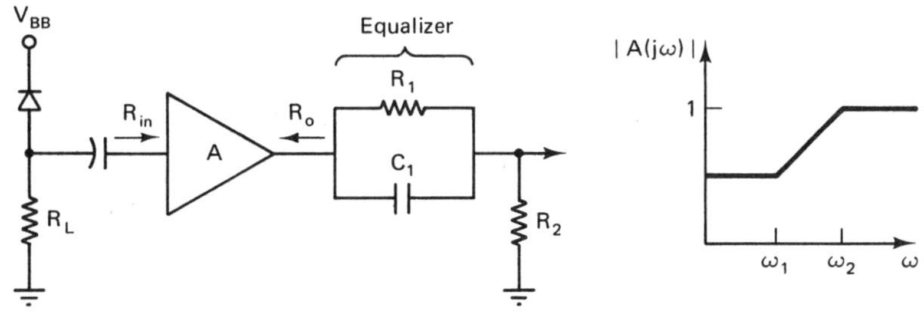

10.7. The rise time of a low-pass system is defined as the time required for the output to go from 10% to 90% of its final value when the input is a unit step. Find the rise time for (a) a single time constant *RC* low-pass filter with half-power bandwidth $W_{1/2}$ and (b) an ideal low-pass filter with bandwidth W.

10.8. Show that

$$NA = \sqrt{n_1^2 - n_2^2}$$

is approximately equal to

$$NA \approx n_1 \sqrt{2\Delta}$$

where

$$\Delta = \frac{n_1 - n_2}{n_1}$$

10.9. In the circuit of Problem 10.6, the effective input resistance of the op-amp is 6 MΩ, and the detector bias resister is 10 MΩ. The total capacitance of the amplifier and the detector is 2 pF.

(a) Compute the 3-dB bandwidth.

(b) Find the thermal noise voltage referred to the amplifier input. Neglect the equalizer for this computation. Assume an operating temperature of 290°K.

(c) The same circuit is now connected as in Problem 10.4. The feedback components are 270 KΩ and 3 pF. Gain of the amplifier is 100. Repeat parts (a) and (b).

10.10. What is the maximum possible transmission distance without a repeater for the following optical fiber system? Assume: **(a)** PIN diode; **(b)** APD diode.

Bit rate = 1 megabits/s.

P_s = 1.5 mW.

Input and output coupling loss is 4 dB combined.

Average splice losses are 1.25 dB/km.

Attenuation loss in the fiber is 5 dB/km.

(c) The fiber to be used has indices of refraction $n_1 = 1.48$ and $n_2 = 1.46$. The diameter is 50 μm. Calculate the fiber bandwidth for parts (a) and (b).

10.11. The BER for binary PCM reception is expressed by (see Figure 10.14)

$$\text{BER} = \frac{1}{2}\,\text{erfc}\left(\frac{Q}{\sqrt{2}}\right)$$

Express the BER in terms of the analog SNR.

10.12. An optical fiber system utilizes an LED transmitter that launches an average optical power of 1 mW at a wavelength of 0.8 μm into a fiber. The cable attenuation is 2 dB/km. If the receiver requires 1000 photons in order to register a binary one with a 10^{-10} BER rate, determine the following.

(a) Maximum transmission distance (without repeaters) when the transmission rate is 2 megabits/s.

(b) Maximum transmission distance when the transmission is 2 gigabits/s.

10.13. An optical fiber PFM-IM system as shown in Figure 10.17 is in use. The detector is a PIN diode with a responsivity of 0.6 A/W. The following system parameters prevail:

$$\text{Peak-to-peak frequency deviation} = 6 \text{ MHz}$$

$$\text{Rest pulse rate} = 30 \text{ MHz}$$

(a) A ratio of peak-to-peak signal power to rms noise power of 60 dB is desired. Calculate the required incident optical power.

(b) Assuming that the detector and source have rise times of 5 ns and 3 ns, respectively, estimate the maximum permissible rise time for the fiber cable such that satisfactory operation of the system will continue.

10.14. In Example 10.4, estimate the maximum bit rate if a Manchester code is used.

10.15. An SNR of 45 dB is required at the detector of an analog system. A postdetection bandwidth of 25 MHz is required. Calculate the minimum optical power incident on the photodetector if it operates at a wavelength of 1.3 μM with a quantum efficiency of 60%.

11

SATELLITE COMMUNICATION NETWORKS

11.0 INTRODUCTION

The first operational satellite system was established in 1965. Today, domestic satellites such as RCA's Satcom, SBS, and Telstar provide communications (voice, data, image) over the continental United States. The International Telecommunications satellites provide services to over 100 countries. These satellites link the world together in business, in major sports broadcasts such as the Olympics and in news broadcasting. Communication satellites for the military play an ever-greater role in strategic and tactical programs. With the availability of the space shuttle, heavier and more powerful satellites are being placed in geosynchronous orbit at more reasonable cost.

The INTELSAT system provides an example of the continuing evolution of satellite systems. Over the past two decades, this evolution has progressed from simple point-to-point links with unique subcarrier defined paths between points to today's multipoint, satellite-switched, multisatellite networks, which tie together hundreds of Earth stations. The INTELSAT satellite system carries approximately two-thirds of the world's international telephony communications. The first operational satellite, INTELSAT 1, was launched in April 1965 and is now up to the sixth-generation satellite, INTELSAT VI. Services provided range from video transmission to digital transmissions for such things as telegraphy, telex, and computer-to-computer. In addition, maritime service is provided by leasing L-band transponder capacity to INMARSAT, the International Maritime Satellite Organization.

INTELSAT introduced, in late 1983, its new business services. This traffic consists of digital transmissions that are intended to originate and be received near the customers' premises. This implies the use of small, low-cost Earth-station antennas. Also planned are such services as video teleconferencing and video text.

A unique feature of satellite communications is the ability to interlink many geographically dispersed users and thus provide multiple access. This is in contrast to submarine cables, which provide only point-to-point communications. Multiple access with a single satellite is provided by two fundamental modulation techniques: frequency-division multiple access (FDMA) and time-division multiple access (TDMA). A typical Earth station in the FDMA mode may transmit 3 to 10 carriers and in return receive 15 to 30 carriers in order to complete its connectivity requirements. Increases in the number of satellite antenna beams require an increase in the number of carriers in order to achieve connectivity among the diverse coverages produced by the satellites. A point is reached where it is both more convenient and cost effective to switch to TDMA. The latest generation of INTELSAT satellites (INTELSAT VI), introduced satellite-switched TDMA (SS/TDMA) in order to access efficiently six separate antenna beams. In this mode, a microwave matrix switch changes the interconnection of each of the six receive antenna beams to each of the corresponding transmit beams in such a way that all interconnections are established within one frame of the TDMA system. Before continuing our discussion further, it is of interest to review particular features of the INTELSAT system.

11.1 THE INTELSAT NETWORK

A communications satellite system consist of a satellite, a ground-based station to control its operation, and a network of Earth stations to provide transmission, up-link and down-link, for communications traffic relayed through the satellite. The satellite itself has two major parts: a payload and a bus. Under the heading of payload, we have antennas, receivers, and transmitters, whose function is to perform the communications transponding function. The bus provides the necessary housekeeping functions needed for payload operation. These include such items as attitude control, power management, and command and telemetry subsystems. After the communications satellite is positioned in space, it occupies an assigned orbital position, or slot, above the equation at geostationary altitude. These slots are boxes with 0.1° to 0.2° sides at 2° to 4° spacing. Since the earth is not a perfect sphere and also because of solar and lunar gravity, the geostationary satellite wanders. The maneuvers that keep the satellite in its alligned slot are referred to as *stationkeeping*, which is effected by firing on-board thrusters. Hydrazine (N_2H_4) is the propellant used in many satellites. Typically, propellant usage is 20 lb to 30 lb per year. A communications satellite may contain on the order of 1300 lb of propellant.

11.2 SATELLITE CLASSES AND STATIONKEEPING

There are two main classes of communication satellites (Figure 11.1). These are spinners and three-axis stabilized satellites. A spinner uses the angular momentum of its spinning body to provide roll and yaw attitude stabilization. The body of a three-axis stabilized satellite remains fixed relative to the earth's surface while an internal subsystem provides roll and yaw stabilization.

Each communications satellite is assigned a longitude in the geostationary arc located approximately 22,300 mi above the equator (See Figure 11.2). Note the slot to which the satellite is assigned. The slot tolerance depends on the communications frequency band. For C-band systems (6 GHz up-link, 4 GHz down-link), the tolerances are typically ±0.1° in longitude and latitude. This decreases to ±0.05° for Ku-Band (14 GHz up-link 12 GHz down-link). Ku-Band antennas exhibit much smaller beamwidth than C-band antennas. Generally speaking, orbital corrections are made every 2 to 6 weeks, as required.

11.2.1 Orbit and Spectrum Conservation

The availability of RF spectrum for use within the geostationary orbit is limited. Ideally, all INTELSAT customers would prefer to access each other with a single

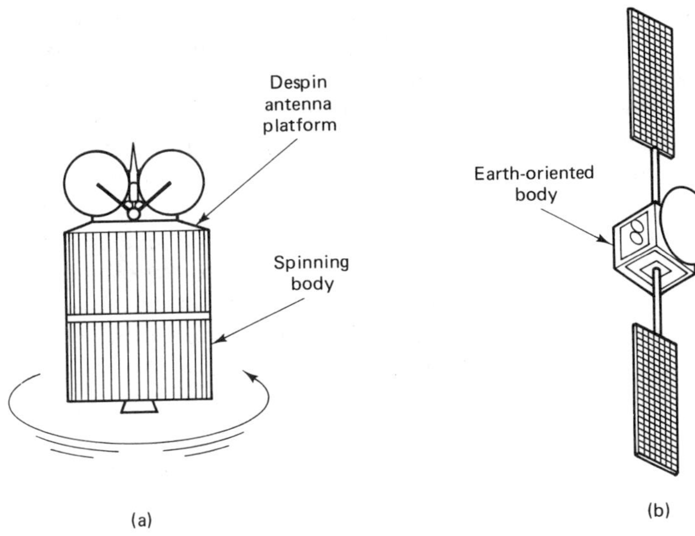

Figure 11.1 Main classes of communications satellites. (a) Spinner. (b) Three-axis stabilized.

Earth station antenna through a single antenna. Practical limitations related to economics, technology, and geometry do not permit this. For transoceanic communications via a geosynchronous satellite, only a relatively narrow range of orbital arc is ideal for interconnecting the largest populated regions with a minimum number of satellites. INTELSAT uses 335.5°E longitude over the Atlantic

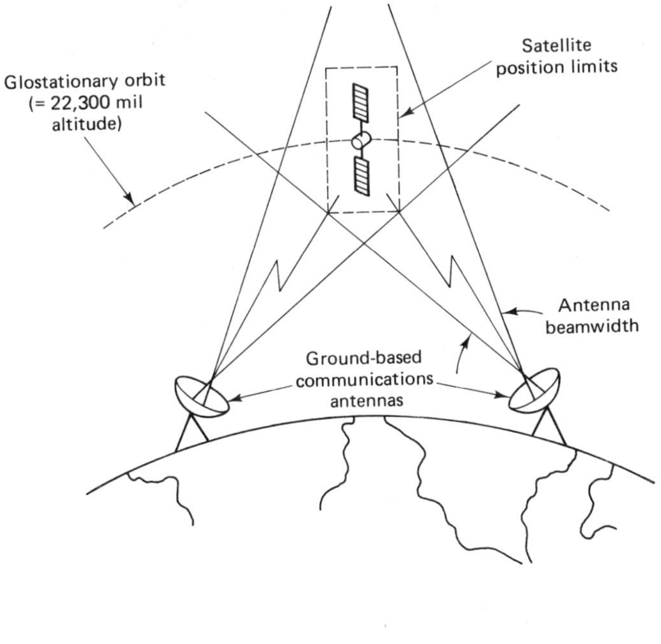

Figure 11.2 Satellite latitude and longitude must be tightly controlled to meet communications system requirements.

Ocean to connect North and South America with Europe, Africa, the Middle East, and with each other. All the traffic demands cannot be met from this one location, so two other satellites are located at 325.5°E and 341.5°E, respectively. Other satellites are located still further east and west.

An important method of conserving the orbit/spectrum resource is frequency reuse. For example, the INTELSAT V series of spacecraft has four types of antenna patterns. A global beam provides coverage over the entire earth's disk from a geosynchronous orbit in the 6/4-GHz frequency band. Four times frequency reuse is obtained in this bandwidth by the use of both east and west hemispheric beams in one polarization sense and zonal beams in the opposite sense. Two times frequency reuse in the 14/11 GHz band is obtained by the use of an east and west steerable spot beam. An example of this type of coverage is shown in Figure 11.3.

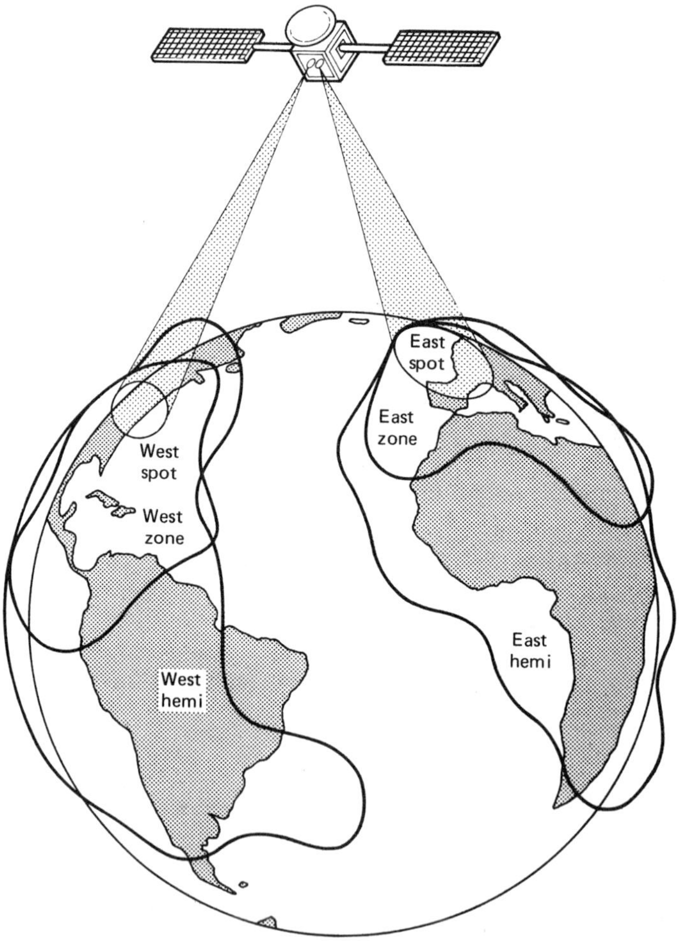

Figure 11.3 Hemispheric and spot converge.

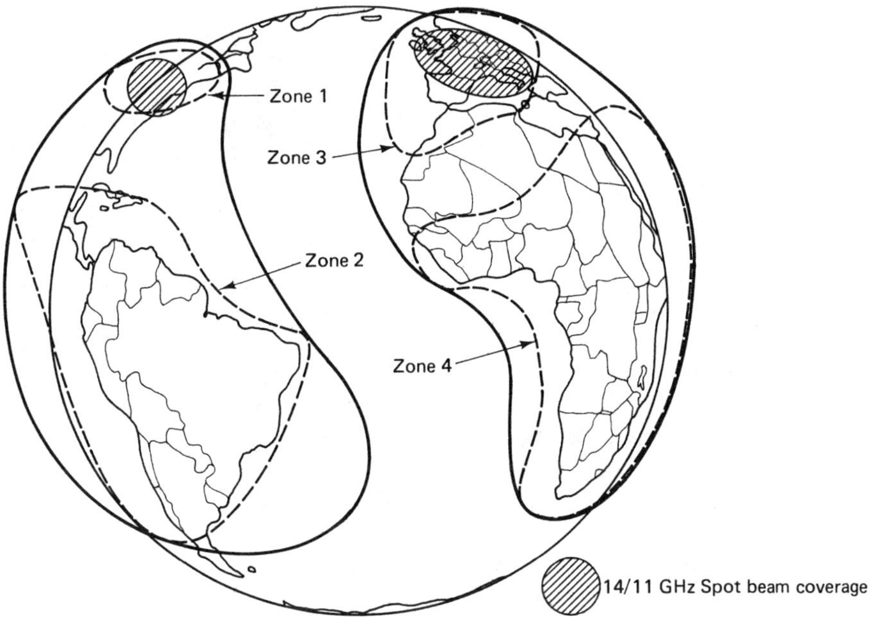

(a)

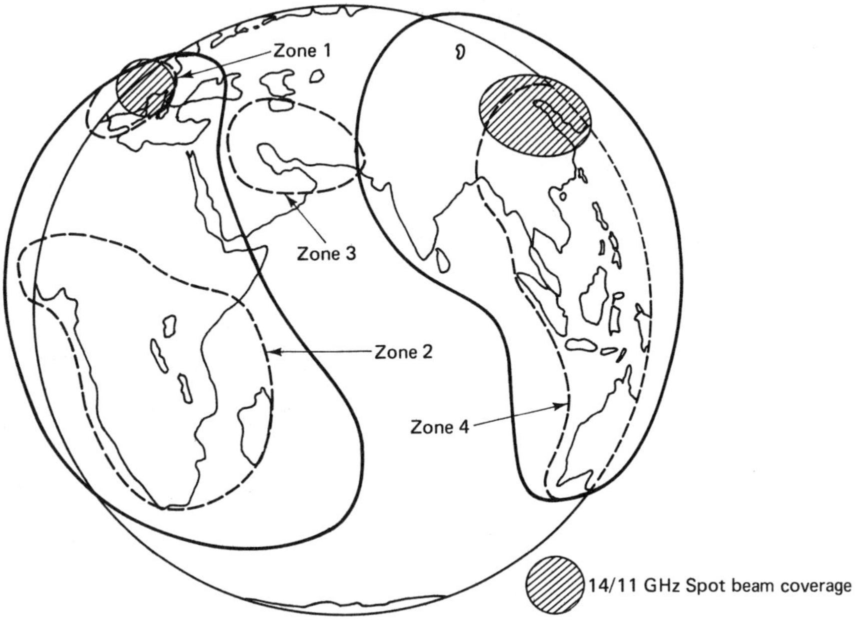

(b)

Figure 11.4 (a) INTELSAT VI coverager contours for the Atlantic Ocean region from 333.5°E longitude. (b) INTELSAT VI coverage contours for the Indian Ocean region from 63°E longitude.

More detailed figures are available with the antenna gain contours, which indicate how the communication performance varies depending on the location of the earth station relative to the satellite. These satellites also added RF spectrum by introducing the use of 14/11 GHz (Ku-Band). Two spatially separated narrow (spot) beams permit an effective doubling of the frequency spectrum. INTELSAT VI satellites extend the principles to achieve sixfold usage of the RF spectrum at 6/4 GHz. Frequency reuse of the 14/11-GHz band is continued and also greater bandwidth is provided over that of INTELSAT V. Contour coverages are illustrated in Figure 11.4.

The two hemispheric beams are copolarized and spatially isolated. A single pair of beams provides coverage requirements for the Atlantic Ocean region, Indian Ocean region, and Pacific Ocean region satellite deployments. The four zone beams shown are copolarized and spatially isolated. These beams overlap the hemispheric beams and are orthogonally polarized to provide hemispheric and zonal isolation. Three separate sets of zone beams are required because the need for zone coverage differs in the three ocean regions. Two spot beam antennas provide twofold reuse of the 14/11 GHz Ku-band frequency spectrum. Each of these spot beam antennas may be steered independently over the full earth disc. The INTELSAT VI communication system consists of 10 independent communication repeaters, 1 repeater for each communications antenna beam. Within the 10 repeaters, there are a total of 50 transponders. The frequency plan for these 50 transponders is shown in Figure 11.5. The 10 antenna beams form a natural

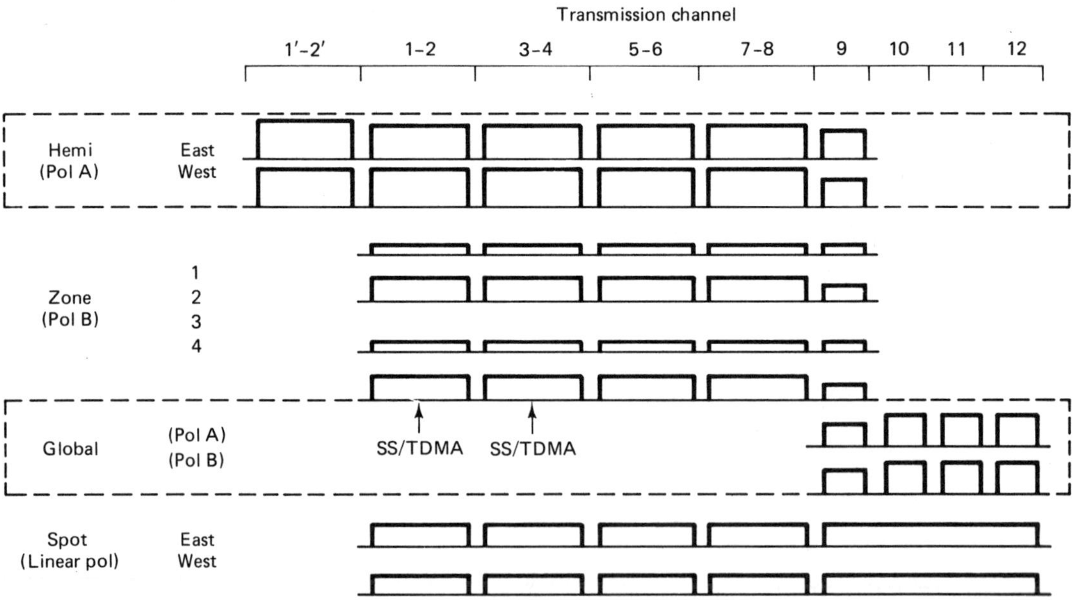

Figure 11.5 INTELSAT VI frequency plan.
(© 1984, IEEE)

partition of the communication system into 10 separate and independent repeaters. Interconnectivity between the 10 repeaters is necessary to realize the full potential of the sixfold reuse of the 6/4 GHz frequencies and the two-fold reuse of the 14/11 GHz frequencies. On INTELSAT VI this interconnectivity is provided both statically and dynamically.

11.2.2 TDMA

Prior to 1984, INTELSAT used FDMA for most of its services. In an FDMA system, each carrier frequency occupies a discrete frequency allocation and is assigned a specific location within a multicarrier transponder. Two FDMA techniques are employed:

1. FDM/FM/FDMA, where the transmitting earth station frequency-division multiplexes (FDM) several single-sideband carrier telephone channels onto one carrier baseband assembly. This in turn FMs a carrier and is then applied to an FDMA satellite network.
2. Single channel per carrier, where each telephone channel modulates a separate radio frequency carrier.

The operation of transponders at maximum power results in cross talk between channels. Beginning in 1984, INTELSAT began to implement TDMA. This eliminates the crosstalk problem, since only one station uses that transponder at a time.

In the TDMA mode, each user accesses the satellite using the same carrier frequency during specified time slots relative to a reference time. A TDMA frame is established, during which all users within the network transmits in burst during their designated time slots. The INTELSAT V TDMA frame is shown in Figure 11.6. The frame is 2 ms, or 120,832 symbols long. Each user is assigned burst time slots within the TDMA frame based on the amount of traffic to be transmitted and received during the frame.

A TDMA satellite switch is contained in INTELSAT VI. Refer to Figure 11.5. Channels 1-2 and 3-4 are switched within the six hemi/zone beams. The satellite switch supplies interconnections between these six receive and transmit

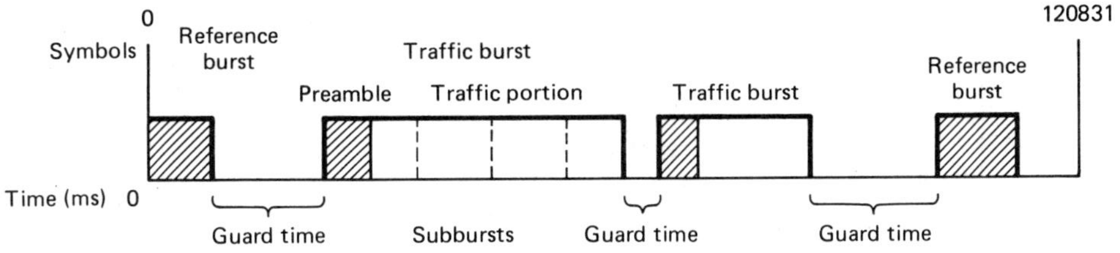

Figure 11.6 Frame format for INTELSAT V TDMA.
(© 1983, IEEE)

beams so that the earth stations in one beam may simultaneously access, via TDMA, the earth stations in all six beams. The six up-link hemi/zone beams are cyclically interconnected through the satellite switch to the six down-link beams. This switching pattern repeats every 2 ms (see Figure 11.6). Time for both TDMA switches is from a common source, so that separate synchronization of the two TDMA transmission channels is unnecessary. Techniques for satellite network synchronization are discussed in the following section.

11.3 TDMA SYNCHRONIZATION TECHNIQUES

SS/TDMA technology is used in recent high-capacity satellites that employ spot beam zones. Thus it is necessary for all earth stations to synchronize their time bases to a common reference. The reference in this particular case is the sync connection window of the satellite. A number of schemes have been proposed to accomplish this.

A TDMA satellite-switched communication system is illustrated in Figure 11.7. This particular system, employing one satellite with *N* transponders along with a number of stations in *N* different spot beams, has great flexibility. A switch operating at a microwave IF frequency permits storage, routing, and switching of signals between multiple spot beam zones. The use of highly directive antennas on the satellite is needed to produce the multiple spot beams. A number of ad-

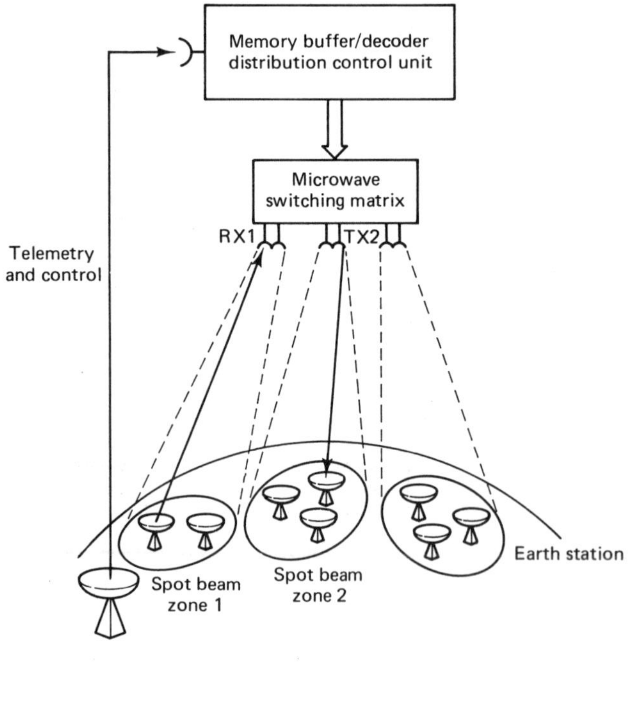

Figure 11.7 TDMA satellite-switched communications system.

vantages are derived from an SS/TDMA configuration. First, the frequency spectrum allocation for each transponder is the same two passbands. We have a passband B for the up-link and a passband B for the down-link. Hence, for N transponders, the total passband is 2 B, not 2 NB. An N-fold saving in bandwidth is realized. Secondly, the effective isotropic radiated power (EIRP) is increased due to the higher gain of spot beam antennas relative to global coverage antennas. This permits an increase in the bit rate, R, while maintaining the same BER. Finally, space is saved in the crowded geostationary area by having a switching satellite take the place of several nonswitching satellites.

Earth stations that exist in the same spot beam must operate in either the FDMA or TDMA mode. We noted previously that with TDMA, only one carrier is amplified in each transponder. The main disadvantage is that all Earth stations require timing information concerning the switching sequence of the satellite. The switching satellite consists of a number of receiver-transmitter pairs (10 for INTELSAT VI), all interconnected by a computer-controlled diode network. An Earth station in spot beam zone 1 communicates with a station in spot beam zone 2 by transmitting a signal which is received by receiver RX1 at the satellite. The satellite switches the signal to transmitter TX2, which radiates to the station in zone 2. This connection lasts for the duration of one time slot (typically 10 μs). Several stations in the same spot beam zone can share a time slot, provided that the total length of transmission does not exceed the time slot.

Refer to Figure 11.8. This represents the normal TDMA synchronization sequence. The simplest TDMA system is one in which all participating stations can see the retransmission from the satellite of the burst from all stations. To establish the beginning of a frame, a reference burst is transmitted by a station designated as the reference station. This reference burst is distributed by the transponder to all the participating stations. The participating stations use the received reference burst as a timing reference. After the application of a delay, D_N, station N transmits its burst so that it arrives at its assigned location in the TDMA frame. A typical computation for D_N is illustrated in Figure 11.9.

The time delay, D_N, is selected so that the time elapsed between departure of the burst from the satellite and the return of a response to the satellite from a ground station is an integer number of TDMA frames. As an example, the frame period for the INTELSAT V satellite is 2 ms (see Figure 11.6). For purposes of traffic management, a multiframe, T_M, is introduced. This consists of 16 TDMA frames and, for INTELSAT V, is 32 ms in length, that is, $T_M = 16\ T_F$. Refer to Figure 11.9. The time line at the top of the figure refers to the timing of events at the satellite. The reference burst, R, occurs with a multiframe period, T_M. At the satellite, all traffic bursts are timed relative to these pulses. The dashed line at the bottom of the figure is the time line for the station located the furtherest distance from the satellite. The round trip propagation time for this path is MT_M, where M is an integer. For a satellite in stationary orbit, the greatest propagation time is 283 ms. Thus the minimum value of M that causes MT_M to exceed this delay is 9; that is, $MT_M = 288$ ms. Note from the figure that a station located

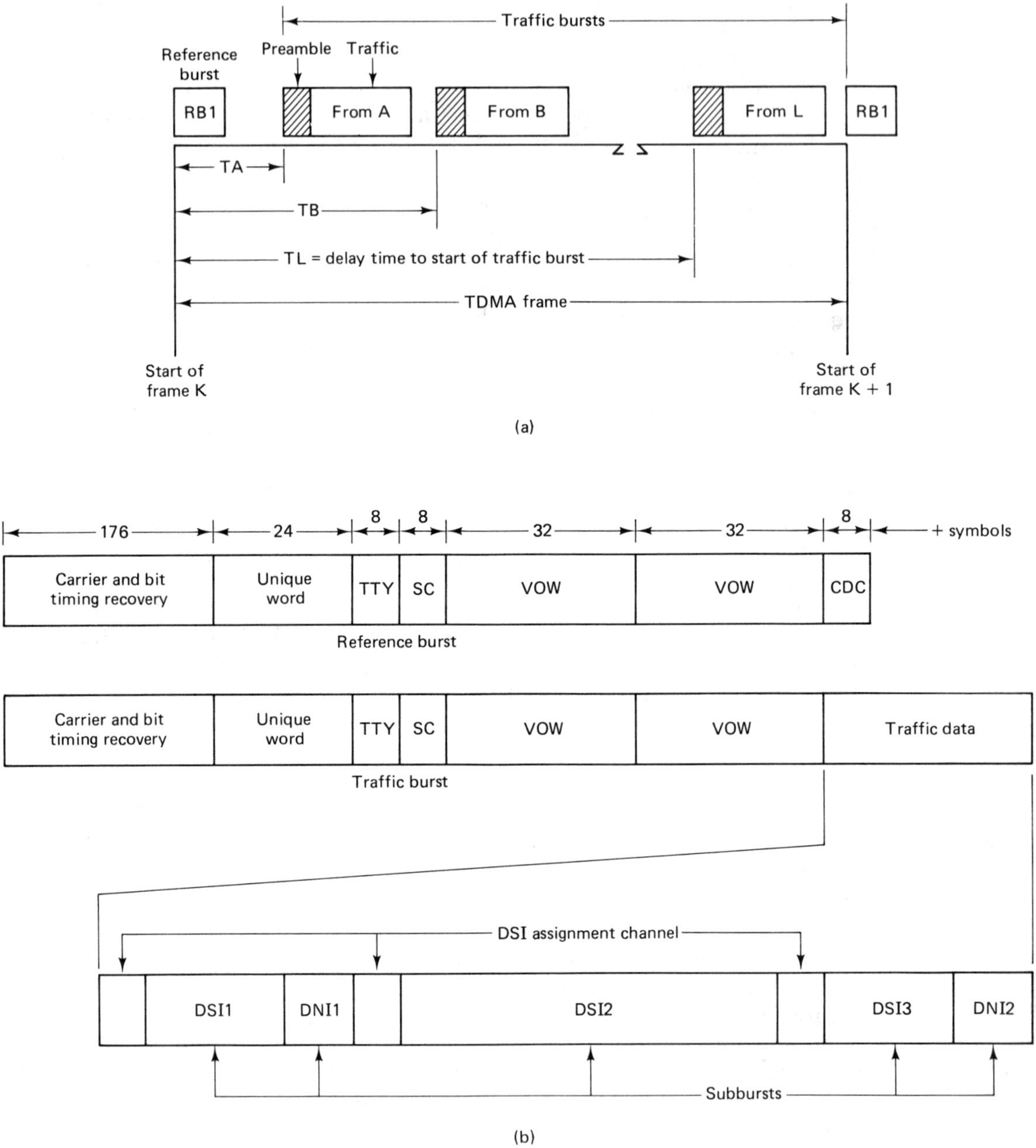

Figure 11.8 (a) TDMA frame structure. (b) Burst format.

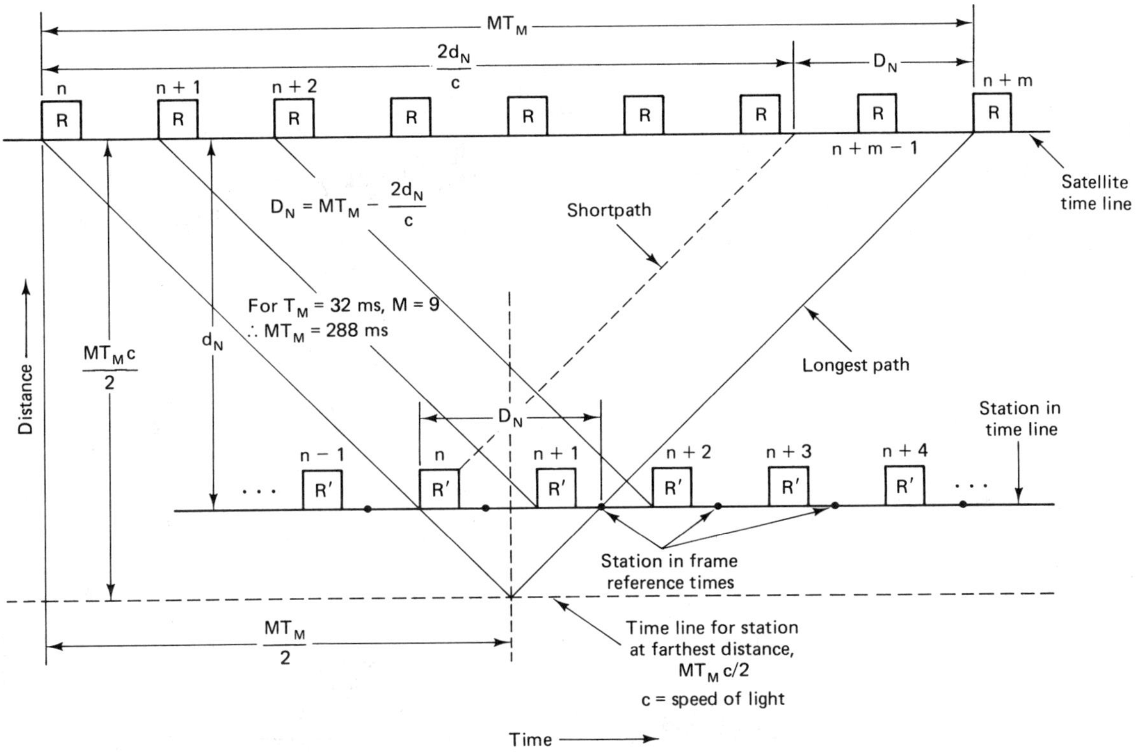

Figure 11.9 Determination of D_N for frame synchronization.

on this time line needs to introduce zero delay ($D_N = 0$) in order to align its local start of TDMA frame. For stations nearer the satellite, a delay D_N equal to the difference between the actual round trip propagation time and MT_M must be introduced. A traffic burst transmitted at this instant will arrive at the beginning of a TDMA frame at the satellite. For a particular station, in order to position its traffice burst at its assigned location in the TDMA frame, a delay Δt must be added to D_N. An SS/TDMA system is illustrated in Figure 11.10(a). In a system such as this the TDMA frame of the Earth stations must be synchronized with the frame of the satellite switch. Timing for the satellite switch is determined by an on-board crystal reference clock. This clock establishes the TDMA frame reference for the entire network and controls the timing of the switch states within the frame so that they occur synchronously with the traffic bursts emitted by traffic stations in all beams.

The first switch state of the sequence provides a time slot into which a special acquisition and synchronization burst is placed by a unit located on the earth known as the acquisition and synchronization unit (ASU). This ASU first performs a special search procedure that homes in on the ASU switch-state slot, selected for the shortest duration of any switch state existing in the switch-state sequence.

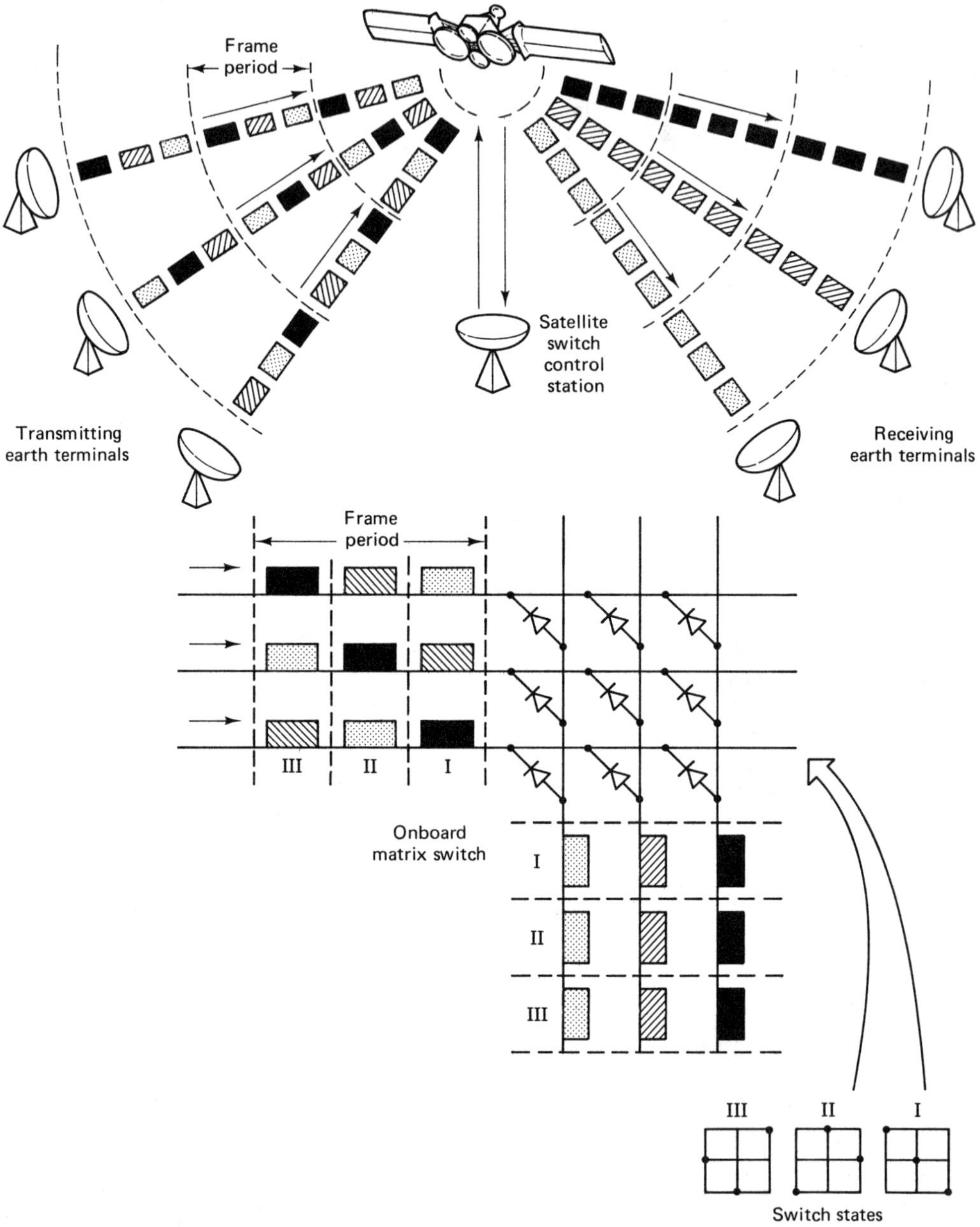

Figure 11.10 Satellite-switched TDMA.

Once it locates the slot, the ASU transmits a special metering burst into the slot, keeping the burst positioned within an accuracy of ±1 symbol of the TDMA transmission symbol rate. In the INTELSAT system, which operates at a 60-megasymbol/s rate, this accuracy is ±16.67 ns. After the ASU has been synchronized, it can be used to establish the TDMA frame timing reference for the reference bursts, which are transmitted by the reference stations of the network. A typical satellite switch-state pattern that provides a facility for performing these later functions is illustrated in Figure 11.11. This figure shows the time slots for

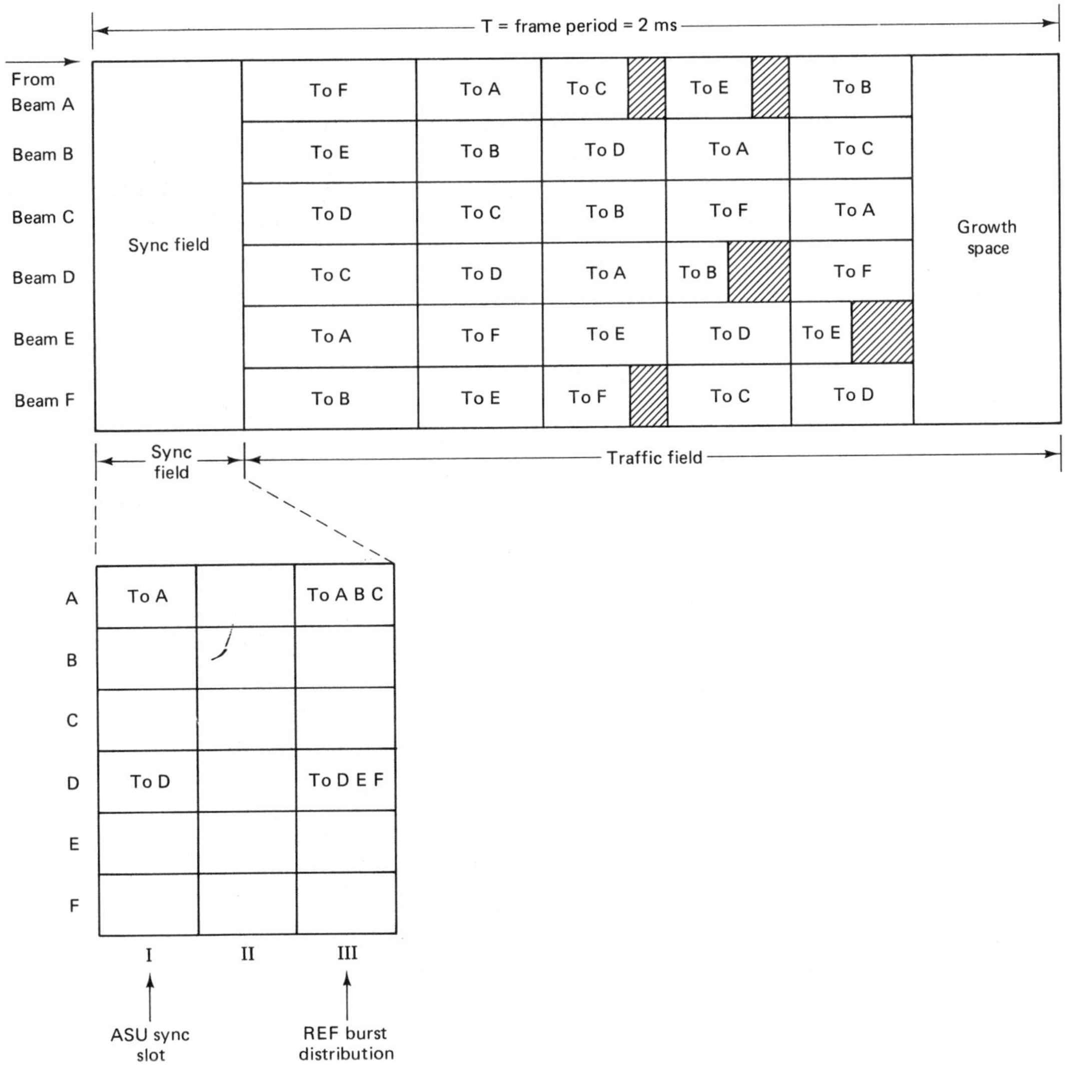

Figure 11.11 SS/TDMA frame formats showing various switch states.

acquisition and synchronization, reference burst distribution, and traffic burst distribution.

The discussion thus far has been general in nature but very important. In the following sections we wish to relate more directly to the thrust of the text, namely, digital communication. In common with the design of any digital communication system, an important first step is an understanding of system parameters. We introduce this subject in the following section.

11.4 SYSTEM PARAMETERS

In discussing digital modulation techniques in previous chapters we observed that a parameter of supreme interest was the BER. Recall that the BER in a digital communication system is a function of the SNR. More specifically, the BER is usually expressed in terms of the energy per bit. The average energy of a transmitted bit is given by

$$E_b = P_o T_b \tag{11.1}$$

where P_o is the power in the bit in watts and T_b is the time duration of the bit in seconds. Obviously, a high energy per bit would be expected to improve the BER.

One of the key elements, then, in system design is to ensure an adequate energy per bit. This is reflected in the required transmitter power, antenna gains, system losses, and so on. The gain of an antenna may be expressed in terms of the gain of an isotropic antenna that has a gain equal to 1. An isotropic antenna radiates constant power in all directions. Specifically, at a distance R from the antenna

$$\text{Isotropic antenna radiation intensity} = \frac{P}{4\pi R^2} \text{ W/m}^2 \tag{11.2}$$

where P is the input power in watts.

The gain of an antenna, G, may be defined as

$$G = \frac{\text{maximum radiation intensity}}{\text{isotropic antenna radiation intensity}} \tag{11.3}$$

where the same input power is assumed for both antennas. At a distance, R, from the transmitting antenna, then, the power density is given by

$$\begin{aligned} P_d &= \frac{P}{4\pi R^2} G_T \\ &= \frac{PG_T}{4\pi R^2} \text{ W/m}^2 \end{aligned} \tag{11.4}$$

where G_T is the gain of the transmitting antennas.

A receiving antenna placed in the path of the propagating energy will intercept a portion of the energy directly proportional to the aperture, A_r, of the antenna. The aperture of the receiving antenna may be expressed as

$$A_r = \frac{\eta \lambda^2 G_r}{4\pi} \text{ m}^2 \tag{11.5}$$

where

λ = wavelength
η = efficiency of the receiving antenna
G_r = gain of the receiving antenna

Using equations (11.4) and (11.5), the total received power, P_r, is given by

$$P_r = \eta \frac{P_T G_T G_r \lambda^2}{(4\pi R)^2} \text{ W} \tag{11.6}$$

EXAMPLE 11.1

An Earth station transmits 2 kW of power to a geosynchronous satellite at a frequency of 6 GHz. Antenna gains at the Earth station and the satellite are 64 dB and 25 dB, respectively. The receiver antenna efficiency is 0.55. Calculate (a) the power density at the satellite and (b) the total received power.

Solution

(a) $$P_d = \frac{PG_T}{4\pi R^2}$$

$$= \frac{(2000)(25{,}118{,}886.4)}{(4\pi)[(22{,}300)(5280)(0.3048)]^2}$$

$$= 3 \times 10^{-7}/\text{m}^2 = 0.3\ \mu\text{W/m}^2$$

(b) $$P_r = \eta P_d G_r \lambda^2$$

$$= (0.055)(90.3 \times 10^{-7})(316.22)\frac{300^2}{6000}$$

$$= 0.130 \times 10^{-7} = 0.013\ \mu\text{W}$$

11.4.1 Link Equations

Equation (11.6) provides the basis for arriving at a power budget for the satellite link. By taking 10 $\log_{10}$ of this equation, we have a form of what is often referred to as the *link equation*; that is,

$$P_r(\text{dB}) = 10 \log P_T G_T - 20 \log \frac{4\pi R}{\lambda} + 10 \log \eta G_r \qquad (11.7)$$

To increase the usefulness of the link equation, additional terms are usually defined. One of these is the noise density, the noise power that is present in a normalized 1 Hz of bandwidth. That is,

$$N_0 = kT_e \qquad (11.8)$$

where k is Boltzmann's constant (−198.6 dBm/°K) and T_e is the equipment system noise temperature in degrees kelvin.

Another commonly used term is the carrier-to-noise density ratio, C/kT_e. Specifically, this relation is expressed as

$$\frac{C}{N_0} = \frac{C}{kT_e} \qquad (11.9)$$

where C is the average wideband carrier power and N_0 is as defined previously. A third parameter is the ratio of gain to equivalent noise temperature. This is a figure of merit used to specify the efficiency of a satellite and of an earth station. It is expressed as

$$\text{Figure of merit} = \frac{G}{T_e} \text{ dB/°K} \qquad (11.10)$$

A final parameter is the ratio of energy per bit to noise density, E_b/N_0. This particular parameter allows comparison of different systems based on such factors as transmission rate, type modulation, and different encoding systems. Specifically, it is expressed as

$$\frac{E_b}{N_0} = \frac{CT_b}{kT_e} = \frac{C}{N_0} T_b \qquad (11.11)$$

Referring to equation (11.11), if C/N_0 is known, we can readily obtain E_b/N_0. For a satellite link, we are interested in both the up-link and the down-link. Using equation (11.6), we have

$$\frac{C_u}{N_{0u}} = \frac{P_{ru}}{kT_e} = \frac{\eta P_T G_T G_r \lambda^2}{(4\pi R)^2 kT_e} \qquad (11.12a)$$

where C_u/N_{0u} is the up-link carrier-to-noise ratio. Operating on both sides of this equation by 10 $\log_{10}$ results in

$$\left(\frac{C_u}{N_{0u}}\right)_{\text{dB}} = 10 \log P_T G_T - 20 \log \frac{4\pi R_u}{\lambda_u} + 10 \log \frac{G_{ru}}{T_e} - 10 \log k + 10 \log \eta \qquad (11.12b)$$

The antenna efficiency term, η, is less than 1, so that it actually represents a loss. In a practical system, additional losses occur due to the cables, connectors, and so on, between the antenna and the microwave system. It is common practice to lump the losses. For the up-link, equation (11.12b) may be expressed as

$$\left(\frac{C_u}{N_{0u}}\right)_{\text{dB}} = \underbrace{10 \log P_T G_T}_{\substack{\text{Earth}\\ \text{station}\\ \text{EIRP}}} - \underbrace{20 \log \frac{4\pi R_u}{\lambda_u}}_{\substack{\text{Free-space}\\ \text{up-link loss}}} + \underbrace{10 \log \frac{G_{ru}}{T_e}}_{\substack{\text{satellite}\\ G/T_e}} + \underbrace{10 \log L_u}_{\substack{\text{Additional}\\ \text{up-link}\\ \text{losses}}} - 10 \log k \tag{11.13}$$

One of the terms in this form of the range equation requires an explanation. EIRP is the product of the transmitted power and the antenna gain. That is,

$$\text{EIRP} = P_t G_t$$

or

$$\text{EIRP(dB)} = P_t(\text{dBW}) + G_t(\text{dB}) \tag{11.14}$$

For the down-link, the basic range equation is

$$\left(\frac{C_d}{N_{0d}}\right)_{\text{dB}} = \underbrace{10 \log(P_s G_{sd})}_{\substack{\text{Satellite}\\ \text{EIRP}}} - \underbrace{20 \log \frac{4\pi R_d}{\lambda_d}}_{\substack{\text{Free-space}\\ \text{down-link}\\ \text{loss}}} + \underbrace{10 \log \frac{G_d}{T_d}}_{\substack{\text{Earth}\\ \text{station}\\ \text{G/T}}} + \underbrace{10 \log L_d}_{\substack{\text{Additional}\\ \text{down-link}\\ \text{losses}}} - 10 \log k \tag{11.15}$$

Comparing equations (11.13) and (11.15), note that they are of the same form. Remember that the satellite is actually a transponder. Consequently, the C/N_0 of the uplink will carry through and contribute to the C/N_0 of the received downlink. In a frequency translating satellite, the overall end-to-end carrier-to-noise ratio,

$(C/N_0)_T$, at the receive earth station is given by

$$\left(\frac{C}{N_0}\right)_T = \frac{1}{N_{0u}/C_u + N_{0d}/C_d} \tag{11.16}$$

The bit energy can be obtained using equation (11.1); that is,

$$E_b = CT_b \tag{11.17}$$

Using this expression, equation (11.16) may be written by

$$\left(\frac{E_b}{N_0}\right)_T = \frac{1}{N_{0u}/E_{bu} + N_{0d}/E_{bd}} \tag{11.18}$$

This expression is very useful in systems calculations for a digital communication system. The probability of error is a function of the total energy per bit-to-noise density ratio $(E_b/N_0)_T$. A noise-link budget calculation for a frequency translating satellite is shown in Table 11.1.

TABLE 11.1 LINK BUDGET OF THE 14/11-GHz EUROPEAN COMMUNICATION SATELLITE (ECS) SYSTEM*

	Up-link	
1.	Transmitter output power at saturation, 2 kW	33 dBW
2.	Backoff and combining losses	7 dB
3.	Transmit antenna gain (15 m, 15 GHz)	64 dB
4.	EIRP	90 dBW
5.	Free-space loss (14 GHz)	207.5 dB
6.	Atmospheric loss (14 GHz, clear weather)	0.6 dB
7.	Satellite G/T	−5.3 dBW/°K
8.	C/T at repeater input	−123.4 dBW/°K
9.	Boltzmann's constant	−228.6 dBW/°K/Hz
10.	Bit rate, 120 megabits/s	80.8 dBHz
11.	E_b/N_0 at repeater input	24.4 dB
	Down-link	
1.	EIRP at beam edge (unmodulated carrier, saturation)	40.8 dBW
2.	Modulation backoff and bandlimiting losses	0.6 dB
3.	Free-space loss (11.7 GHz)	205.6 dB
4.	Atmospheric loss (11.7 GHz, clear weather)	0.4 dB
5.	Power at receive antenna	−165.8 dBW
6.	Receive antenna gain (15 m, 11.7 GHz)	62 dB
7.	Receive system noise temperature (clear weather) 270 °K	24.3 dB°K
8.	Earth station G/T	37.7 dB/°K
9.	C/T at receiver input	−128.1 dBW/°K
10.	Boltzmann's constant	−228.6 dBW/°K/Hz
11.	Bit rate, 120 megabits/s	80.6 dBHz
12.	E_b/N_0 at receiver input	19.7 dB

* This system will provide 17,000 two-way digital telephone circuits by 1990. Individual satellite transponders will carry 120 megabits/s of traffic.

EXAMPLE 11.2

The output power of an earth station transmitter is 2.2 kW and the diameter of its antenna is 15 m. If the transmitting frequency is 14 GHz, calculate the EIRP of the station. Assume losses of 7 dB between the transmitter and the antenna.

Solution Using equation (11.3), it may be shown that the antenna gain is given by

$$G = K\left(\frac{4D^2}{\lambda}\right) = 0.55 \frac{4 \times 15^2}{300/14{,}000}$$

$$= 431{,}200 = 66.3 \text{ dB}$$

In this expression $K = 0.55$ is the antenna efficiency. Using equation (11.14), we have

$$\text{EIRP} = P(\text{dBW}) + G(\text{dB})$$

$$= (33.42 - 7) + 66.3 = 92.72 \text{ dBW}$$

EXAMPLE 11.3

Using results from Example 11.2 and Table 11.1, find E_b/N_0 at the satellite's repeater input if the bit rate is 100 megabits/s (80 dB·Hz).

Solution Using equation (11.15), we have

$$\left(\frac{C_u}{N_{0u}}\right)_{\text{dB}} = 92.72 - 207.5 + (-5.3) + (-0.6) - (-228.6)$$

$$= 107.92 \text{ dBW/Hz}$$

$$= 137.92 \text{ dBm/Hz}$$

The up-link bit energy is given by equation (11.17):

$$E_{bu} = CT_b$$

or

$$(E_{bu}) \text{ dBm} = C_u \text{ (dBW)} + 10 \log T_b(\text{s})$$

Then

$$\frac{E_b}{N_0} = \frac{C_u T_b}{N_0}$$

$$= 107.92 \text{ dBW/Hz} + 10 \log \tfrac{1}{100} \times 10^{-6}$$

$$= 107.92 - 80 = 27.92 \text{ dB}$$

11.5 SYSTEM CONSIDERATIONS

When considering the establishment of a satellite link, a number of factors need to be taken into account. If the communication distance involved is short—for example, intrastate—we would want to consider a microwave or fiber optic link. For distances less than 700 mi to 800 mi, the economics of fiber optics may be the viable option in terms of the cost factor. The two key performance parameters in an all digital transmission network are

- BER
- System availability

We are already familiar with the implications of the BER. Under system availability we consider the combined effects of equipment reliability and transmission path reliability. Availability is defined as the percentage of time that the system BER performance is equal to or greater than the threshold BER objective established. For example, for high-quality data services, the BER under fading conditions must be on the order of 10^{-7}. Thus a system availability of 0.9999 implies that 99.99 percent of the time, the BER performance is 10^{-7} or better. Some typical satellite transmission parameters are itemized in Table 11.2. In the following we consider the design of a satellite link.

11.5.1 Satellite Link Design

Assume that a *C*-Band satellite system with the performance parameters of Table 11.2 is to be designed. A BER of 10^{-9} during clear weather is feasible if a forward-

TABLE 11.2 SATELLITE TRANSMISSION SYSTEM CONSIDERATIONS

Performance	
Clear weather	Coded BER 1×10^{-9}
Faded BER	Coded BER 1×10^{-7}
System availability	0.99995
Rainy weather and sun transit degradation	
Transmission relay	
Inherent delay may be objectionable to voice service users	
Impact on throughput efficiency on data transmission, particularly when using half duplex protocols.	
Entrance/exit links	
C-Band earth stations typically cannot be colocated with service point or technical operating center because of interference problems	
Microwave entrance or exit links generally required	
System capacity expansion may be constrained by a lack of interference-free channels on entrance or exit links.	

error-correcting code is used (See Chapter 4). We wish to design the C-Band satellite network with 13.0-m earth stations. The question, then, is how much link margin is required to meet the faded performance objectives (BER $= 10^{-7}$).

One of the first parameters that must be determined is the amount of fading to be expected during rain storms. Figure 11.12 illustrates the electromagnetic attenuation in decibels per kilometer as a function of rainfall rate and transmission frequency. For our example, assume that the earth station is to be located in a region subject to very heavy rainfall (100 mm/h). This is represented by curve *E*. For an up-link (transmit) frequency of 6.425 GHz and a down-link (receive) frequency of 4.2 GHz, the attenuations are then 0.7 dB/km and 0.13 dB/km, respectively. In general, the size of a rainfall cell is inversely proportional to the severity (neglecting hurricanes). A typical rainstorm cell is on the order of 16 km across and 5 km in altitude. For a cell this size and with an antenna elevation of 14.5°, the rain-path length is 16.5 km. The minimum rain margins required for a propagation availability of 0.9995 are 11.6 dB at 6.425 GHz and 2.2 dB at 4.2 GHz. Normal design practice is to include the down-link rain margin (2.2 dB) in the system-operating margin. The *G*/*T* (receive antenna gain to system noise tem-

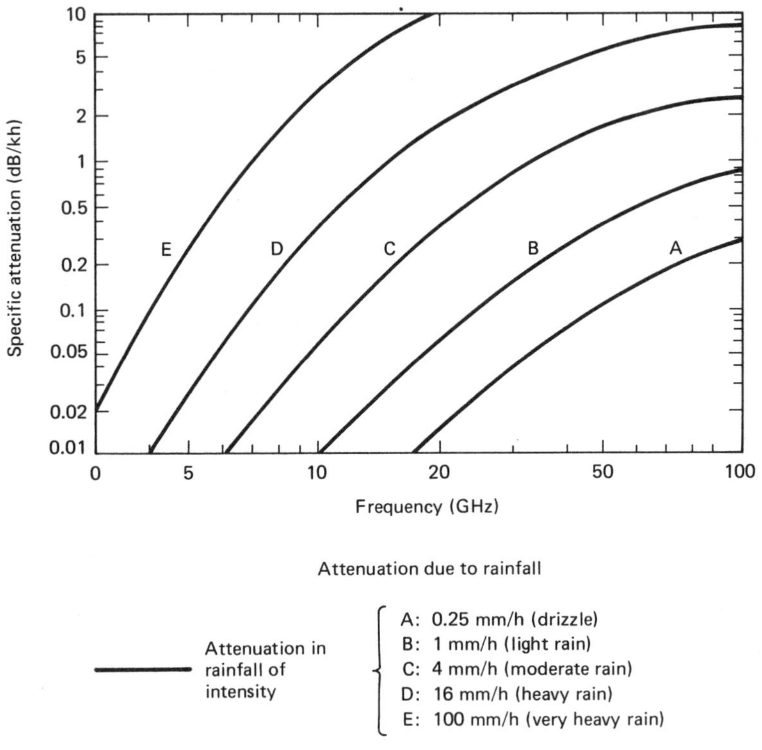

Figure 11.12 Attenuation due to rainfall.
(© 1985, IEEE)

perature) of a 13-m antenna is on the order of 32.7 dB/K. This antenna in conjunction with a 3.3 kW high-power amplifier (HPA) is sufficient to provide the needed up-link margin. Normal operating practice for the up-link is to adjust the earth station EIRP to provide the design down-link operating parameters. Thus when rain occurs on the up-link, transmitter (HPA) power is increased to offset the rain attenuation losses. For our example, this ranges up to 11.6 dB increase in transmitter power.

In terms of typical numbers, a 60-megabit/s TDMA system equipped with a rate $-\frac{7}{8}$ convolutional encoder (see Chapter 4) and a hard-decision decoder, requires an E_b/N_0 ratio of 12.8 dB in order to meet the clear-weather objective of BER $= 10^{-9}$. Using equation (11.11) the system carrier-to-noise ratio, CNR_s, can be calculated. That is,

$$\text{CNR}_0 = \frac{E_b}{N_0} T_b = \frac{E_b R}{N_0} \tag{11.19}$$

The CNR_s is then

$$\begin{aligned} \text{CNR}_s &= \frac{C}{N_0 B} = \frac{E_b R}{N_0 B} \\ &= \frac{E_b}{N_0} + 10 \frac{\log R}{B} \end{aligned} \tag{11.20}$$

where, for our example,

$$\frac{E_b}{N_0} = 12.8 \text{ dB}$$

$$R = 60 \text{ megabits/s}$$

$$B = 36 \text{ MHz}$$

Then

$$\begin{aligned} \text{CNR}_s &= 12.8 + 10 \log \frac{60 \times 10^6}{36 \times 10^6} \\ &= 12.8 + 77.8 - 75.6 = 15.0 \text{ dB} \end{aligned}$$

Refer to equation (11.16). CNR_s consists of contributions from both the up-link and the down-link. In addition, if interference exists, there is a third term: carrier-to-interference ratio (C/I). In order to compute the down-link carrier-to-noise ratio, a number of other parameters must be specified. The EIRP of a C-Band satellite (the Westar satellite, for example), is on the order of 32 dBW. Remember from the contour plots of Figure 11.4 this can vary with geographic location of the earth station. For our 60-megabit/s TDMA system, the satellite transponder can be nominally operated at the saturation level. In practice, the BER performance begins to degrade as a TWT transponder is driven into saturation. This occurs because of AM/PM and AM/AM conversion. Typically, the optimum BER is found with the output power backed off by 0.5 dB. On the other hand, solid-state transponders can generally be operated at saturation. With these parameters defined we can compute the down-link carrier-to-thermal noise density (C/N_0) using equation 11.15. Thus, in terms of dB

$$\frac{C}{N_0} = \mathrm{EIRP}_{\mathrm{sat}} - Lp + k^{-1} + G/T - M \tag{11.21}$$

where:

$\mathrm{EIRP} = 31.5$ dBW
$Lp =$ down-link path loss at 4.2 GHz $= 196.6$ dB
$k^{-1} =$ Boltzman's constant $= 228.6$ dBW/K-Hz
$\frac{G}{T} =$ Earth station $G/T = 32.7$ dB/K
$M =$ margin $-$(2.6-dB operating/rain margin + 0.5 dB miscellaneous losses)

Then,

$$\frac{C}{N_0} = 31.5 - 196.6 + 228.6 + 32.7 - 3.1$$

$$= 93.0 \text{ dB}\cdot\text{Hz}$$

The system down-link CNR_{DL} is:

$$\mathrm{CNR}_{DL} = \frac{C}{N_0} - 10 \log B = 93.0 - 75.6 = 17.5 \text{ dB.}$$

With the down-link parameters satisfied, we can perform a similar computation for the up-link and then determine the overall CNR_s for the entire link. Remember for a BER $= 10^{-9}$, $\mathrm{CNR}_s = 15$ dB is required. We can determine the EIRP required for the Earth station by starting with the saturation flux density at the satellite's receive antenna and adding to this the propagation losses. A nominal figure for saturation flux density is -82.0 dBW/m^2. Also, from the transfer characteristics of a C-band TWT transponder, the input power backoff at a 0.5-dB output power backoff is on the order of 3 dB. Thus, with backoff, the operating flux density at the satellite's receive antenna is -85 dBW/m^2. The EIRP for the transmitter's Earth station is then:

$$\mathrm{EIRP}_{es} = \psi_0 + \mathrm{BSL} + M \tag{11.22}$$

where

$\psi_0 =$ operating flux density $= -85$ dBW/m^2
$\mathrm{BSL} =$ beam spreading loss $= 162.6$ dB
$M =$ atmospheric loss $= 0.5$ dB

Thus

$$\text{EIRP}_{es} = -85.0 + 162.6 + 0.5 = 78.1 \text{ dBW}$$

and the up-link carrier-to-thermal-noise density ratio is

$$\begin{aligned} C/N_0 &= \text{EIRP}_{es} - Lp + k^{-1} + \frac{G}{T} \\ &= 78.1 - 200.1 + 228.6 - 6.0 \qquad (11.23) \\ &= 100.6 \text{ dBW}\cdot\text{Hz} \end{aligned}$$

where

$\text{EIRP}_{es} = 78.1$ dBW
Lp = up-link path loss at 6 GHz = 200.1 dB
k^{-1} = Boltzman's constants = 228.6 $(\text{dBW/K}\cdot\text{Hz})^{-1}$
$\frac{G}{T}$ = satellite $\frac{G}{T} = -6.0$ dB/K.

After factoring in the bandwidth, the up-link carrier-to-noise ratio is

$$\begin{aligned} \text{CNR}_{\text{up}} &= \frac{C}{N_0} - 10 \log B \\ &= 100.6 - 75.6 = 25 \text{ dB} \end{aligned}$$

By an extension of equation (11.16), the system CNR is

$$\text{CNR}_s = \frac{1}{1/\text{CNR}_{DL} + 1/\text{CNR}_{UL} + 1/C/I}$$

or (11.24)

$$\text{CNR}_s \text{ (dB)} = \text{CNR}_{DL} \oplus \text{CNR}_{UL} \oplus \frac{C}{I}$$

where $\oplus$ indicates a power addition. For $C/I = 20.8$ dB, we have

$$\text{CNR}_s = 17.5 \oplus 25.0 \oplus 20.8 \Rightarrow 15.3 \text{ dB}$$

Note that this result exceeds the required CNR_s required for a clear weather BER $= 10^{-9}$ by 0.3 dB.

We can adjust the transmitting Earth station HPA output power to meet the EIRP requirement of 78.1 dBW. The actual output power level may be calculated by

$$P_{HPA} = \text{EIRP}_{es} - G_T + L_{\text{mux}} \qquad (11.25)$$

where

$$\begin{aligned} \text{EIRP}_{es} &= 78.1 \text{ dBW} \\ G_T &= 56.5 \text{ dB} \\ L_{\text{mux}} &= \text{RF multiplexer loss} = 7.0 \text{ dB} \end{aligned}$$

TABLE 11.3 C-BAND LINK ANALYSIS SUMMARY

Coding rate	7/8R HD
Modulation	QPSK
Antenna size	13.0 m
CNR system required	15.0 dB for BER = 10^{-9}
Propagation availability	99.995%
C/I	20.8 dB

	Description	Transmission rate (kilobits/s) 60 megabits/s
Down-link		
EIRP	satellite saturated (dBW)	32.0
EIRP	operating (dBW)	31.5
OPBO	(ouput power backoff) (dB)	0.5
G/I	ES (dB/K)	32.7
LD/L	Down-link path loss (dB)	196.6
M_R	(operating/rain margin) (dB)	2.6
M_I	(miscellaneous losses) (dB)	0.5
K	Boltzmann's constant (dBW/K/Hz)	228.6
C/N_0	down-link (dB·Hz)	93.0
B	System noise bandwidth (dB·Hz)	75.6
CNR *D/L* (dB)		17.5
Up-link		
φ_s	saturation flux density (dBW/M^2)	−82.0
IPBO	input power backoff (dB)	3.0
ϕ_0	operation—operation flux density (dBW/M^2)	−85.0
BSL	beam spreading loss (4 R^2) (dB)	162.6
M	margin (dB)	0.5
EIRP	*E/S* (dBW)	78.1
L U/L	up-link path loss (dB)	200.1
G/T	satellite (dB/K)	−6.0
K	Boltzmann's constant (dBW/K·Hz)	228.6
C/N_0	up-link (dB·Hz)	100.6
B	system noise bandwidth (dB·Hz)	75.6
CNR	up-link (dB)	25.0
EIRP_{ES}	78.1 dBW	
RF mux loss	7.0 dB	
HPA output power	28.6 dBW (724 W)	

$$\begin{aligned} \text{CNR system} &= \text{CNR}_{D/T} \oplus \text{CNR} \oplus C/I \text{ (dB)} \\ &= 25.0 \oplus 17.5 \oplus 20.8 \\ &= 15.3 \text{ dB} \end{aligned}$$

$\oplus$ denotes power addition

Thus

$$P_{PHA} = 78.1 - 56.5 + 7.0 = 28.6 \text{ dBW}$$
$$= 724 \text{ W}$$

The RF multiplexer loss assumes multiple transponders and includes such things as filter, circulators, waveguide losses, and hybrid combiners. These system calculators are itemized in Table 11.3.

Both C-Band and Ku-Band satellites suffer from another form of performance degradation that arises from sun transit. Two times each year, the sun moves across the equator. During the spring equinox it moves northward across the equator. Conversely, during the fall equinox, it moves southward across the equator. For a period of time on the order of 5 da during both transits, it is in the field of view of the earth station antenna for a few minutes each day. For both equinoxes, the total outage for a 13-m C-band system is on the order of 18 min. An additional consideration for a C-Band system is locating an interference-free site for the earth station. It may be necessary to locate the station out of an urban area and then engineer an interference-free microwave path to the terminating office facilities. This adds considerably to the cost. This problem does not exist at Ku-band, since there are no corresponding terrestrial based microwave systems at Ku-Band. Obviously, trade-off studies are needed in an initial design.

11.6 LOCATING THE SATELLITES

At early ages, we all become familiar with the gridlike system used to describe any location on earth. The angular measurements called latitude and longitude were established years ago by the British to aid in map-making and navigation. The globe was divided into 360° of longitude with 0° passing through the Greenwich observatory in England. If one travels west from the Greenwich meridian (0° longitude), we have west longitude, which increases from 0° to 180°. Similarly, eastward travel represents an increase from 0° to 180° of east longitude. In a similar manner, the earth was divided north and south of the equator with parallel lines called degrees of latitude. The position of the equator is represented by 0°.

As discussed previously, the satellites are located along the equatorial belt at an altitude of 22,300 mi. In order to locate a satellite, it is necessary first to locate due south. This can be done by locating magnetic north with a compass and then correcting for magnetic deviation in the immediate area. Once true north is located, a 180° reversal yields true south. In the northern hemisphere this is the direction that the antenna must point. For a polar-mounted antenna (Figure 11.13), the angles of azimuth and elevation must be determined. This can be accomplished using the following equations:

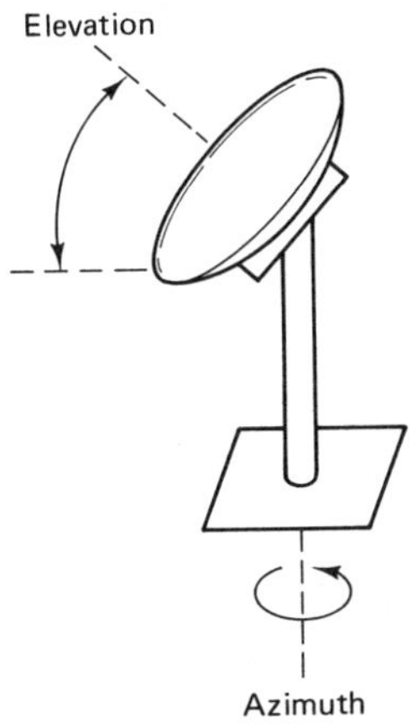

Figure 11.13 AZ-EL antenna mount.

Azimuth:

$$\mathrm{Az} = 180 + \tan^{-1}\frac{\tan B}{\sin A} \qquad \text{northern hemisphere} \tag{11.26}$$

$$\mathrm{Az} = \tan^{-1}\frac{\tan B}{\sin A} \qquad \text{southern hemisphere} \tag{11.27}$$

where

$$A = \text{latitude}$$
$$B = \text{longitude}$$

Elevation:

$$\mathrm{EL} = \cos^{-1}\left[\frac{D^2 + R^2 - (R - H)^2}{2DR}\right] - 90° \tag{11.28}$$

where D is the line of sight distance between the earth station and the satellite. It may be expressed as

$$D = [R^2 + (R + H)^2 - 2R(R + H)\cos C]^{1/2} \tag{11.29}$$

where

R = earth radius (3974 mi)
H = satellite height (22,300 mi)
C = central angle
= $\cos^{-1}[\cos A \cdot \cos B]$

Tables of satellite locations are given in the appendix.

The most popular antenna is the dish antenna because of its high gain (Figure 11.14). For example, for antenna diameters between 3 m and 11 m, the corre-

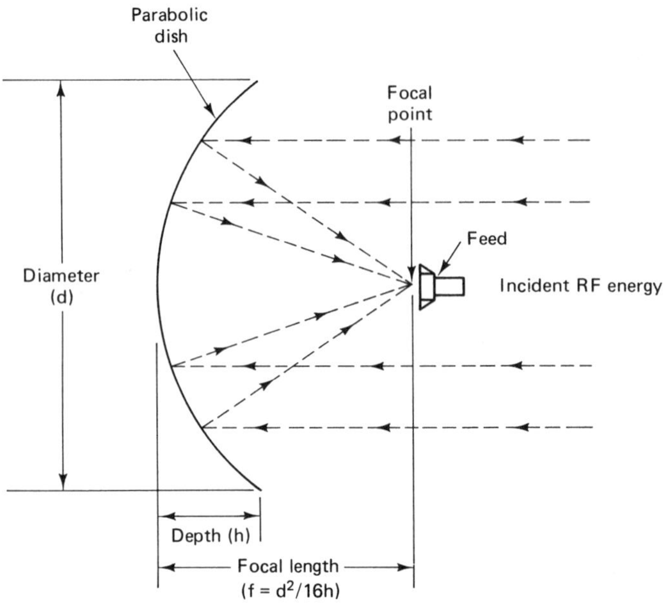

Figure 11.14 Parabolic dish antenna.

sponding gains range from 39.4 dB to 54 dB. The exact expression for gain of this antenna is given in Example 11.3.

PROBLEMS

11.1. Calculate the power picked up from a 25-W transmitter at a distance of 10^{14} m operating at a frequency of 12 GHz. The transmitting antenna is a dish with effective area 6 m^2 and the receiving antenna is a dish with effective area 2000 m^2.

11.2. A geostationary satellite is a satellite in orbit over the equator, which rotates with the earth once every 24 h, so that it apparently always stays in the same place in the sky. Calculate the height of such a satellite above the earth's surface and the time it takes a signal to travel from the nearest point on the earth to the satellite and back. (Acceleration due to gravity at distance r from the center of the earth is gR^2/r^2, where $g = 9.81$ ms^{-2} and where $R = 6.37 \times 10^6$ m is the radius of the earth. This must be equal to the centrifugal acceleration W^2r, where W is the angular velocity of the satellite in radians per second corresponding to one revolution every 24 h.)

11.3. The noise at the terminals of a receiving antenna corresponds to a temperature of 300 K. The receiving antenna has an effective area of 1.5 m^2 and is connected to a preamp with a 15-MHz bandwidth. Calculate the power that must be transmitted by a satellite such that the SNR at the antenna termals is 50 dB. The coverage from the satellite is an area of 0.5×10^6 km^2.

11.4. The up-link distance from the Earth station to a satellite is the same as the down-link distance. Why is the path loss greater on the up-link than the down-link?

11.5. Would the communicator be more concerned with the jamming of the satellite up-link or the satellite down-link? Explain.

11.6. A satellite in geostationary orbit is to be used for communication to an Earth station. The following parameters apply:

$$P_t = 10 \text{ W}$$

$$\text{Frequency} = 300 \text{ MHz}$$

$$G_t = 10 \text{ dB}; \qquad G_r = 30 \text{ dB}$$

An error probability of 10^{-5} is desired in the presence of noise $N_0 = 10^{-20}$ W/Hz.

(a) Find the maximum data rate assuming ideal PSK modulation.

(b) What is the maximum data rate if 32-ary modulation is used?

(c) If an arbitrarily large bandwidth is allowed along with an arbitrarily complex coding scheme, what is the maximum data rate permitted? How does this compare to parts (a) and (b)?

11.7. A digital signal using FSK modulation is received from a satellite. The SNR is 20 dB.

(a) Assuming a BER of 10^{-6}, what is the maximum data rate?

(b) Assume a system bandwidth of 200 Hz is available. What is the capacity of the system and the corresponding E_b/N_0 ratio?

(c) What is the bandwidth efficiency in bits per second per hertz of the system in part (b)?

11.8. A satellite in synchronous orbit is transmitting to a ground station over a slant range of 40,000 km. The satellite antenna has a gain of 6 dB and the operating frequency is 4 GHz. The ground station has a parabolic antenna 15 m in diameter with an aperture efficiency of 0.54. What transmitted power at the satellite is required to achieve a received signal power of 10^{-12} W?

11.9. How does the systems designer determine the link margin required in order to satisfy a specified link availability?

11.10. What is the difference between spot beam and global beam coverage? Assuming geostationary orbit, calculate the beam angle to give full global coverage. How many satellites are required for complete earth coverage?

11.11. Assume a 7° elevation for transmission to and from a satellite in geostationary orbit. Calculate the round-trip propagation delay.

11.12. If the spacing between satellites is changed from 3° to 2°, how close to each other will the satellites be? What problems do you foresee this will entail for antenna designers?

11.13. In Table 11.1, the C/T at the input of the Earth station receiver is -128.1 dBW/K. If the transmission rate is 70 Mb/s, what is the E_b/N_0 ratio at the receiver input?

11.14. The up-link frequency of an earth station is 6 GHz and the antenna has a diameter of 30 m. The satellite G/T is -5.3 dB/K. For a transmission rate of 80 megabits/s, calculate E_b/N_0 at the satellite input.

11.15. In Figure 11.9, show that the synchronization delay, D_N, for the Nth satellite is given by

$$D_N = MT_M - \frac{2\, d_N}{C}$$

11.16. Show the following power addition (see equation 11.24): 17.5 ⊕ 25.0 ⊕ 20.8 = 15.3 dB.

11.17. A satellite transponder is to be designed for QPSK modulation. The received bit rate is 120 megabit/s.

(a) Determine the required E_b/N_0 ratio and C/N ratio if $P_e = 10^{-7}$ is required.

(b) What is the required C/N ratio and required RF bandwidth if the 4-ϕ unit is replaced by a two-phase BPSK unit?

11.18. Refer to the TDMA frame shown in Figure 11.6. A Bell system T1 frame (1.544 megabit/s) is to be transmitted over the satellite link.

(a) How many bits are transmitted per frame?

(b) Assuming QPSK modulation, what is the symbol rate?

(c) What is the duration of a symbol period?

11.19. Replace the C-band satellite link described in Section 11.5.1 with a Ku-band system (12.2 GHz down-link and 14.5 GHz up-link). For the same performance objectives, what down-link rain margin is required? Do you consider a Ku-band design to be feasible for this particular orbit slot?

APPENDIX

List of Abbreviations

AO	Atlantic Ocean
BOL	Bolivia
BSS	Broadcasting Satellite Service
CL	Colombia
DOD	Department of Defense
DOM	Domestic
EC	Ecuador
ESA	European Space Agency
EXP	Experimental
Ford A.S.S. Corp	Ford Aerospace Satellite Services Corporation
FSS	Fixed Satellite Service
GOVT	Government
IO	Indian Ocean
INDON	Indonesia
INT or INT'L	International
ISRO	Indian Space Research Organisation
KDD	Kokusai Denshin Denwa Co., Ltd.
MARIT	Maritime Mobile Satellite Service
MET or M	Meteorological
MOBILE	Mobile Satellite Service
MOD	Ministry of Defense
MOPT	Ministry of Posts and Telegraphs
MP	Major Path
NASDA	National Aeronautics and Space Development Agency
NOAA	National Oceanographic and Atmospheric Agency
NTT	Nippon Telephone and Telegraph
PO	Pacific Ocean
PR	Peru
PR China	Peoples Republic of China
PRI	Primary
PT & T	Postal Telephone and Telegraph
RITE	Research Institute of Telecommunications and Economics
RRL	Radio Research Laboratory
SBS	Satellite Business Systems
SP COMM CO	Southern Pacific Communications Co.
VN	Venezuela
*	Application pending at the F.C.C.
**	Temporary location

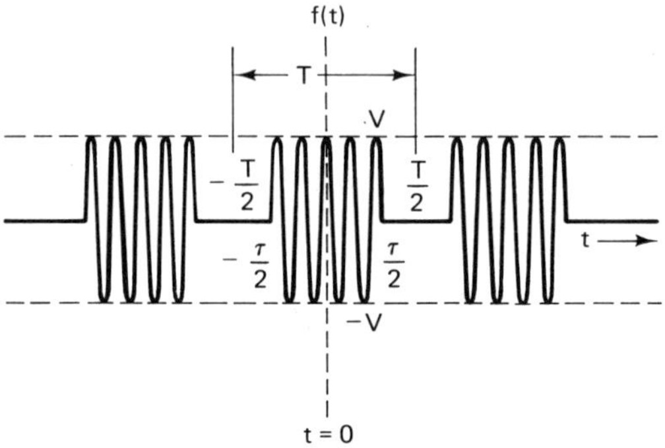

Figure A.1 Pulsed RF wave.

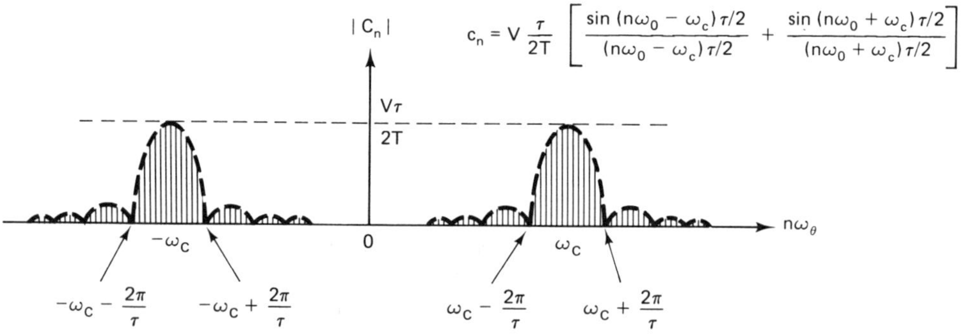

Figure A.2 Frequency spectrum of pulsed RF wave.

Table 1 Geosynchronous Satellites in the Eastern Hemisphere (© 1984, IEEE)

Location	Up (GHz)	Down	Service	Name of Satellite	Operator	Transponders	EIRP	Launch	Life
E000	8	7	GOVT	SKYNET IVA	UK MOD			86	0
E001	11,12.5	6.5–7,14	BUSINESS	GDL-5	LUXEMBOURG			85	10
E001-7	none	none	FSS	INTELSAT RETIRED	INTELSAT	none	none	0	0
E002.7	0.15,0.4,2.1	0.14,0.5,1.7	MET	METEOSAT-2	EURO MET SAT	2		81	3
E002.7	0.15,0.4,2.1	0.14,0.5,1.7	MET	METEOSAT-OPN'L	EURO MET SAT	2		87	3
E003	6, 8, and 14	4, 7, and 12	DOM, GOVT	TELECOM-1C	FRANCE	12	26–47	85	7
E005	17	12	REGIONAL	TELE-X	NORDIC NATIONS	4		87	6
E006.6	14	11	BSS & FSS	OTS-2	EURO SPC AGY	6	35 to 45	78	5
E007	6 and 30	4,20,40,90	REGIONAL	ATHOS	CNES & FR. PTT	8 and 4	37.5 and 51	87	5
E007	14	11 and 12.5	REGIONAL	EUTELSAT I-3	EUTELSAT COUNCIL	11	39.8	85	10
E007	2,6,30	2,4,20		F-SAT-1	FRANCE			87	0
E010	2,6,30,90	2,4,20,40	EXP	APEX	FRANCE			86	0
E010	14	11 and 12.5	REGIONAL	EUTELSAT I-2	EUTELSAT COUNCIL	11	39.8	84	10
E012	3	2	GOVT	PROGNOZ-2	USSR	9		0	0
E013	30 and 50	20 and 40	DOM	ITALSAT	ITALY	13		88	5
E013.1				COSMOS-0775	USSR			75	0
E014	6	4	DOM	NIGERIAN NAT SAT-1	NIGERIA			0	0
E015	6 and 14	4 and 11	FSS	AMS-1	ISREAL			86	0
E015	6 and 14	4 and 11	FSS	AMS-2	ISREAL			86	0
E016	0.4,8,14,44	0.3,7,12,20	GOVT	SICRAL 1A	ITALY PTT			87	0
E017	14.2–14.6	11.7–12.5	BSS	SABS	SAUDI ARABIA		63	0	0
E019	6	2.5 and 4	FSS & BSS	ARABSAT-1	ARAB SAT COMM	25	41 and 31	84	7
E019	11,12.5	6.5–7,14	BUSINESS	GDL-6	LUXEMBOURG			85	10
E020	6	4	DOM	NIGERIAN NAT SAT-2	NIGERIA			0	0
E022	0.4,8,14,44	0.3,7,12,20	GOVT	SICRAL 1B	ITALY PTT			87	0
E023.5	14,30	11,12,20	DOM & BSS	DFS-1	GERMANY	11	48,49	87	10
E026	6	2.5 and 4	FSS & BSS	ARABSAT-2	ARAB SAT COMM	25	41 and 31	85	7
E028.5	14,30	11,12,20	DOM & BSS	DFS-2	GERMANY	11	48 and 49	87	10
E032	14	12	VIDEO?	VIDEOSAT-1	FRANCE			87	0
E035	8	7	GOVT	GALS-06	USSR PT&T	10		84	0
E035	3	2	GOVT	PROGNOZ-3	USSR	9		0	0
E035	5.93–6.23	3.6–3.9	DOM	STATSIONAR-02	USSR PT&T	6	19 to 30	0	0
E035.8	5.75–6.25	3.4–3.9	DOM	RADUGA-09	USSR PT&T	10	30	81	0
E037.2	6	4	FSS	RADUGA-11	USSR PT&T	10	30	82	0
E038	14	12	DOM	PAKSAT I	PAKISTAN			87	0
E040	5.73–6.28	3.4–3.95	DOM	STATSIONAR-12	USSR PT&T	11	19 to 30	0	0
E041	14	12	DOM	PAKSAT-2	PAKISTAN			0	0
E041	8	7	GOVT	SKYNET IVB	UK MOD			86	0
E043.5	6	4	FSS	RADUGA-13	USSR PT&T	10	30	83	0
E045	8	7	GOVT	GALS-02	USSR PT&T	10		0	0
E045	14	11	DOM	LOUTCH-P02	USSR PT&T	10		0	0
E045	5.73–6.03	3.4–3.7	DOM	STATSIONAR-09	USSR PT&T	6	22 to 31	0	0
E045	0.3–0.4,1.6	0.2–0.3,1.5	MOBILE	VOLNA-03	USSR PT&T	2		0	0
E050.7	6.2	0.714	BSS	EKRAN-03	USSR PT&T	1	49	79	1
E051.5				COSMOS-0637	USSR PT&T			74	0
E052.1	6 and 8	4 and 7	GOVT	GORIZONT-09	USSR PT&T			84	0
E053	14	11	DOM	LOUTCH-02	USSR PT&T	10	39	0	0
E053	5.98–6.28	3.65–3.95	DOM	STATSIONAR-05	USSR PT&T	6	23 to 30	0	0
E053	0.4,1.6,1.6	0.3,1.5,1.5	MOBILE	VOLNA-04	USSR PT&T	3		0	0
E054.8	1.4, 6, 8 and 14	1.5, 4, 7, and 11	DOM	GORIZONT-05	USSR PT&T			82	0
E055.5	14	11	DOM	LOUTCH-02 LATER	USSR PT&T	10	39	0	0

Table 1 *(continued)*

Location	Up (GHz)	Down	Service	Name of Satellite	Operator	Transponders	EIRP	Launch	Life
E056.8	6	4	INTL	INTELSAT IVA-F2	INTELSAT	20	20 to 29	76	7
E057	1.6,6,14	1.5,4,11	INT&MARIT	INTELSAT IO-SPARE-1	INTELSAT			0	0
E057	6 and 14	4 and 11	INTL	INTELSAT V-F1	INTELSAT	27	21 to 42	81	7
E059.8	14	12	BSS	YURI (BS-1)	MOPT JAPAN	2	46 to 55	78	3
E060	2,8	2,7	GOVT	DSCS-2-F04	DOD USA	4	28,43	73	5
E060	2,8	2,7	GOVT	DSCS-3-F04	DOD USA	7	23–44	84	10
E060	1.6,6,14	1.5,4,11	INT&MARIT	INTELSAT IO-MP/PRI	INTELSAT			0	0
E060	1.6, 6, and 14	1.5, 4, and 11	INT&MARIT	INTELSAT V-F7	INTELSAT	28	21 to 42	83	7
E063	1.6,6,14	1.5,4,11	INT&MARIT	INTELSAT IO-PRI/MP	INTELSAT			0	0
E063	1.6, 6, and 14	1.5, 4, and 11	INTL	INTELSAT V-F5	INTELSAT	28	21 to 42	82	7
E064.5	1.6,6	1.5,4	MARIT	MARECS-C	INMARSAT	2	18 to 30	0	7
E065	6,17	4,12	DOM, BSS	PRC-1	CHINA B/C SAT			87	5
E065.9	0.15,18	0.14,12	EXP	SIRIO-1	ITALY	2	23	77	2
E066	1.6,6,14	1.5,4,11	INT&MARIT	INTELSAT IO-SPARE-2	INTELSAT			0	0
E070	6	4	DOM	STW-2	PR CHINA	1	50	85	3
E071.9	6	0.714 and 3	BSS	EKRAN-09	USSR PT&T	1	49	82	2
E072.5	1.6 and 6	1.5 and 4	MARIT	MARISAT-INDIAN-103	COMSAT GENERAL	8	29 and 19	76	8
E073.3	0.3,8	0.25	GOVT	FLEETSATCOM-2	US NAVY	23	16 to 28	79	5
E074	0.46,6	2.5,4	DOM/BSS/M	INSAT-IB	ISRO-INDIA	14	32 to 42	83	7
E074.8	6,8	4,7	DOM, GOVT	GORIZONT-02	USSR			79	0
E075	20	44	GOVT	FLTSATCOM-B INDIAN O	US NAVY			86	0
E075	0.3 and 8	0.25	GOVT	LEASAT-1	HUGHES COMM SVC	13	17 to 28	84	10
E076	0.149,2	0.137,1.7	MET	GOMS	USSR	2		0	0
E076.4	6,8	4,7	DOM, GOVT	GORIZONT-03	USSR PT&T			79	0
E079.5	SHF	SHF	EXP	COSMOS-1366	USSR			82	0
E079.9	6	4	FSS	RADUGA-12	USSR PT&T	10	30	83	0
E080	4.40–4.68	3.95–4.00	DOM	POTOK-2	USSR PT&T	3		0	0
E080	6,17	4,12	DOM, BSS	PRC-2	CHINA B/C SAT			88	5
E080	3	2	GOVT	PROGNOZ-4	USSR			0	0
E080	5.93–6.23	3.6–3.9	DOM	STATSIONAR-01	USSR PT&T	6	19 to 30	0	0
E080	5.73–6.28	3.4–3.95	DOM	STATSIONAR-13	USSR PT&T	11	19 to 30	0	0
E080.2	6	4	DOM	PALAPA-A1	PERUMTEL (INDON)	12	32	76	7
E080.2	6	4	DOM	PALAPA-A2	PERUMTEL (INDON)	12	32	77	7
E083.7	6	4 and 7	FSS	RADUGA-14	USSR PT&T	10	30	84	0
E084.5	5.75–6.25	3.4–3.9	DOM	RADUGA-04	USSR PT&T	10	30	78	0
E085	8	7	GOVT	GALS-03	USSR PT&T	10		0	0
E085	14	11	DOM	LOUTCH-P03	USSR PT&T	10		0	0
E085	5.93–6.23	3.6–3.9	DOM	STATSIONAR-03	USSR PT&T	6	19 to 30	0	0
E085	0.3 and 1.6	0.2 and 1.5	MOBILE	VOLNA-05	USSR PT&T	2		0	0
E085.4	6	0.714 and 3	BSS	EKRAN-08	USSR PT&T	1	49	82	2
E085.8	5.75–6.25	3.4–3.9	DOM	RADUGA-10	USSR PT&T	10	30	81	0
E086.2	6.2	0.714	DBS	EKRAN-07	USSR PT&T	1	49	81	1
E089.4	6 and 8	4 and 7	GOVT	GORIZONT-08	USSR			83	0
E090	14	11	DOM	LOUTCH-03	USSR PT&T	10	39	0	0
E090	5.98–6.28	3.65–3.95	DOM, BSS	STATSIONAR-06	USSR PT&T	6	23 to 30	0	0
E090	0.4,1.5,1.5	0.3,1.6,1.6	MOBILE	VOLNA-08	USSR PT&T	3		84	0
E092	17	12	DOM	PRC-3	PR CHINA			88	5
E092.5	5.75–6.25	3.4–3.9	DOM	RADUGA-03	USSR PT&T	10	30	77	0
E094	1.7 and 6	2, 2.5 and 4	DOM, BSS	INSAT-IC	INDIA	14	32 to 42	86	0
E095	14 and 15	11 and 13	TRACKING	CSDRN	USSR			0	0
E095	5.73–6.28	3.4–3.95	DOM	STATSIONAR-14	USSR PT&T	11	19 to 30	0	0

Table 1 (*continued*)

Location	Up (GHz)	Down	Service	Name of Satellite	Operator	Transponders	EIRP	Launch	Life
E098.2	6.2	0.714	BSS	EKRAN-06	USSR PT&T	1	49	80	1
E099	6.2	0.714	BSS	STATSIONAR-T	USSR PT&T	1	49	0	0
E099	6.2	0.714	BSS	STATSIONSAR-T2	USSR PT&T	1	49	0	0
E100.5	6.000	0.714	TV	EKRAN-10	USSR	1	49	83	0
E102	0.148 and 6	0.136 and 4	DOM	APPLE	INDIA	1	30.5	81	2
E102.2	14 and 2	12 and 2	BSS	YURI-2A	MOPT JAPAN	2	46 to 55	84	3
E102.3	6.000	0.714	TV	EKRAN-11	USSR	1	49	83	0
E108	6	4	REGIONAL	PALAPA-B1	INDONESIA	24	34	83	8
E110	14 and 2	12 and 2	BSS	BS-2B (SPARE)	MOPT JAPAN	2	46–55	85	3
E110	14	12	BSS	BS-3	MOPT JAPAN			87	5
E113	6	4	REGIONAL	PALAPA-B2 REPLACEMENT	INDONESIA	24	34	86	8
E116.0	6.000	0.714	TV	EKRAN-12	USSR	1	49	84	0
E118	6	4	REGIONAL	PALAPA-B3	INDONESIA	24	34	0	8
E120	30	20	DOM	NTT-1	NIPPON TEL & TEL			87	10
E120	30	20	DOM	NTT-2	NIPPON TEL & TEL			95	10
E120.8	6 and 30	4 and 20	DOM	CS-2B	NASDA & NTT	8	29 and 37	83	5
E125.8	6.225–0.425	4.00–4.200	DOM	STW-1	PR CHINA	2	35	84	3
E128	5.73–6.28	3.4–3.95	DOM	STATSIONAR-15	USSR PT&T	11	19 to 30	0	0
E130	EXP	EXP	EXP	ETS-5	RRL-JAPAN			87	0
E130	8	7	GOVT	GALS-05	USSR PT&T	10		84	0
E132.0	6 and 30	4 and 20	DOM	CS-2A (SAKURA)	NASDA & NTT	8	29 and 37	83	5
E135	30 and 50	20 and 40	DOM	CS-3A	NASDA-JAPAN			88	0
E135	30 and 50	20 and 40	DOM	CS-3B	NASDA-JAPAN			88	0
E140	0.4, 2.0, 2.2	0.5, 1.7, 2.3	MET	GMS-3A (HIMIWARI)	KDD	4	4 to 16	84	7
E140	14	11	DOM	LOUTCH-04	USSR PT&T	10	39	0	0
E140	5.98–6.28	3.65–3.95	DOM	STATSIONAR-07	USSR PT&T	6	23 to 30	0	0
E140	0.4, 1.5, 1.5	0.3, 1.6, 1.6	MOBILE	VOLNA-06	USSR PT&T	3		0	0
E140.8	0.2, 0.4, 2	0.1, 0.5, 1.7	MET	GMS-1 (HIMIWARI-1)	JAPAN	4	4 to 16	77	3
E142.6	6, 8, and 14	4, 7, and 11	DOM	GORIZONT-06	USSR PT&T			82	0
E145	6	4	DOM	STATSIONAR-16	USSR			0	0
E146.9	0.2, 0.4, 2	0.1, 0.5, 1.7	MET	GMS-2 (HIMIWARI-2)	JAPAN	4	4 to 16	81	3
E151.1	6 and 30	4 and 20	DOM	CS (SAKURA)	NASDA-JAPAN	8	29 and 37	77	3
E156	14.0–14.5	12.25–0.75	DOM/BSS	AUSSAT-1	AUSTRALIA	15	36 to 48	85	7
E160	14.0–14.5	12.25–0.75	DOM, BSS	AUSSAT-2	AUSTRALIA	15	36 to 48	85	7
E160	0.4, 2.9, 2.1	0.5, 1.7, 2.3	MET	GMS-3B (HIMIWARI)	KDD	4	4 to 16	85	7
E164	14.0–14.5	12.25–0.75	DOM, BSS	AUSSAT-3	AUSTRALIA	15	36 to 48	88	7
E164.8	0.149, 6	0.136 + , 5	EXP	ATS-1	NASA	1	17.8	66	15
E170	6	4	INTL	PAC. AREA SAT SYS-A	RITE-JAPAN	12?	87	7	
E170	6 and 14	4 and 12	INTL	PAC. AREA SAT SYS-B	RITE-JAPAN			87	7
E172	20	44	GOVT	FLTSATCOM-B W. PAC O	US NAVY			86	0
E173.1	0.3, 8	0.25	GOVT	FLEETSATCOM-4	US NAVY	23	16.5–28	80	5
E174	6	4	INTL	INTELSAT IVA-F6	INTELSAT	20	20 to 29	78	7
E174	6, 14	4, 11	INTL	INTELSAT PO-PRI	INTELSAT		0	0	7
E175	2, 8	2, 7	GOVT	DSCS-3-F2	DOD USA	7	23–44	84	10
E175	2, 8	2, 7	GOVT	DSCS-WP	DOD USA			0	0
E176.5	0.3 and 8	0.25	GOVT	LEASAT-2	HUGHES COMM SVC	13	17 to 28	84	10
E176.7	1.6 and 6	1.5 and 4	MARIT	MARISAT-INDIAN-103	COMSAT GENERAL	8	23, 29, and 19	77	8
E177.5	1.6, 6	1.5, 4	MARITIME	MARECS-B2	INMARSAT	2	17 to 35	84	7
E179	6	4	INTL	INTELSAT IVA-F3	INTELSAT	20	20 to 29	78	7
E179	6, 14	4, 11	INTL	INTELSAT PO-SPARE	INTELSAT		0	0	0
E179	6 and 14	4 and 11	INT'L	INTELSAT V, VA & MCS	INTELSAT			0	7
E179.6	2, 8	2, 7	GOVT	DSCS-2-F08	DOD USA	4	28, 434S77	5	

Table 2 Geosynchronous Satellites in the Western Hemisphere
Note: * Location request pending at the Federal Communications Commission
**Temporary location until Gstar A2 is operational (© 1984, IEEE)

Location	Up (GHz)	Down	Service	Name of Satellite	Operator	Transponders	EIRP	Launch	Life
W001	6	4	INT'L	INTELSAT AO-SPARE 4	INTELSAT			0	0
W001	6	4	INTL	INTELSAT IV-F8	INTELSAT	12	20 and 31	74	7
W001	6 and 14	4 and 11	INTL	INTELSAT VA-F1	INTELSAT	28	21 to 42	84	7
W004	6	4	INT'L	INTELSAT AO-SPARE 5	INTELSAT			0	0
W005	6, 8, and 14	4, 7, and 12	DOM, GOVT	TELECOM-1B	FRENCH PTT	12	26–47	85	7
W008	6, 8, and 14	4, 7, and 12	DOM, GOVT	TELECOM-1A	FRENCH PTT	12	26–47	84	77
W010.6	6,8	4,7	DOM, GOVT	GORIZONT-04	USSR PT&T			80	0
W011	2,14,30	2,12,20		F-SAT-2	FRANCE			86	0
W011	5.73–6.28	3.4–3.95	DOM	STATSIONAR-11	USSR PT&T	11	19 to 30	0	0
W011.33	6	4	EXP	SYMPHONIE-2	FRANCE & GERMANY	2	29	75	5
W011.5	2,8	2,7	GOVT	DSCS-2-F14	DOD USA	4	28,43	79	5
W013	2,8	2,7	GOVT	DSCS-AO	DOD USA			0	0
W013.5	4.40–4.68	3.95–4.00	DOM	POTOK-1	USSR PT&T	3		0	0
W014	14	11	DOM	LOUTCH-01	USSR PT&T	10	39	0	0
W014	5.98–6.28	3.65–3.95	DOM	STATSIONAR-04	USSR PT&T	6	23 to 30	0	0
W014	1.6	1.5	MOBILE	VOLNA-02	USSR PT&T	3		0	0
W014.5	6 and 8	4 and 7	GOVT	GORIZONT-07	USSR			83	0
W014.9	1.6 and 6	1.5 and 4	MARIT	MARITSAT-ATLANTIC-101	COMSAT GENERAL	8	29 and 19	76	8
W016	14 and 15	11 and 13	TRACKING	WSDRN	USSR			0	0
W016.2	2,8	2,7	GOVT	DSCS-3-F1	DOD USA	7	23–44	82	10
W017.3	2,8	2,7	GOVT	NATO IIIB	NATO	3	31 and 36	77	7
W018	8	7	GOVT	NATO IV	NATO	3	31 and 36	0	7
W018.5	1.6,6,14	1.5,4,11	INT&MARIT	INTELSAT AO-MAJ-2	INTELSAT			0	0
W018.5	1.6, 6, and 14	1.5, 4, and 11	INTL	INTELSAT V-F6	INTELSAT	28	21 to 42	83	7
W019	17	12	BSS	HELVESAT-1	SWISS TEL-TEC	5	64	87	7
W019	17	12	BSS	HELVESAT-2	SWISS TEL-TEC	3	64	88	7
W019	17.3–17.7	11.7–12.1	BSS	LUXSAT	RADIO TELE LUX	5	63	88	7
W019	13,18,30	12,12,20	EXP, BSS	OLYMPUS 1	ESA	9	46,53,63	87	5
W019	17.8–18.1	11.7–12.5	BSS	SARIT	ITALY	5	86	0	
W019	17	12	BSS	TDF-1A	TELE-DIF-FRANCE	3	63.9	85	7
W019	17	12	BSS	TDF-1B	TELE-DIF-FRANCE	5	63.9	87	7
W019	17	12	BSS	TDF-2	TELE-DIF-FRANCE	5	63.9	90	10
W019	17	12	BSS	TV-SAT-A3 F1	WEST GERMANY	3	65.6	85	7
W019	2.1,17	2.3,12	BSS	TV-SAT-A5	GERMANY	5	65.6	90	10
W019.1	2,8	2,7	GOVT	NATO IIIC	NATO	3	31 and 36	78	7
W020	11,12.5	6.5–7,14	BUSINESS	GDL-4	LUXEMBOURG			85	10
W021.5	6	4	INT & MAR	INTELSAT AO-DOM&MCS	INTELSAT			0	0
W021.5	6	4	INTL	INTELSAT IVA-F4	INTELSAT	20	20 to 29	77	7
W022.0	0.3,8	0.25	GOVT	FLEETSATCOM-3	US NAVY	23	16.5–28	80	5
W023	20	44	GOVT	FLTSATCOM-B E. ATL O	US NAVY			86	0
W023	0.3 and 8	0.25	GOVT	LEASAT-4	HUGHES COMM SVC	13	17 to 28	85	10
W024	3	2	GOVT	PROGNOZ-1	USSR	9		0	0
W024.5	6,14	4,11	INTL	INTELSAT AO-PRI	INTELSAT			0	0
W024.5	6 and 14	4 and 11	INTL	INTELSAT V-F3	INTELSAT	27	21 to 42	81	7
W024.5	6 and 14	4 and 11	INT & MARIT	INTELSAT VA-F3	INTELSAT	28	21 to 42	84	7
W024.7	5.75–6.25	3.4–3.9	DOM	RADUGA-07	USSR PT&T	10	30	80	
W025	14	11	DOM	LOUTCH-P01	USSR PT&T	10		0	0
W025	0.148 and 2.1	0.137 and 1.7	MET&EXP	SIRIO-2 REPLACEMENT	ESA			85	5

Table 2 *(continued)*

Location	Up (GHz)	Down	Serive	Name of Satellite	Operator	Transponders	EIRP	Launch	Life
W025	5.73–5.93	3.4–3.58	DOM	STATSIONAR-08	USSR PT&T	4	22 to 31	0	0
W025	1.5,0.3	1.5,1.6,0.2	MOBILE	VOLNA-01	USSR PT&T	2		0	0
W026	1.6,6	1.5,4	MARIT	MARECS-ATL1	INMARSAT	2	17 to 30		
W026.1	1.6,6	1.5,4	MARIT	MARECS-A	INMARSAT	2	18 to 30	81	7
W026.5	8	7	GOVT	GALS-01	USSR PT&T	10		0	0
W027.5	1.6,6,14	1.5,4,11	INT & MARIT	INTELSAT AO-SPARE 2	INTELSAT			0	0
W027.5	6 and 14	4 and 11	INTL	INTELSAT V-F4	INTELSAT	27	21 to 42	82	7
W027.5	1.6, 6, and 14	1.5, 4, and 11	INTL	INTELSAT VA-F2	INTELSAT	28	21 to 42	84	7
W029.6	2,8	2,7	GOVT	NATO IIIA	NATO	3	31 and 36	76	7
W031	17	12.2–12.7	BSS	FALKLANDS	U.K.			0	0
W031	6	4	INTL	INTELSAT AO-SPARE 1	INTELSAT			0	0
W031	6	4	INTL	INTELSAT IVA-F1	INTELSAT	20	20 to 29	75	7
W031	14 and 18	11 and 12	BSS & FSS	UNISAT-1	UNITED SATS LTD	6	45 and 63	86	10
W031	14 and 18	11 and 12	BSS & FSS	UNISAT-2	UNITED SAT LTD	6	45 and 63	86	10
W031	14 and 18	11 and 12	BSS & FSS	UNISAT-3	UNITED SAT LTD	6	45 and 63	87	10
W034	17	12.2–12.7	BSS	GUYANA & JAMAICA	GUYANA & JAMAICA			0	0
W034.5	6,14	4,11	INTL	INTELSAT AO-MAJ-1	INTELSAT			0	0
W034.5	6 and 14	4 and 11	INTL	INTELSAT V-F2	INTELSAT	27	21 to 42	80	7
W037.5	14	12	VIDEO?	VIDEOSAT-2	FRANCE			87	0
W037.5*	14	11	INTL	ORION-1	ORION SAT CORP	22	44–47.4	87	7
W038-44	none	none		INTELSAT RETIRED.	INTELSAT		none	0	0
W040.2	2 and 15	2.2 and 13	TRACKING	TDRS-F1	SPACECOM	5	26 to 52	83	10
W041	2,15	2+,13	TRACKING	TDRS SPARE	SPACECOM		53	86	10
W041	2 and 15	2.2 and 13	TRACKING	TDRS-F2	SPACECOM	5	26 to 52	85	10
W042	17	12.2–12.7	BSS	GRENADA	GRENADA			0	0
W043	14	11 and 12.5	INTL	CYGNUS-43	CYGNUS SAT CORP	16	46–48	0	10
W045	17	12.2–12.7	BSS	BRAZIL	BRAZIL 1-4			0	0
W045	14	11 and 12.5	INTL	CYGNUS-45	CYGNUS SAT CORP	16	46–48	0	10
W045.2	0.3,8	0.25	GOVT	FLEETSATCOM-5	US NAVY	23	17–27	81	5
W047*	14	11	INTL	ORION-2	ORION SAT CORP	22	44–47.4	87	7
W050	6	4	INTL	INTELSAT IV-F1	INTELSAT	12	20 and 31	75	7
W050*	14	11	INTL	ORION-3	ORION SAT CORP	22	44–47.4	87	7
W053	6	4	INT'L	INTELSAT AO-SPARE 3	INTELSAT			0	0
W053	1.6, 6, and 14	1.5, 4, and 11	INT & MARIT	INTELSAT V-F8	INTELSAT	28	21 to 42	84	7
W055	17	12.2–12.7	BSS	ARGENTINA	ARGENTINA #2			0	0
W056*	14	11 and 12	INTL	ISI-1	INT'LSAT INC	16	42–44	0	10
W057	17	12.2–12.7	BSS	FALKLANDS	ARGENTINA			0	0
W057	UHF,8	UHF,8	GOVT	TACSAT-2	DOD USA	24		0	0
W057*	6	4	DOM	DIGISAT 1B	DIGITAL TELESAT	24	37	88	10
W057*	14	12	DOM	DIGISAT-1A	DIGITAL TELESAT	24	44	88	10
W057*	6.4 and 14	10.7–11.2	INT'L	PANAMSAT-1	PAN AM SAT CORP	12 (6N,6 S bound)		87	10
W058*	14	11 and 12	INTL	ISI-2	INT'L SAT INC	16	42 and 44	0	10
W061.5	17	12.2–12.7	BSS	USA-BSS	EASTERN-1	32		0	0
W062.5*	6 and 14	4 and 12	DOM	COL COMM CORP-62.5	COL COMM CORP	48	34 and 47	0	10
W064	17	12.2–12.7	BSS	BRAZIL	BRAZIL 3-5			0	0
W065	6	4	DOM	SBTS A-2	BRAZIL	24	38	85	8
W065*	14	12	DOM	SATCOM K4	RCA AMERICOM	16	38–39	0	10
W067	14	12	DOM	SATCOM K3	RCA AMERICOM	16	38–39	86	10
W067	6	4	DOM–INTL	SATCOM VI	RCA AMERICOM	24	37	87	10

Table 2 (*continued*)

Location	Up (GHz)	Down	Service	Name of Satellite	Operator	Transponders	EIRP	Launch	Life
W069	6 and 14	4 and 12	DOM	SPACENET-2	GTE-SPACENET	18	34, 36, and 39	84	7
W070	20	44	GOVT	FLTSATCOM-B W. ATL O	US NAVY			86	0
W070	6	4	DOM	SBTS A-1	BRAZIL	24	35	85	8
W070*	1.6 and 5.2	5.1	MOBILE	GEOSTAR CORP.	GEOSTAR-70	2	31	87	0
W070.5	17	12.2–12.7	BSS	CANADA-BSS	CANADA			0	0
W071.5	17	12.2–12.7	BSS	URAGUAY	URAGUAY			0	0
W072	6	4	DOM	SATCOM IIR	RCA AMERICOM	24	34–35	83	8
W072.5	17	12.2–12.7	BSS	CANADA-BSS	CANADA			0	0
W073*	6 and 14	4 and 12	FSS	FORDSAT-3	FORD A.S.S. CORP	48	35–50	88	10
W073*	14	12	DOM	GALAXY H5	HUGHES COMM INC	16	50	86	10
W073*	14	12	DOM	M.M.-73	MART. MARIETTA	28	47–53	88	10
W074	17	12.2–12.7	BSS	BRAZIL	BRAZIL			0	0
W074.0	6	4	DOM	GALAXY II	HUGHES COMM INC	24	34	83	9
W075	6	4	DOM	SATCOL II	COLOMBIA	24		89	10
W075*	14	12	DOM	GALAXY H6	HUGHES COMM INC	16	50	86	10
W075*	14	12	DOM	M.M.-75	MART. MARIETTA	28	47–53	88	10
W075*	14	12	DOM	RAINBOW-3	RAINBOW SAT INC	24	44–47	87	10
W075*	6	4	DOM	SPOTNET-C2	NAT EXCHANGE	24	34	87	9
W075*	14	12	DOM	SPOTNET-K3	NAT EXCHANGE	24	57–60	87	10
W075*	14	12	DOM	SPOTNET-K4	NAT EXCHANGE	24	57–60	87	10
W075*	14	12	DOM	USAT-3	US SAT SVCS INC	20	43–46	88	8
W075.1	0.4, 1.7, and 2	0.4, 1.7, and 2	MET	GOES-5	NOAA	4		81	5
W075.4	6	4	DOM	SATCOL IA	COLUMBIA	24	34.6	87	10
W075.4	6	4	DOM	SATCOL IB	COLUMBIA	24		87	10
W076	6	4	DOM	TELSTAR 302	AT&T COMM.	24	32–34	84	10
W076.0	6	4	DOM	COMSTAR D1	COMSAT GENERAL	24	33	76	7
W076.0	6	4	DOM	COMSTAR D2	COMSAT GENERAL	24	33	76	7
W077	14	12	DOM	SATCOM K1	RCA AMERICOM	16	38–39	85	10
W078	17	12.2–12.7	BSS	AMIGO 3	MEXICO			0	0
W079	14	12	DOM	RAINBOW-2	RAINBOW SAT INC	24	44–47	88	10
W079	6	4	DOM	WESTAR II	WU TELEGRAPH	12	33	74	7
W080.7			EXP	COSMOS-1540	USSR			84	0
W081	6 and 14	4 and 12	DOM	ASC-2	AMER SAT CO	24	34, 36, and 42	86	8
W081	17	12.2–12.7	BSS	BRAZIL	BRAZIL			0	0
W081*	6	4	DOM	WESTAR VII	WESTERN UNION	24	34+	88	10
W082	17	12.2–12.7	BSS	CANADA-BSS	CANADA			0	0
W083	14	12	DOM	ABCI-1	ADV BUS COMM	16	41–53	87	10
W083	6	4	DOM	SATCOM IV	RCA AMERICOM	24	32	82	10
W083	6	4	REGIONAL	STSC-1	CUBA			88	0
W083*	14	12	DOM	USAT-4	US SAT SVCS INC	20	43–46	88	8
W083.1	14	12	EXP	CTS (HERMES)	CANADA & NASA	2	46 and 56	76	2
W083.5	17	12.2–12.7	BSS	HAITI & DOMINICAN RP	HAITI & DOM. REP			0	0
W084.5	17	12.2–12.7	BSS	NE SOUTH AMERICA	TRIN/SUR/GUYANA			0	0
W085	14	12	DOM	USAT-1	US SAT SVCS INC	20	43–46	86	8
W085*	0.8, 1.6, 14	0.8, 1.5, 12	MOBILE	MOBILESAT 1	MOBILE SAT CORP	2	50, 53, 45	87	7
W086	17	12.2–12.7	BSS	PERU	PERU			0	0
W086	6	4	DOM	WESTAR VIII	WESTERN UNION	24	35	88	10
W087	14	12	DOM	SATCOM K2	RCA AMERICOM	16	38–394S85	10	
W087*	14	12	DOM	WESTAR C	WESTERN UNION	16	46+	89	10

Table 2 (*continued*)

Location	Up (GHz)	Down	Service	Name of Satellite	Operator	Transponders	EIRP	Launch	Life
W087.0	6	4	DOM	COMSTAR D3	COMSAT GENERAL	24	33	78	7
W088.5	6	4	DOM	TELSTAR 303	AT&T COMM.	24	32–34	85	10
W088.75	6 and 14	4 and 12	DOM	SPACENET-3	GTE-SPACENET	24	34, 36, and 39	85	7
W089	17	12.2–12.7	BSS	CUBA	CUBA			0	0
W091	17	12.2–12.7	BSS	CANADA-BSS	CANADA			0	0
W091*	30	20	30/20 GHZ	HUGHES KA-1	HUGHES GALAXY	32	58	88	0
W091	14	12	DOM	SBS-4	SAT BUS SYST	10	37–44 and 53	84	7
W091	6	4	DOM	WESTAR III	WESTERN UNION	12	33	79	7
W091	6	4	DOM	WESTAR VI-S	WESTERN UNION	24	34+	84	10
W091*	14	12	DOM	WESTAR B	WESTERN UNION	16	46+	88	10
W091.4	0.402, 2	0.14, 0.4, 1.7	MET	GOES-3	NOAA	4		78	5
W091.4	5.75–6.25	3.4–3.9	DOM	RADUGA-01	USSR PT&T	10	30	75	0
W092.5	17	12.2–12.7	BSS	CARIBBEAN	CARIBBEAN			0	0
W093*	30	20	30/20 GHZ	HUGHES KA-2	HUGHES GALAXY	32	58	89	0
W093*	6, 14, and 30	4, 12, and 20	DOM	ASC-4	AMER SAT CO	46	34, 42, and 52	89	8
W093*	14	12	DOM	COMSTAR K1	COMSAT GENERAL	16	47–50	88	10
W093*	6	4	DOM	EQUASTAR-1	EQUATORIAL COMM	24	36	87	10
W093*	6 and 14	4 and 12	DOM	FORDSAT-2	FORD A.S.S. CORP	48	35–50	88	10
W093*	14	12	DOM	GALAXY H7	HUGHES COMM INC	16	50	86	10
W093*	14	12	DOM	RAINBOW-4	RAINBOW SAT INC	24	44–47	87	10
W093*	14	12	DOM	WESTAR A	WESTERN UNION	16	46+	88	10
W093.5	6	4	DOM	GALAXY III	HUGHES COMM INC	24	34	85	9
W094	17	12.2–12.7	BSS	ARGENTINA	ARGENTINA			0	0
W095	17	12.2–12.7	BSS	ECUADOR	ECUADOR			0	0
W095.0	14	12	DOM	SBS-3	SAT BUS SYST	10	37–44	82	7
W096	17	12.2–12.7	BSS	BERMUDA	BERMUDA			0	0
W097.0	14	12	DOM	SBS-2	SAT BUS SYST	10	37–44	81	7
W099	17	12.2–12.7	BSS	PARAGUAY	PARAGUAY			0	0
W099	6	4	DOM	WESTAR IV	WESTERN UNION	24	34	82	10
W099*	14	12	DOM	SBS-6	SBS	14	47–49	87	10
W099.9	0.24–0.4, 36	0.24–0.4, 38	GOVT	LES-9	US DEPT OF DEF	3		76	5
W100	30	20	EXP	ACTS-1	NASA			88	3
W100	20	44	GOVT	FLTSATCOM-B E. PAC O	US NAVY			86	0
W100	0.3 and 8	0.25	GOVT	LEASAT-3	HUGHES COMM SVC	13	17 to 28	85	10
W100*	1.6 and 5.2	5.1	MOBILE	GEOSTAR-100	GEOSTAR CORP.	2	31	87	0
W100.0	14	12	DOM	SBS-1	SAT BUS SYST	10	37–44	80	7
W100.2	0.3, 8	0.25	GOVT	FLEETSATCOM-1	US NAVY	23	17–27	78	5
W101	17	12.2–12.7	BSS	USA-BSS	EASTERN-2	32		0	0
W101*	18	12.2–12.7	BSS	ACC-1	ADV COMM CORP	6		0	0
W101*	6 and 14	4 and 12	DOM	ASC-3	AMER SAT CO	24	34, 36, and 42	87	8
W101*	17	12.2–12.7	BSS	CBS-101	CBS INC	6	61	87	7
W101*	14	12	DOM	COMSTAR K2	COMSAT GENERAL	16	47–50	88	10
W101*	17	12.2–12.7	BSS	DBSC-1	DBS CORP	10		87	7
W101*	6 and 14	4 and 12	DOM	FORDSAT-1	FORD A.S.S. CORP	48	35–50	88	10
W101*	14	12	DOM	GSTAR A-3	GTE SATELLITE	16	40–46	854S10	
W101*	17	12.2–12.7	BSS	RCA-DBS-1	RCA AMERICOM	6	58	86	7
W101*	6	4	DOM	SPOTNET-C1	NAT EXCHANGE	24	34	87	9
W101*	14	12	DOM	SPOTNET-K1	NAT EXCHANGE	24	57–60	87	10
W101*	14	12	DOM	SPOTNET-K2	NAT EXCHANGE	24	57–60	87	10

Table 2 (*continued*)

Location	Up (GHz)	Down	Service	Name of Satellite	Operator	Transponders	EIRP	Launch	Life
W101	17	12.2–12.7	DBS	STC-101A	SAT TV CORP	6	57	86	7
W101	17	12.2–12.7	DBS	STC-101B	SAT TV CORP	6	57	86	7
W101*	17.3–17.8	12.2–12.7	BSS	USSB-1	HUBBARD B/C	6	57	87	7
W102	17	12.2–12.7	BSS	BRAZIL	BRAZIL			0	0
W103	17	12.2–12.7	BSS	COLOMBIA	COLOMBIA			0	0
W103	14	12	DOM	GSTAR A-1	GTE SATELLITE	16	42	84	10
W104	17	12.2–12.7	BSS	VENEZUELA	VENEZUELA			0	0
W104.5	6	4	DOM	ANIK D1	TELESAT CANADA	24	36	82	9
W104.8	8	7	GOVT	NATO II	NATO		19.5	71	5
W105	14	12	DOM	GSTAR A-2	GTE SATELLITE	16	42	85	10
W105*	0.868	0.823	MOBILE	SKYLINK 1	SKYLINK AMERICA	2	56	87	0
W105**	14	12	DOM	ANIK C2	TELESAT CANADA	16	48	83	8
W105.2	0.149,6	0.136/7,4	EXP	ATS-3	NASA	1	17.8	67	14
W105.4	0.24–0.4,36	0.24–0.4,38	GOVT	LES-8	US DEPT OF DEF	3		76	5
W106	17	12.2–12.7	BSS	CHILE	CHILE			0	0
W106.2	0.4,2	0.14,0.5,1.7	MET	GOES-2	NOAA	4		77	5
W106.5	0.868 and 14	0.823 and 12	MOBILE	M-SAT	CANADA/USA	5	37–38	87	7
W107.5	14	12	DOM	ANIK C1	TELESAT CANADA	16	48	84	8
W108	0.8,1.6,8	0.8,1.5,7	GOVT	MUSAT	CANADA			0	0
W108.3	2,8	2,7	GOVT	DSCS-2-F01	DOD USA	4	28,43	71	5
W109	6 and 14	4 and 12	DOM	ANIK B	TELESAT CANADA	18	36 and 47	78	7
W110	6	4	DOM	ANIK D2	TELESAT CANADA	24	36	85	9
W110	17	12.2–12.7	BSS	USA-BSS	EASTERN	32		0	0
W110*	17.3–17.8	12.2–12.7	BSS	USSB-1 ALTERNATE	HUBBARD B/C	6	57	87	7
W110*	17	12.2–12.7	BSS	WU-DBS-F1	WU TELEGRAPH CO	4	55.5	86	0
W110*	17	12.2–12.7	BSS	WU-DBS-SPARE-110	WU TELEGRAPH CO	4	55.5	0	0
W111.0	6	4	DOM	ANIK A3	TELESAT CANADA	12	33	75	7
W112.6	6	4	DOM	SATCOM II	RCA AMERICOM	24	32	76	8
W113.5	6 and 14	4 and 12	DOM	MORELOS-1	MEXICO	22 to 24	39 and 45	85	9
W115	17	12.2–12.7	BSS	ANDEAN BSS	VN/CL/PR/EC/BOL			0	0
W116.5	6 and 14	4 and 12	DOM	MORELOS-2	MEXICO	22 to 24	39 and 45	85	9
W117.5	14	12	DOM	ANIK C3	TELESAT CANADA	16	48	82	8
W119	17	12.2–12.7	BSS	USA-BSS	EASTERN	32		0	0
W119*	17	12	BSS	GSC-2	GRAPHIC SCANNIN	2	54.1	0	0
W119*	17	12.2–12.7	BSS	RCA-DBS-2	RCA AMERICOM	6	58	86	7
W119*	17	12.2–12.7	BSS	VSS-1	VIDEO SAT SYST	6	56	86	8
W119*	17	12.2–12.7	BSS	WU-DBS-F3	WU TELEGRAPH CO	4	55.5	0	0
W120	6 and 14	4 and 12	DOM	SPACENET-1	GTE-SPACENET	18	34, 36, and 39	84	7
W122	14	12	DOM	USAT-2	US SAT SVCS INC	20	43–46	87	8
W122*	6	4	DOM	EQUASTAR-2	EQUATORIAL COMM	24	36	87	10
W123.0	6	4	DOM	WESTAR V	WESTERN UNION	24	34	82	10
W124*	14	12	DOM	SBS-5	SBS	14	47–49	87	10
W125*	0.8,1.6,14	0.8,1.5,12	MOBILE	MOBILESAT 2	MOBILE SAT CORP	2	50,53,45	87	7
W125*	6	4	FSS	TELSTAR 304	AT&T	24	32–34	88	10
W127	17	12.2–12.7	BSS	AMIGO 4	MEXICO			0	0
W127.0	6	4	DOM	COMSTAR D4	COMSAT GENERAL	24	33	81	7
W128	6 and 14	4 and 12	DOM	ASC-1	AMER SAT CO	24	34, 36, and 42	85	9
W129	17	12.2–12.7	BSS	CANADA-BSS	CANADA			0	0
W130	14	12	DOM	ABCI-2	ADV BUS COMM	16	41–53	87	10

Table 2 (*continued*)

Location	Up (GHz)	Down	Service	Name of Satellite	Operator	Transponders	EIRP	Launch	Life
W130*	1.6 and 5.2	5.1	MOBILE	GEOSTAR-130	GEOSTAR CORP.	2	31	87	0
W130.1	2,8	2,7	GOVT	DSCS-2-F11	DOD USA	4	28,43	78	5
W130.8	0.4,2	0.13,0.5,1.7	MET	GOES-1	NOAA	4		75	5
W131	6	4	DOM	SATCOM IIIR	RCA AMERICOM	24	32,34	81	9
W132	14	12	DOM	RAINBOW-1	RAINBOW SAT INC	24	44–47	87	10
W134*	14	12	DOM	DIGISAT II	DIGITAL TELESAT	24	44	88	10
W134.0	6	4	DOM	GALAXY I	HUGHES COMM INC	24	37	83	10
W135	2,8	2,7	GOVT	DSCS-3-F3	DOD USA	7	23–44	83	10
W135	2,8	2,7	GOVT	DSCS-EP	DOD USA			0	0
W135	2,8	2,7	GOVT	NATO IIID	NATO	3	31 and 36	0	7
W135*	0.868	0.823	MOBILE	SKYLINK 2	SKYLINK AMERICA	4	53.4	87	0
W135.2	0.4, 1.7, and 2	0.4, 1.7, and 2	MET	GOES-6	NOAA	4		83	5
W136	17	12.2–12.7	BSS	AMIGO 1	MEXICO			0	0
W137*	6	4	DOM	AURORA II	ALASCOM, INC.	24	34	89	10
W137*	6	4	DOM	GALAXY IV	HUGHES COMM INC	24	34.5	86	9
W138	17	12.2–12.7	BSS	CANADA-BSS	CANADA			0	0
W139	6	4	DOM SPARE	SATCOM IR	RCA AMERICOM	24	36	83	10
W139.2	0.4,2	0.13,0.5,1.7	MET	GOES-4	NOAA	4		80	5
W141	6 and 14	4 and 12	DOM	MORELOS-3	MEXICO	22 to 24	39 and 45	0	9
W141*	6	4	DOM	AURORA III	ALASCOM, INC.	24	34	91	10
W142.4	8	7	FSS	DSCS-2-F15	USAF	4	28,43	82	5
W143	6	4	DOM	AURORA I	ALASCOM, INC.	24	34	82	10
W145	6 and 14	4 and 12	DOM	MORELOS-4	MEXICO	22 to 24	39 and 45	0	9
W146	17	12.2–12.7	BSS	AMIGO 2	MEXICO			0	0
W147.5*	6 and 14	4 and 12	DOM	COL COMM CORP-147.5	COL COMM CORP	48	34 and 47	0	10
W148	17	12.2–12.7	BSS	USA-BSS	WESTERN-1	32		0	0
W148*	18	12.2–12.7	BSS	ACC-2	ADV COMM CORP	6		0	0
W148*	17	12.2–12.7	BSS	CBS-148	CBS INC	6	61	88	7
W148*	17	12.2–12.7	BSS	DBSC-2	DBS CORP	10		87	7
W148*	17	12	BSS	GSC-1	GRAPHIC SCANNIN	2	54.1	0	0
W148*	17	12.2–12.7	BSS	RCA-DBS-3	RCA AMERICOM	6	58	87	7
W148	17	12.2–12.7	DBS	STC-148A	SAT TV CORP	6	57	87	7
W148*	17.3–17.8	12.2–12.7	BSS	USSB-2	HUBBARD B/C	6	57	87	7
W148*	17	12.2–12.7	BSS	VSS-2	VIDEO SAT SYST	6	56	86	8
W148*	17	12.2–12.7	BSS	WU-DBS-F2	WU TELEGRAPH CO	4	55.5	85	0
W148*	17	12.2–12.7	BSS	WU-DBS-F6	WU TELEGRAPH CO	4	55.5	0	0
W148*	17	12.2–12.7	BSS	WU-DBS-SPARE-148	WU TELEGRAPH CO	@	55.5	0	0
W157	17	12.2–12.7	BSS	USA-BSS	WESTERN	32		0	0
W157*	17	12.2–12.7	BSS	RCA-DBS-4	RCA AMERICOM	6	58	87	7
W157*	17	12.2–12.7	BSS	WU-DBS-F4	WU TELEGRAPH CO	4	55.5	0	0
W160	14 and 15	11 and 13	TRACKING	ESDRN	USSR			0	0
W166	17	12.2–12.7	BSS	USA-BSS	PACIFIC-1	32		0	0
W168	4.40–4.68	3.95–4.00	DOM	POTOK-3	USSR PT&T	3		0	0
W170	8	7	GOVT	GALS-04	USSR PT&T	10		0	0
W170	14	11	DOM	LOUTCH-P04	USSR PT&T	10		0	0
W170	5.73–6.03	3.4–3.7	DOM	STATSIONAR-10	USSR PT&T	6	19 to 30	0	0
W170	0.3,1.6	0.24,1.5	MOBILE	VOLNA-07	USSR PT&T	2		0	0
W171	2 and 15	2 and 14	TRACKING	TDRS-F3	SPACECOM	5	26 to 52	85	10
W175	17	12.2–12.7	BSS	USA-BSS	PACIFIC-2	32		0	0

INDEX

A

Acceptance angle, 266,270
Acceptance cone, 269
Acquisition (PLL), 142–145
Acquisition time, 176,241
Adaptive delta modulation, 29–30
Adaptive equalization, 227–229
A-Law companding, 20
Aliasing, 21–23
AM/AM conversion, 321
Amplitude modulation, 154,192,193
  comparison, 194–199
Amplitude shift keying (ASK), 42–43,192
AM/PM, 321
Analog-to-digital conversion, 15–18
Angle modulation, 45–61,153–154,193–199 *See also* modulation
Angular momentum, 300
Antenna gain, 219–224,312–317
Aperture (numerical), 269–274
Asynchronous systems, 10–11,166
Atmosphere path, 197–198
Autocorrelation function, 117–119,246
  properties of, 118–125
AWGN (additive white gaussian noise), 83,166,219
  power spectral density, 169

B

Backoff loss, 321
Balanced modulator, 154
Bandlimiting, 199–208
Bandwidth (definition), 85–87,271
  conservation of, 194
  compression, 73–75
  efficiency of, 53,71–73,153,196
  Nyquist, 6–7,85,199
Baseband, 7,67
Baud, 31,64
Binary, 5,16
Binary cyclic codes, 95–99
Biphase, 39–41,55,235
Bipolar, 11,37
  modulator, 128
Bit, 4–5
  error rate (BER), 35,41,47,73,90–91,196,312
Bit synchronization, 116,164–179
Block code, 91–95
  check code, 91–95
Burst, 116,308–312
Bus (satellite), 300

C

Capture range, 143–145
Carrier extraction, 55–59
  Costas loop, 56–57,161
  multiply-filter-divide, 57–59
  remodulator, 57
Carrier recovery and synchronization, 153–164
Carrier sense multiple access/collision detection (CSMA), 40
Carrier-to-noise density ratio, 64,213,214,314–324
Channel capacity, 8,83–85
Characteristic equation, 146
Charge transfer device, 244
Chip synchronization, 240–241,245–247
Companding, 18–20,64–65,83–85
Code rate, 88,92
Codes:
  Block, 91–95
  concatenated, 91
  convolutional, 100–104
  cyclic, 91,95–99
  Golay, 255
  Gold, 123–126
  Pseudonoise, 99–100
  systematic, 91
Codeword, 91
Coding (channel), 83,87–88
  efficiency, 88
  error control, 83
  error detection, 92
  forward acting, 92
Coding gain, 87
Coherent detection, 46–47,227
Constraint length, 101
Continous phase, 128–130,193

Correlation coefficient, 118–119
Correlative level encoding, 67–69
Cosine filter, 203–204
Costas loop, 56–57,158–164
  extended, 187
Coupling loss, 283

D

Decision directed loop, 157–158
Delay modulation, 41–42
Delta modulation, 26–29
  adaptive, 29–30
Differential PCM, 30
Differential PSK, 60–61
Digital modulation techniques, 40–71,192–197
  comparison, 64–65,194–197
Digital radio systems, 191–229
  bandlimited, 199–208
  channel simulation, 225–227
  fading, 197–199
  interference effects, 196–197
  link analysis, 218–225
  receiving system design parameters, 212–218
  spectrum measurements, 210–212
  third order intercept point, 215–218
Digital synchronization circuits, 153–185
  TDMA, 306–312
Digital-to-analog conversion, 15–16
  quantizing noise, 16–18,28
Direct sequence spread spectrum, 126,237–240,245–251
Doppler, 227,239–240
Dynamic range (receiver), 215–218

E

Earth stations, 306–307,321
Effective aperture area (antennas), 219,313
Encoding formats, 38
  biphase, 39
  miller, 41
  NRZ, 38–39
  RZ, 39
Eight phase PSK, 51,62–64
Eight QAM, 66
Effective isotropic radiated power (EIRP), 315–316
Energy-to-noise ratio (E_b/N_0), 69–73,83–85,208,224,251–252,314–317,320
Equalization, 20–21,227–229
Equatorial belt, 324
Equinox, 324
Equivalent noise temperature, 213
Error control and correction, 84–105
Error function, 69
Error rate probability, 69–71
Error voltage (PLL), 48–49,135
Ethernet, 40–41
Eye diagram, 24

F

Fade margin, 135,156
Fast hopping, 257
Filter, 135,156
  active, 138
  bandpass, 255
  brickwall, 205
  impulse response (raised cosine), 204
  loop, 146
  raised cosine, 203
  receiver, 205
  transmitter, 206
  transversal, 228,242
FDM/FM/FDMA, 305
Fiber guide, 266–272
Fiber losses, 270,283–285
Fiber optics codes, 288–289
Flux density, 321
Forward error correcting, 92,255,318–319
Frame synchronization, 182–185
Free running frequency, 10–11,164
Free space loss, 220
Frequency devision multiple access (FDMA), 305
Frequency hopping, 113–115,237–239,249
Frequency plans (satellite), 300–305

Frequency range (VCO), 143
Frequency shift keying (FSK), 43–50,126–128

G

Gain-to-equivalent noise temperature ratio (G/T_e), 314
Gaussian (white) noise, 178
Generator matrix, 93–95
Generator polynomial, 96–99
Geostationary satellite, 326
Geosynchronous orbit, 299
Golay code, 255
Gold code, 123–126
Guard space, 23

H

Hamming distance, 89
Hard limiting decoder, 320
Hemispheric beams, 303–304
Heterodyne detection, 268
Hopping pattern (frequency), 328
Horn type antenna, 219

I

I-channel, 192,201
Index of refraction, 273
Index profile (refractive), 272
Injection laser diodes (ILD's), 276
INTELSAT, 300–312
Intensity modulation systems, 281–288
Interference signal (CW), 198
Intermediate frequency (IF), 226,306
Intermodulation product distortion (IMD), 215
Intersatellite link, 265
Intersymbol interference (ISI), 201,204
ISDN H1 channel, 291–293
Isolator (circulator), 201
Isotropic antenna, 219

J

Jitter, 12,55
Jitter amplitude, 179
Jitter frequency, 179–180
Jitter reduction techniques,168–171

L

Lambertian source, 274–276
Lens, 265
Light emitting diodes (LED's), 274–279
  lasers (ILD's), 276
Linear phase (filter), 203
Link analysis, 218–225
  optical, 274,281–286
  satellite, 312–324
Look angles (satellite), 324
  azimuth and elevation, 324–325
Low pass filter, 49,286
  step response, 286–287

M

Manchester code, 36,39–41
Mark, 10,255
M-ary phase shift keying (MPSK), 50–55,62
Matched filter, 158,241–245
Mean output power level, 210–212
Merseene prime, 120
Midtread quantization, 20
Minimum shift keying (MSK), 196–197
Modulation, 41–45,192–197
  amplitude phase keying (APK), 66
  amplitude shift keying (ASK), 42
  Binary phase modulation (PSK), 193
  continous phase modulation (CPSM), 128–130
  delay (miller), 41
  delta, 26–30
  differential (DPCM), 30–31
  differential PSK, 60–61,193–194
  frequency shift keying (FSK), 43–50,193

Modulation (*cont.*)
multiple APK (MAPK), 69–71
offset-key (QAM), 195
On-off keying, 281
On-off keyed (QPSK), 195
pulse amplitude modulation (PAM), 286
pulse frequency modulation (PFM), 289–293
pulse modulation, 285–286
pulse position modulation (PPM), 286
quadrature amplitude modulation (QAM), 43,192
staggered QPSK (SQPSK), 61,127
Modulation index, 193
Multimode step index fiber, 271–274
Multipath fading, 196–198
Multiple access, 305–312
CDMA, 256–258
FDMA, 305–306
TDMA, 306–312

N

Narrowband noise, 170–171
Natural frequency, 137–139,179
Noise, 219
Noise (thermal), 212
Noise bandwidth, 87,175–177
Noise density, 169,178
Noise effects, 68–71
Noise figure, 213–214
Noise margin, 223
Noise temperature (effective), 221
Noncoherent FSK, 45–47
Nonlinear amplifier, 200
Nonlinear receiver gain, 215
Nonreturn-to-zero (NRZ), 38–39,288
Numerical aperture, 270–274
Nyquist bandwidth (*see* bandwidth)
Nyquist sampling rate, 21–23

O

Offset-keyed (*see* modulation)
On-Off keying (*see* modulation)
Optical transmitters and receivers, 224–280
Orbit, 300
Orthogonal, 119
Oscillator (VCO), 135–137,159

P

Parity, 93
Partial response signaling (PRS), 68
PCM codes, 36–42,288–289
Performance comparison, 64–70,192–199
Phase ambiguity, 57,156,231
Phase detector, 135–137,160
Phase locked loops, 47–50,135–139,175–182,227
Phase reference, 47,154
Phase shift (filter), 203
Phase shift keying (*see* modulation)
Photons, 266
Pilot carrier synchronization, 154–156
PIN diode, 277–278
Primitive, 120
Probability of error (*see* error rate probability)
Propagation time, 307
Pulse amplitude modulation (PAM-*see* modulation)
Pulse code modulation (PCM-*see* modulation)
Pulse frequency modulation (PFM-*see* modulation)
Pulse position modulation (PPM-*see* modulation)
Pulse shaping, 128
Pulse width modulation (PWM-*see* modulation)

Q

Q-channel, 53,192,201
Quadrature amplitude modulation (QAM-*see* modulation)
Quadrature components, 128–130
Quadrature PSK (QPSK-*see* modulation)

Quantization error (noise), 16–18
mean square value, 27–28
step size, 27
Quantizing range, 18

R

Raised cosine (*see* filters)
Random fading loss, 224–225
Rayleigh distribution, 224–225
Reference burst, 307–312
Refraction, 272–273
Refractive index, 273
Relative intensity noise, 267–268
Repeaters, 287
Resolution, 124
Responsivity, 278
Return-to-zero (RZ), 39,288
Rise time, 286–288
Roll-off factor, 203–204

S

Sampling rate, 21–22
Sampling time, 6,204
Satellite communications, 297–326
Satellite locations, 324–326
Satellite switched TDMA, 184–185,306–312
Satellite system parameters, 312–314
Satellite transponder, 304
Saturation (TWT), 320–321
Shannon, 64,83–84
Shot noise, 266–268
Signal state space diagram, 51
Signal-to-quantization noise ratio, 28–29
Spectral dispersion, 273
Spectrum measurements (digital radio), 210–213
Spectral source width (fiber), 273
Spread spectrum, 113–130,235–258
Stationkeeping, 300
Synchronization, 153–185,306–312
bit, 164–179
carrier, 153–164
carrier (pilot), 154–157
chip, 240–241,245–247
frame, 183–185
Syndrome, 95
System design parameters (digital radio), 212–218

T

T1-Pcm frame, 184
Time delay (DN), 307–309
Time division multiple access (TDMA), 185,306–312
Timing recovery, 55
Timing signal (clock recovery), 164
Threshold comparitor, 11
Tracking loop (PLL), 142–147
Traffic burst, 309–312

U

UART, 168
Up-link, 314

W

White noise, 83,166,219

Z

Zero dispersion wavelength, 273